Imagery and GIS

Best Practices for Extracting Information from Imagery

Kass Green
Russell G. Congalton
Mark Tukman

Esri Press
REDLANDS | CALIFORNIA

Cover credits: NASA/Earth Observatory; Garry Killian/Shutterstock

Esri Press, 380 New York Street, Redlands, California 92373-8100
Copyright © 2017 Esri
All rights reserved. First edition 2017
Printed in the United States of America
21 20 19 18 17   1 2 3 4 5 6 7 8 9 10

Library of Congress Cataloging-in-Publication Data

Names: Green, Kass.
Title: Imagery and GIS : best practices for extracting information from
  imagery / Kass Green, Russell G. Congalton, and Mark Tukman.
Description: Redlands, California : Esri Press, [2017] | Includes
  bibliographical references and index.
Identifiers: LCCN 2017016441 | ISBN 9781589484542 (pbk. : alk. paper)
Subjects: LCSH: Remote sensing. | Geographic information systems. | Image
  processing.
Classification: LCC G70.4 .G743 2017 | DDC 621.36/78--dc23 LC record
available at
https://urldefense.proofpoint.com/v2/url?u=https-3A__lccn.loc.gov_2017016441&d=DwIFAg&c=n6-cguzQvX_tUIrZOS_4Og&r=RhmcbAxStnbJpr06
ef1onNDeVX-gjVopdqeQ8i7DbIY&m=D46jekE3rLqAedyzhCR6CiiAjGL2PtB677NAry_6e8E&s=w10SupBXRhN_tmghKOJHZR5cg_2T3Fm6g8l3KW9
nzM8&e=

The information contained in this document is the exclusive property of Esri unless otherwise noted. This work is protected under United States copyright law and the copyright laws of the given countries of origin and applicable international laws, treaties, and/or conventions. No part of this work may be reproduced or transmitted in any form or by any means, electronic or mechanical, including photocopying or recording, or by any information storage or retrieval system, except as expressly permitted in writing by Esri. All requests should be sent to Attention: Contracts and Legal Services Manager, Esri, 380 New York Street, Redlands, California 92373-8100, USA.

The information contained in this document is subject to change without notice.

US Government Restricted/Limited Rights: Any software, documentation, and/or data delivered hereunder is subject to the terms of the License Agreement. The commercial license rights in the License Agreement strictly govern Licensee's use, reproduction, or disclosure of the software, data, and documentation. In no event shall the US Government acquire greater than RESTRICTED/LIMITED RIGHTS. At a minimum, use, duplication, or disclosure by the US Government is subject to restrictions as set forth in FAR §52.227-14 Alternates I, II, and III (DEC 2007); FAR §52.227-19(b) (DEC 2007) and/or FAR §12.211/12.212 (Commercial Technical Data/Computer Software); and DFARS §252.227-7015 (DEC 2011) (Technical Data–Commercial Items) and/or DFARS §227.7202 (Commercial Computer Software and Commercial Computer Software Documentation), as applicable. Contractor/Manufacturer is Esri, 380 New York Street, Redlands, CA 92373-8100, USA.

@esri.com, 3D Analyst, ACORN, Address Coder, ADF, AML, ArcAtlas, ArcCAD, ArcCatalog, ArcCOGO, ArcData, ArcDoc, ArcEdit, ArcEditor, ArcEurope, ArcExplorer, ArcExpress, ArcGIS, arcgis.com, ArcGlobe, ArcGrid, ArcIMS, ARC/INFO, ArcInfo, ArcInfo Librarian, ArcLessons, ArcLocation, ArcLogistics, ArcMap, ArcNetwork, *ArcNews*, ArcObjects, ArcOpen, ArcPad, ArcPlot, ArcPress, ArcPy, ArcReader, ArcScan, ArcScene, ArcSchool, ArcScripts, ArcSDE, ArcSdl, ArcSketch, ArcStorm, ArcSurvey, ArcTIN, ArcToolbox, ArcTools, ArcUSA, *ArcUser*, ArcView, ArcVoyager, *ArcWatch*, ArcWeb, ArcWorld, ArcXML, Atlas GIS, AtlasWare, Avenue, BAO, Business Analyst, Business Analyst Online, BusinessMAP, CityEngine, CommunityInfo, Database Integrator, DBI Kit, EDN, Esri, esri.com, Esri—Team GIS, Esri— *The GIS Company*, Esri—The GIS People, Esri—The GIS Software Leader, FormEdit, GeoCollector, Geographic Design System, Geography Matters, Geography Network, geographynetwork.com, Geoloqi, Geotrigger, GIS by Esri, gis.com, GISData Server, GIS Day, gisday.com, GIS for Everyone, JTX, MapIt, Maplex, MapObjects, MapStudio, ModelBuilder, MOLE, MPS—Atlas, PLTS, Rent-a-Tech, SDE, SML, Sourcebook•America, SpatiaLABS, Spatial Database Engine, StreetMap, Tapestry, the ARC/INFO logo, the ArcGIS Explorer logo, the ArcGIS logo, the ArcPad logo, the Esri globe logo, the Esri Press logo, The Geographic Advantage, The Geographic Approach, the GIS Day logo, the MapIt logo, The World's Leading Desktop GIS, *Water Writes*, and Your Personal Geographic Information System are trademarks, service marks, or registered marks of Esri in the United States, the European Community, or certain other jurisdictions. CityEngine is a registered trademark of Procedural AG and is distributed under license by Esri. Other companies and products or services mentioned herein may be trademarks, service marks, or registered marks of their respective mark owners.

Ask for Esri Press titles at your local bookstore or order by calling 800-447-9778, or shop online at esri.com/esripress. Outside the United States, contact your local Esri distributor or shop online at eurospanbookstore.com/esri.

Esri Press titles are distributed to the trade by the following:

In North America:
Ingram Publisher Services
Toll-free telephone: 800-648-3104
Toll-free fax: 800-838-1149
E-mail: customerservice@ingrampublisherservices.com

In the United Kingdom, Europe, Middle East and Africa, Asia, and Australia:
Eurospan Group
3 Henrietta Street   Telephone: 44(0) 1767 604972
London WC2E 8LU   Fax: 44(0) 1767 601640
United Kingdom   E-mail: eurospan@turpin-distribution.com

For
Gene Forsburg and Jack McDevitt
Janet and Robert Congalton
Mel and Lois Tukman

Contents

Acknowledgments   ix

Section 1   Discovering Imagery
Chapter 1   Introduction   3
Chapter 2   Thinking About Imagery   11
Chapter 3   Imagery Fundamentals   27
Chapter 4   Choosing and Accessing the Right Imagery   69

Section 2   Using Imagery
Chapter 5   Working with Imagery   107
Chapter 6   Imagery Processing: Controlling Unwanted Variation in the Imagery   145

Section 3   Extracting Information from Imagery
Chapter 7   Understanding Variation on the Ground—the Importance of the Classification Scheme   179
Chapter 8   Digital Elevation Models   219
Chapter 9   Data Exploration: Tools for Linking Variation in the Imagery to Variation on the Ground   229
Chapter 10   Image Classification   267
Chapter 11   Change Analysis   313

Section 4   Managing Imagery and GIS Data
Chapter 12   Accuracy Assessment   337
Chapter 13   Managing and Serving Imagery   361
Chapter 14   Concluding Thoughts   385

Acronyms   393
Glossary   399
References   411
Image Credits   415
Index   417

Acknowledgments

This book would not have been possible without the generous contributions of our colleagues who reviewed the text, permitted us to include their graphics, provided case studies, and offered valuable insights. In particular we would like to thank Maggi Kelly, Gerald Kinn, Jarlath O'Neil-Dunne, and Cassandra Pallai. We would also like to express our gratitude to the Sonoma County Agricultural Preservation and Open Space District, and the California Department of Fish and Wildlife—our project with them has enriched this book with many real world examples. Esri personnel have made this book truly professional and we thank their editors, product managers, application developers, and graphic artists. We are especially grateful to Esri's Claudia Naber, who shepherded the book from inspiration to completion, and Peter Becker whose commitment, persistent questions, and leadership continually raised our standards and improved the content. Finally, we thank our families whose patience and support gave us the freedom and energy to pursue our passion for imagery and to share it with others in this book.

Section 1
Discovering Imagery

Chapter 1
Introduction

Why Imagery and GIS?

Imagery—it allures and fascinates us; its measurements inform us. It draws us in to explore, analyze, and understand our world. First comes the astonishment of its raw beauty—the enormity of a hurricane, the stark glaciers in Greenland, the delicate branching of a redwood's lidar profile, a jagged edge of a fault line in radar, the vivid greens of the tropics, the determined lines of human impact, the rebirth of Mount Saint Helens' forests, the jiggly wiggly croplands of Asia and Africa, the lost snows of Kilimanjaro. Each image entices us to discover more, to look again and again.

Then we start to ask questions. Why do trees no longer grow here? Can trees grow here again? How much has this city expanded? Will the transportation corridors support emergency relief? Why did this house burn while the one next door is untouched by flames? What crops flourish here? Will they produce enough food to feed the people of this region? Why has this landscape changed so dramatically? Who changed it? When we bring imagery and GIS together we can answer these questions and many more. By combining imagery and GIS, we can inventory our resources, monitor change over time, and predict the possible impacts of natural and human activities on our communities and the world.

This book teaches readers about the many ways that imagery brings value to GIS projects and how GIS can be used to derive value from imagery. Imagery forms the foundation of most GIS data. Whether it be a map of transportation networks, elevation contours, building footprints, facility locations, vegetation type, or land use, the information in most GIS datasets is derived primarily from imagery. Alternatively, GIS allows us to more efficiently and effectively derive information from imagery. Organizing imagery in a GIS brings the power of spatial information management and analysis to imagery.

The purpose of this book is to unlock the mysteries of imagery, to make it readily usable by providing you with the knowledge required to make *informed decisions* about imagery. More than just an overview of remote sensing technology, this book takes a hands-on, decision-focused approach. Each chapter evaluates practical considerations and links to online interactive examples. The book also includes multiple real-world case studies that highlight the most effective use of imagery and provide advice on deciding between alternative image sources and approaches. The book provides guidance on

1. choosing the best imagery to meet your needs;
2. effectively working with and processing imagery;
3. efficiently extracting information from imagery; and
4. assessing, publishing, and serving imagery datasets and products.

Why Now?

Humans have always coveted a bird's-eye view. The resulting knowledge of where we are relative to others and the resources we need has long been treasured and is necessary for survival. Remote sensing, the science of measuring the attributes of an object from a distance, provides us with imagery. Offering valuable insights into how humans interact with the earth, imagery and GIS allow citizens, governments, corporations, and nonprofits to fundamentally understand patterns of resource status, use, and change.

It took thousands of years for humans to invent cameras and aircraft, but within 30 years of their invention they were combined, and remote sensing was born. In the late 1800s and early 1900s, early remote sensing systems consisted of cameras placed first on balloons and kites, and then on airplanes. Later, the military operations of World Wars I and II as well as the Cold War spurred remote sensing into a field of science, resulting in methods and technologies that allow us to analyze and measure features from a distance. Remote sensors are now everywhere—from your cell phone camera, to the video camera above your bank teller machine, to satellites hundreds of miles in space. Imagery and GIS support a broad array of applications including weather prediction, disaster response, military reconnaissance, flood planning, forest management, habitat conservation, wetland preservation, mineral exploration, famine early warning, agriculture yield estimates, urban planning, wildfire prevention and control, fisheries management, transportation planning, humanitarian aid, climate monitoring, and change detection.

Precision agriculture

Information gathered during harvest, including yield at any given location, helps growers track their results and provides valuable input for calculating seeding and soil amendment rates for the following year. The images on this page and the next one are interactive at thearcgisimagerybook.com.

Humanitarian aid

Access to up-to-date imagery shows tha creation of the Zaatari refugee camp over a nine-day period in July 2012. Designed to hold over 60,000 people, its population skyrocketed to over 150,000 before new camps relieved some of the pressure. The story map *The Uprooted* tells the tale.

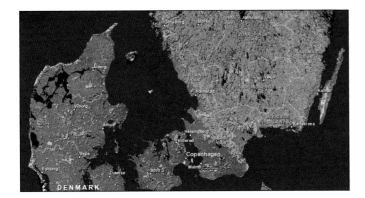

Forestry

Dynamic access to data on forests in Europe is derived from the Corine Land Cover 2006 inventory. Corine means "coordination of information on the environment".

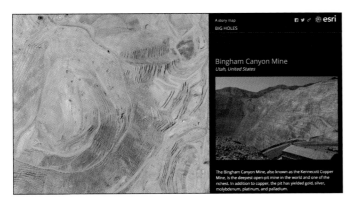

Mining

The geologic nature of the landscape comes to life using earth-orbiting satelites.

Natural disaster assessment

This scene shows the destruction of Hurricane Sandy's storm surge in Seaside, New Jersey. The active swipe map compares pre- and postevent imagery from the National Oceanic and Atmospheric Administration (NOAA).

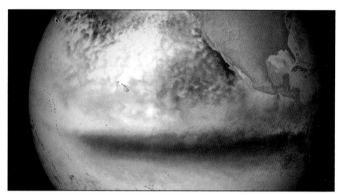

Climate and weather study

This short map presentation from NOAA answers many of the quations about the effects of El Niño. Scroll down to learn more about this climate feature and its characteristics.

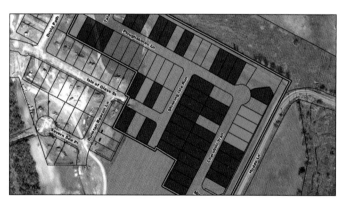

Engineering and construction

Development projects actively under construction in the City of Pflugerville, Texas, are displayed here.

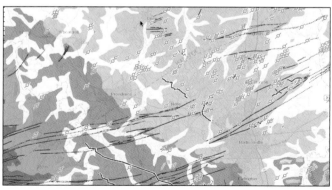

Oil and gas exploration

This geologic map compiled by the Kentucky Geological Survey relates theme of land use, environmental protection, and economic development.

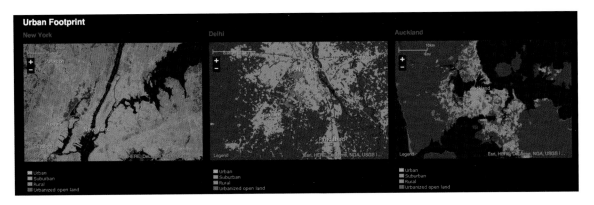

The Urban Observatory is an ambitious project led by TED founder Richard Saul Wurman to compile data that allows comparison of metro areas at common scales.

Remote sensing has always been a rapidly changing field with technologies readily adopted as they become operational and cost effective. However, recently the pace of adoption has quickened. Long a staple of military operations, remotely sensed imagery has recently exploded for civilian use as availability and access have increased while prices have declined. This rapidly quickening pace of change results from

- the evolution of sensors from capturing images on film to capturing them on digital arrays. As a result, storing, accessing, and analyzing imagery have become much easier and faster. As microelectronic performance continues to improve, sensors will continue to become lighter, smaller, more powerful, and less expensive.
- platform improvements resulting in more agile and smaller platforms that are less expensive to operate. Besides airplanes and large satellites, imagery is now collected from unmanned airborne systems (sometimes called drones) and constellations of small satellites.
- increasing accessibility because of growing supply, policy changes, and the ability to quickly serve cached imagery across the web. While much high-resolution satellite imagery is license restricted, both the United States and the European Union offer global imagery at no cost in the public domain from their moderate-resolution systems (Landsat and Sentinel), and high-resolution airborne imagery is freely available from many local, state, or federal agencies across the United States. Additionally, archived high-resolution imagery is readily available for free viewing on many web services, including ArcGIS Online, Google Earth, and Bing.
- improved positional accuracy. GPS and other technologies allow for precise registration of imagery to the ground, which supports the easy integration of imagery with other GIS datasets. Additionally, humans can now easily locate themselves on web-served imagery using the GPS in their cell phones.
- the advent of cloud storage and the plummeting cost of computer disk space and memory. Imagery is Big Data and the files can be very large, but Big Data becomes

less and less of a barrier to use as the cost of data storage continues to decline and accessibility improves.
- spatial information becoming mainstream. Until the turn of the century, few people had the expertise or resources required to manipulate and analyze imagery, and maps remained nonintuitive. Now, with spatial information at our fingertips, many more people are spatially aware, and location has become a commodity. As a result, remote sensing and GIS have attracted a generation of brilliant software engineers who were brought up using computers and who rapidly bring innovations in computer science and database management to the geospatial sciences.

Book Organization

The organization of the book follows the organization of a typical imagery project workflow and is broken into four sections. The first section, Discovering Imagery—four chapters—provides the information needed to choose the best imagery to meet your needs. Chapter 2 introduces the structure of imagery data and presents a construct for thinking about imagery that is the foundation of this book, and also provides a decision framework for all of your work with imagery. Chapter 3 examines the fundamental collection and organizational characteristics of imagery that determine what imagery dataset will bring the most value to your projects. Chapter 4 provides a framework for choosing the best imagery to meet your needs and describes the variety of imagery datasets available.

The second section, Using Imagery in a GIS—two chapters—focuses on how to manipulate imagery to increase its value within a GIS. Chapter 5 discusses imagery storage and formats, displays, mosaicking, and accessing imagery as web services. Chapter 6 reviews the methods used to control unwanted variation in imagery caused by the earth's atmosphere and terrain.

The third section, Extracting Information From Imagery—five chapters—details how to efficiently and accurately extract information from imagery. Chapter 7 introduces the importance of developing a robust classification scheme to characterize variation on the ground. Chapter 8 reviews how digital elevation models are created from imagery. Chapter 9 introduces imagery elements and discusses a variety of techniques and tools for exploring the correlation between imagery variation and variation on the ground. Chapter 10 reviews image classification approaches ranging from manual interpretation to sophisticated semi-automated classification. Chapter 11 discusses the concepts and methods commonly employed for using imagery to monitor change.

The fourth section, Managing Imagery and GIS Data—three chapters—focuses on ensuring the effective management and use of imagery and maps created from imagery.

Chapter 12 introduces concepts and techniques for assessing the positional and thematic accuracy of imagery products and services. Chapter 13 reviews using ArcGIS to publish and serve imagery, imagery products, and imagery services. The book's concluding chapter lists experience-proven tips for successfully deriving the most value from imagery.

This book is illustrated with over 150 figures which clarify many of concepts presented. Over 30 of these figures are linked to interactive applications, which allow you to explore the concepts in more depth. If a figure is linked to an application, you will see a blue Esri url in the figure caption.

Case Study of Sonoma County, California

During the writing of this book, the authors also had the pleasure of creating a high-resolution vegetation type map of Sonoma County, California (approximately 1 million acres) for the Sonoma County Agricultural Preservation and Open Space District and its partners. A map of 85 vegetation types at a 1-acre minimum mapping unit (or smaller for some wetland and riparian features) was created using a variety of imagery and nonimagery sources including Landsat, National Agricultural Imagery Program (NAIP) imagery, hyperspectral imagery, digital elevation models, wildfire history, weather measurements, previously created vegetation maps, and NASA-funded six-inch multispectral imagery and lidar data[1]. Additionally, other GIS layers were created from the imagery including an impervious-surfaces map, a croplands map, building footprints, and many hydrologic data deliverables such as stream centerlines. The project products support decision making for natural resource planning, land conservation, sustainable community and climate protection planning, public works projects, hydrologic evaluations, watershed assessments and planning, and disaster preparedness throughout the county. The timeliness, detail, and richness of the Sonoma vegetation mapping project supported the development of many figures and case studies presented in this book. You can learn more about this project and download its imagery and products at http://sonomavegmap.org/.

---

[1] Lidar data and orthophotography were provided by the University of Maryland under grant NNX13AP69G from NASA's Carbon Monitoring System (Dr. Ralph Dubayah and Dr. George Hurtt, Principal Investigators). This grant also funded the creation of derived forest cover and land-cover information, including a countywide biomass and carbon map, a canopy cover map, and digital elevation models (DEMs). The Sonoma County Vegetation Mapping and LiDAR Program funded lidar-derived products in the California State Plane Coordinate System, such as DEMs, hillshades, building footprints, one-foot contours, and other derived layers. The entirety of this data is freely licensed for unrestricted public use, unless otherwise noted.

Chapter 2
Thinking About Imagery

Introduction

This chapter introduces the fundamental concepts that define imagery—its structure, uses, and classification. More importantly, the chapter introduces the four fundamental steps required to rigorously consider the type of information to be extracted from imagery, and how those considerations will drive all decisions you make about acquiring, using, serving, and classifying imagery. These steps form the foundation of imagery workflows and shape the structure of this text.

What Is Imagery?

Images capture and store data measured about locations. Historically, most imagery was captured on film, and stored and displayed on either film, glass, or paper. Now, nearly all imagery is captured digitally and stored in a gridded form. Even historical paper maps and photos are often now scanned and stored as digital images such as the vegetation maps of Sonoma County from the 1960s shown in figure 2.1.

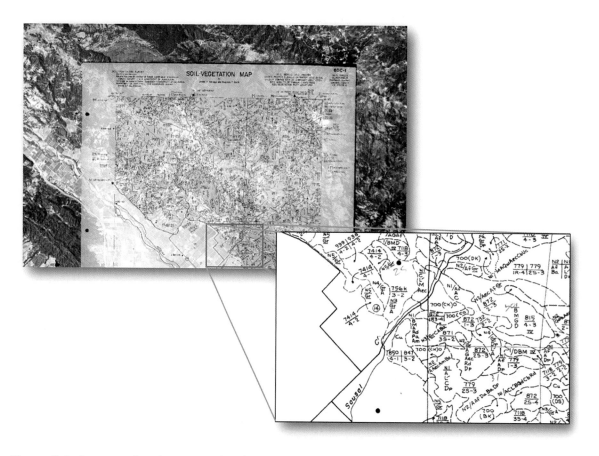

Figure 2.1. A scanned and registered soil vegetation map created from 1960s aerial photography overlaid onto 2013 imagery in Sonoma County, California.

Many images are measurements of reflected or emitted electromagnetic energy (discussed more in chapter 3) captured by a sensor, whether it's the camera on your cell phone, the magnetic resonance imaging device in a medical laboratory, or a sophisticated sensor on an unmanned aerial vehicle, an airplane, or a satellite. Other types of imagery data include scientific measurements of a location's properties, such as its precipitation, temperature, or water depth and flow.

Imagery Data Structure

As measurements, all images are continuous data. Continuous data is measured on a continuum and can be split into finer and finer increments, contingent upon the precision of the sensor making the measurements. Sensors that capture imagery return numerical values within a range defined by the sensing instrument.

Most remote sensors collect data in a rectangular array, and remotely sensed data not captured in a rectangular array is usually resampled into a rectangular array after collection. Examples of image data include imagery collected from optical sensors, such as that collected by USGS's Landsat satellites and by contractors collecting aerial National Agricultural Imagery Program imagery for the USDA.

Discreet point location measurements, such as those from weather stations or buoys, are also represented in a rectangular array. Three-dimensional (x, y, and z) point data, such as that collected by lidar sensors, is referred to as a point cloud, but also often transformed into a rectangular array for both visualization and analysis (figure 2.2).

Figure 2.2. Lidar returns from a Teledyne Optech Titan bathymetric lidar system. The image is color coded by elevation for both topographic height and water depth. Source: Teledyne Optech.

Image files are structured as gridded rectangular arrays or raster data, with each cell representing a measurement value captured by the remote sensing measurement. The data is stored as rows and columns of contiguous rectangular cells laid out in a grid (figure 2.3). Each cell contains a value. The values may be integer or floating point. The cells of an image raster are often referred to as picture elements or pixels and contain data values that measure some characteristic of each cell's location, such as its temperature, elevation, or spectral reflectance.

Because each cell has both a row and column location within the grid, the cells have inherent coordinates even though those coordinates may still need to be converted to map coordinates if the imagery is to be used in a GIS with other GIS layers (see chapter 6 for more information on georeferencing imagery).

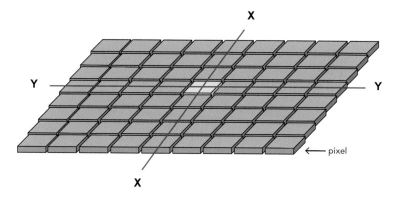

Figure 2.3. A two-dimensional raster grid. Most imagery is stored as a raster where each cell is referred to as a pixel and has an associated x and y coordinate.

Imagery and GIS

If imagery data has geographic coordinates, it can be incorporated into a GIS as a layer and registered with the other geographic layers in the GIS. This overlay capability is the fundamental concept upon which GIS operates. When combined with other GIS data, imagery transcends its status as merely a picture and becomes a true data source that can be combined, compared, analyzed, and classified with other data layers of the same area, as shown in figure 2.4.

Figure 2.4. Imagery as a GIS layer

For use in a GIS, imagery is usually stored as it has been collected: in raster format. Point imagery data can be converted to raster data either by giving the cells between the sample points values of zero or by interpolating between the sample points. Similarly, line and polygon data can also be converted to raster representations and so handled similar to images.

Rasters versus Vectors

GIS data is stored as either rasters or vectors. Vectors represent the world with points, lines, or polygons. A point is one location represented by x, y, and z coordinates. A line is a linear connection between points. Sometimes lines are connected into a network of topologically connected lines. A polygon is a set of lines joined together to enclose an area. Polygons are drawn to outline the shape of an object of interest.

Rasters divide the landscape into a grid of equal-area rectangular cells. The rectangular shape of an individual cell does not represent a specific object on the ground. Rather, the cell is an arbitrary delineation. Lines or polygon shapes on the ground are represented by connected raster cells, as shown in figure 2.A.

Most imagery is collected as raster data, which is why most imagery is captured and stored as rasters. Because of the simple structure of rasters, raster spatial analysis is relatively uncomplicated. However, unlike vectors, rasters do not have meaningful boundaries. In a raster, a lake is a cluster of spatially adjacent cells classified as water. There is no way to analyze the lake as a singular object—it is merely a collection of connected water cells. In a vector system, a lake is a polygon object with a defined boundary, which also carries information about the other objects sharing its boundary. As a result, we can measure the size of the lake, analyze the wildlife habitat next to the lake, and measure the distances from the lake to cabins. Vector spatial analysis is usually more computationally intensive than raster analysis, but vectors also better represent the shapes of the world as they actually exist, with curves and straight lines. As a result, vector maps are more aesthetically pleasing. Fortunately, raster maps can be converted to vector maps and vice versa, which means that map users can thoughtfully choose which data structure will best meet their needs. However, conversion from raster to vector or vector to raster data introduces changes that can potentially create errors, and care must be taken when this process is performed.

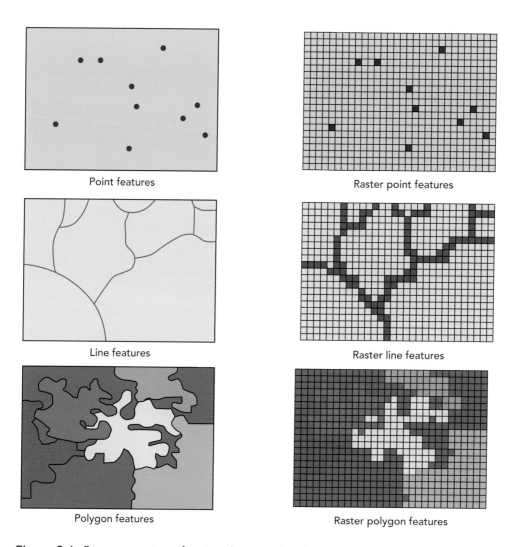

Figure 2.A. Representation of points, lines, and polygons in vector and raster formats

Characteristics of Rasters

Because most imagery is captured and stored as rasters, it is important to understand the characteristics of rasters, such as type, bands, and cell size.

Type

The cells of rasters can contain either continuous or discrete values. As mentioned earlier, image data is continuous. However, when we classify an image into information, the resulting values can be either continuous, as in an elevation model (figure 2.5), or discrete, as in a land use map such as that shown in figure 2.6. Unlike continuous data, discrete data classes cannot be mathematically divided more finely. Discrete information can take on only finite predefined values such as "tank," "lake," "urban," "forest," "building," or "agriculture." Rasters of discrete values represent information that has been classified from image data; they are no longer considered images, but are rather now raster format maps.

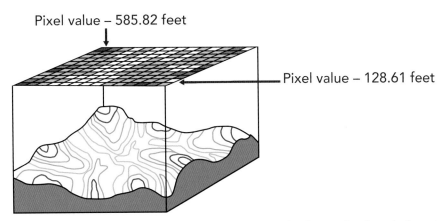

Figure 2.5. An example of a continuous raster in the form of a digital elevation model (DEM)

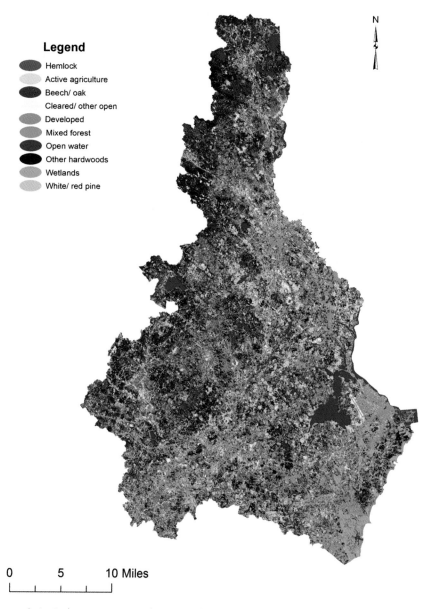

Figure 2.6. A thematic map showing discrete land-cover classes present in the Coastal Watershed in southeastern New Hampshire created from Landsat 8 imagery

Bands

Imagery measurements are collected and stored in raster bands. For example, a panchromatic image raster includes only a single band of measurements, shown as a single layer in figure 2.3, and is typically shown in grayscale. Multispectral imagery contains several bands of measurements, as shown in figure 2.7. Figure 2.8 shows a portion of Landsat imagery over Sonoma County, California. The numerical values of the cells of three bands of the seven-band image are displayed. When the red, green, and blue bands are displayed in the red, green, and blue colors of a computer screen they create the natural-color image of figure 2.8.

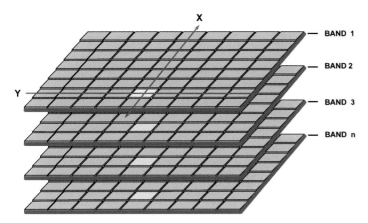

Figure 2.7. Multispectral data. If more than one type of measurement is collected for each cell, the data is called multispectral, and each type of measurement is represented by a separate band.

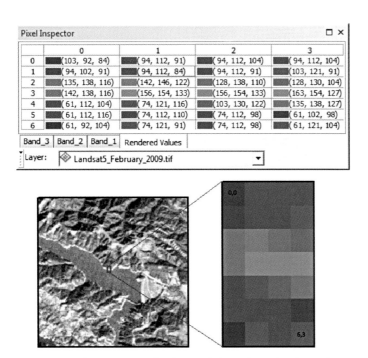

Figure 2.8. The numerical values of three bands of Landsat imagery over a portion of Sonoma County, California.

Hyperspectral data contains 50 to more than 200 bands of measurements and is usually represented as a cube of spectral values over space (figure 2.9). Image cubes are also used to bring the temporal dimension into a set of images, as when multiple Landsat images are analyzed of the same area over time.

Figure 2.9. A hyperspectral data cube captured over NASA's Ames Research Center in California. Hyperspectral data includes 50 to more than 200 bands of measurements. Source: NASA

Figure 2.10. The impact of raster cell size on the level of detail depicted. The larger the cell, the less discernible detail. In this example a car is represented by three different image cell sizes but displayed at the same scale. The smaller the cell, the more information available to identify the rectangle of eight large reddish pixels on the right as a red sedan on the left.

Cell Size

The cell size, or spatial resolution, of a raster will determine the level of spatial detail displayed by the raster. Figure 2.10 illustrates the effect of cell size on spatial resolution. The cell must be small enough to capture the required detail but large enough for computer storage and analysis to be performed efficiently. More features, smaller features, or greater detail in the extent of features can be represented by a raster with a smaller cell size. However, more is not always better. Smaller cell sizes result in larger raster datasets to represent an entire surface; therefore, there is a need for greater storage space, which often results in longer processing time.

Choosing an appropriate cell size is not always simple. You must balance your application's need for spatial resolution with practical requirements for quick display, processing time, and storage. Essentially, in a GIS, your results will only be as accurate as your least accurate dataset. The more homogeneous an area is for critical variables, such as topography and land use, the larger the cell size can be without affecting accuracy.

Determining an adequate cell size is just as important in your GIS application planning stages as determining what datasets to obtain. A raster dataset can always be resampled to have a larger cell size; however, you will not obtain any greater detail by resampling your raster to have a smaller cell size. Chapter 3 discusses cell size and image spatial resolution in more detail.

How Is Imagery Used in a GIS?

The three primary uses of imagery in a GIS are
1. as a base image to aid the visualization of map information, as shown in figure 2.11
2. as an attribute of a feature. For example, an image of vegetation taken from the ground may serve as an attribute of a vegetation survey point displayed on a map, as shown in figure 2.12
3. as a data source from which information is extracted through the process of image classification. For example, imagery may be interpreted by image analysis to determine the current state of situations for disaster response, environmental monitoring, or military planning. Imagery can also be transformed into informational map classes through manual interpretation or semi-automated classification.

The focus of much of this book is on the third use—image classification, which is the process of utilizing imagery in a GIS to produce maps.

Figure 2.11. Imagery as a base image. This figure shows airborne infrared imagery as a base image with parcel boundaries (in yellow) and field data points (in green). (esriurl.com/IG211). Source: Sonoma County Agriculture Preservation and Open Space District

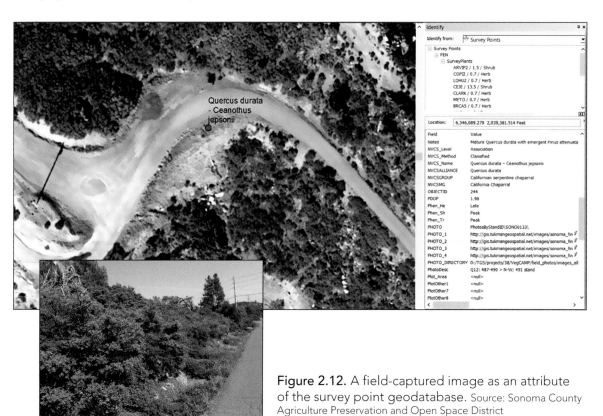

Figure 2.12. A field-captured image as an attribute of the survey point geodatabase. Source: Sonoma County Agriculture Preservation and Open Space District

22 Chapter 2 : Thinking About Imagery

Image Classification—Turning Data into Map Information

To simplify and make sense of our world, humans classify the continuous stream of data received by our sensory system—our eyes, ears, tongue, nose, and skin. We receive the *data* and our brains turn it into *information*. For example, if we see a four-legged animal, shorter than 1 meter, with a long snout and canine teeth, we might identify it as a dog, wolf, or coyote. If we determine it is a dog and the dog is growling, with its hackles up and its teeth bared, we know it is a threatening dog. If the dog is wagging its tail and lowering its body into a submissive posture, we know that it is a friendly dog. Dog, wolf, coyote, threatening, and friendly are all categories of information our brains determine from the data we receive.

When we see an image, our brains immediately start to explore and classify it. We identify features and note how they are related to one another. In a GIS system, when the data of an image is "classified," it is converted from continuous *data* into either continuous or categorical *information* and a map is created. Table 2.1 below provides an overview of the differences between continuous data such as an image, and continuous and categorical information which are derived from imagery.

Table 2.1. Overview of the differences between continuous data, continuous information, and categorical information

Property	Continuous Data	Continuous information	Categorical information
Pixel or point values	Are measurements of something, such as elevation or spectral reflectance	Are interpolations of data points to create a continuous raster model	Represent discreet classes, such as "forest," "grass," or "water." Classes are usually nouns.
Value type	Continuous	Continuous	Thematic
Range of values	Integer or floating point values with hundreds to billions of possible values	Integer or floating point values with hundreds to billions of possible values	Integer values with a range of values large enough to store all classes. Typically, the number of values is between two and several hundred.
Number of bands	One or more	Only one	Only one

Types of Maps Created from Imagery

Three types of maps are produced from the classification of imagery: digital elevation models (DEMs) and their derivatives, thematic raster and vector maps, and maps of feature locations.

Digital Elevation Models

DEMs provide continuous information about the elevation of the earth—either its bare surface without vegetation or structures, or the elevation of its terrain including the height of the vegetation and structures. DEMs can be created from survey point data or from points collected from imagery. The ability to create DEMs across large areas from imagery offers distinct advantages over using much more labor-intensive and expensive ground surveys to produce DEMs. DEMs and their derivatives, such as slope and aspect, are among the most commonly used geospatial data layers.

Thematic Vector and Raster Maps

A thematic map is a vector or raster map of themes such as land-cover types, soil types, land use, or forest types. Thematic map classes are discrete, not continuous. A thematic map covers the entire area of the landscape and labels everything into thematic classes. Figure 2.6 is an example of a thematic map of land-cover types for an area of the Coastal Watershed in southeastern New Hampshire. Thematic maps are created through manual interpretation of imagery or semiautomated image classification.

Feature Maps

A subset of thematic maps is feature maps. Rather than label the entire landscape, feature maps identify only a single object type, resulting in a binary map in which the feature is located and identified, and everything else is mapped as null; not that feature. Often, the feature of interest is a very specific type of object such as an airplane, military vehicle, or other unique entity that is out of place and unexpected in a particular environment. Sometimes, the objects of interest are common objects such as water bodies, roads, or buildings. Feature extraction is usually performed manually, but computer algorithms have also been developed to automatically extract features. Usually, automated feature extraction results in a number of false positives (i.e., the location of points that are not the feature of interest), which are then manually reviewed and corrected.

Imagery Workflows

Incorporating imagery in a GIS requires first deciding how you want to use the imagery. Is it as a base image, as an attribute of a feature, or to make a map? If your goal is to make a map, you must relate the objects on the imagery to features on the ground. To do so, four steps must be completed. You must

1. understand and characterize the variation on the ground that you want to map,
2. control variation in the imagery not related to the variation on the ground,
3. link variation in the imagery to variation on the ground, and
4. capture the variation in the imagery and other data sets as your map information.

First, you must decide how you want to characterize the phenomena on the earth that you want to identify, analyze, and display on the map; i.e., *you need to understand the variation on the ground that you want to capture on the map*. Once you understand the variation on the ground, you will need to create a set of rules that classify the variation on the ground into meaningful categories for your proposed uses of the map. It is the map categories and proposed uses that will drive your choice of what type of imagery to acquire for your project. Knowing how to best make that choice is the objective of chapters 3 and 4. Knowing how to build a rigorous classification scheme is the objective of chapter 7.

Next, you must work with your imagery in your GIS, register it to the ground, and remove or manage any spurious variation in the imagery caused by clouds, cloud shadows, or atmospheric conditions that could likely lead to map errors; i.e., *you need to control unwanted variation in the imagery*. Chapter 5 reviews working with imagery in ArcGIS, and chapter 6 discusses registering imagery to the ground and dealing with unwanted image variation.

Third, you must understand the variation in the imagery and how it relates to the variation you want to map; i.e., *you must link variation in the imagery to variation on the ground*. To do so, you will inspect the imagery to understand how the image object elements of color/tone, shape, size, pattern, shadow, texture, location, context, height, and date vary across the landscape. There are analytics you can perform on the imagery to discover how well the imagery varies with the classes you want to map, and you may decide to manipulate the imagery data to produce indices or derivative bands that help derive more information from the imagery. You may discover that some of the variation on the ground that you want to map cannot be derived from the imagery. In that case, you must discover other data sources (i.e., ancillary data), such as DEMs, that will help you make the map. Creating DEMs and their derivatives is the topic of chapter 8. Understanding how to link variation in the imagery to variation on the ground is the learning objective of chapter 9.

Fourth, you will classify the imagery to create maps of digital elevation, feature locations, or thematic landscape classes by *capturing the variation in the imagery and ancillary data that is related to your map classes*. This work may be performed manually or with the help of a computer. There are many methods of classifying imagery. Explaining those methods and describing how to choose which method to use are the objectives of chapters 10 and 11. Once the image is classified into a map, you will want to assess the map's accuracy, which is the topic of chapter 12. Finally, you may want to publish your imagery and maps, which is the topic of chapter 13.

Chapter 3
Imagery Fundamentals

Introduction

Imagery is collected by remote sensing systems managed by either public or private organizations. It is characterized by a complex set of variables, including
- collection characteristics: image spectral, radiometric, and spatial resolutions, viewing angle, temporal resolution, and extent; and
- organizational characteristics: image price and licensing and accessibility.

The choice of which imagery to use in a project will be determined by matching the project's requirements, budget, and schedule to the characteristics of available imagery. Making this choice requires understanding what factors influence image characteristics. This chapter provides the fundamentals of imagery by first introducing the components and features of remote sensing systems, and then showing how they combine to influence imagery collection characteristics. The chapter ends with a review of the organizational factors that also characterize imagery. The focus of this chapter is to provide an understanding of imagery that will allow the reader to 1) rigorously evaluate different types of imagery within the context of any geospatial application, and 2) derive the most value from the imagery chosen.

Collection Characteristics

Image collection characteristics are affected by the remote sensing system used to collect the imagery. Remote sensing systems comprise *sensors* that capture data about objects from a distance, and *platforms* that support and transport sensors. For example, humans are remote sensing systems because our bodies, which are platforms, support and transport our sensors—our eyes, ears, and noses—which detect visual, audio, and olfactory data about objects from a distance. Our brains then identify/classify this remotely sensed data into information about the objects. This section explores sensors first, and then platforms. It concludes by discussing how sensors and platforms combine to determine imagery collection characteristics.

A platform is defined by the *Glossary of the Mapping Sciences* (ASCE, 1994) as "A vehicle holding a sensor." Platforms include satellites, piloted helicopters and fixed-wing aircraft, unmanned aerial systems (UASs), kites and balloons, and earth-based platforms such as traffic-light poles and boats. Sensors are defined as devices or organisms that respond to stimuli. *Remote sensors* reside on platforms and respond "to a stimulus without being in contact with the source of the stimulus" (ASCE, 1994). Examples of remote sensing systems include our eyes, ears, and noses; the camera in your phone; a video camera recording traffic or ATM activity; sensors on satellites; and cameras on UASs, helicopters, or airplanes.

Imagery is acquired from terrestrial, aircraft, marine, and satellite platforms equipped with either analog (film) or digital sensors that measure and record electromagnetic energy.[1] Because humans rely overwhelmingly on our eyes to perceive and understand our surroundings, most remote sensing systems capture imagery that extends our ability to see by measuring the electromagnetic energy reflected or emitted from an object. Electromagnetic energy is of interest because different types of objects reflect and emit different intensities and wavelengths of electromagnetic energy, as shown in figure 3.1. Therefore, measurements of electromagnetic energy can be used to identify features on the imagery and to differentiate diverse classes of objects from one another to make a map.

---

[1] Most remote sensing systems record electromagnetic energy, but some, such as sonar systems, record sound waves.

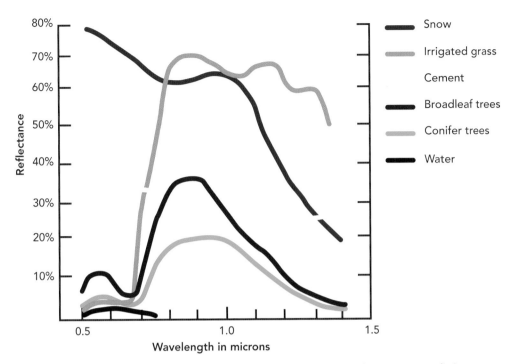

Figure 3.1. Comparison of example percent reflectance of different types of objects across the electromagnetic spectrum (esriurl.com/IG31)

The type of sensor used to capture energy determines which portions of the electromagnetic spectrum the sensor can measure (the imagery's spectral resolution) and how finely it can discriminate between different levels of energy (its radiometric resolution). The type of platform employed influences where the sensor can travel, which will affect the temporal resolution of the imagery. The remote sensing system—the combination of the sensor and the platform—impacts the detail perceivable by the system, the imagery's spatial resolution, the viewing angle of the imagery, and the extent of landscape viewable in each image.

Sensors

This section provides an understanding of remote sensors by examining their components and explaining how different sensors work. As mentioned in chapter 1, a wide variety of remote sensors have been developed over the last century. Starting with glass-plate cameras and evolving into complex active and passive digital systems, remote sensors have allowed us to "see" the world from a superior viewpoint.

All remote sensors are composed of the following components, as shown in figure 3.2:
- Devices that capture either electromagnetic energy or sound, either chemically, electronically, or biologically. The devices may be imaging surfaces (used mostly in electro-optical imaging) or antennas (used in the creation of radar and sonar images).
- Lenses that focus the electromagnetic energy onto the imaging surface.
- Openings that manage the amount of electromagnetic energy reaching the imaging surface.
- Bodies that hold the other components relative to one another.

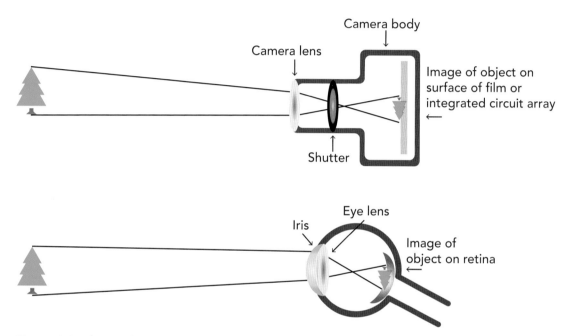

Figure 3.2. The similar components of the human eye and a remote sensor

Our eyes, cameras, and the most advanced passive and active digital sensors fundamentally all work the same way. Electromagnetic energy passes through the opening of the sensor body where it reaches a lens that focuses the energy onto the imaging surface. Our brains turn the data captured by our retinas into information. Similarly, we convert remotely sensed image data into information through either manual interpretation or semi-automated image classification.

Imaging Surfaces

Imaging surfaces measure the electromagnetic energy that is captured by digital sensors such as a charged coupled device (CCD) or a complementary metal-oxide-semiconductor (CMOS) array. The wavelengths of energy measured are determined by either filters or dispersing elements placed between the sensor opening and the imaging surface. The energy is generated either passively by a source (such as the sun) other than the sensor, or actively by the sensor.

The Electromagnetic Spectrum

Most remote sensing imaging surfaces work by responding to photons of electromagnetic energy. Electromagnetic energy is caused by the phenomenon of photons freeing electrons from atoms. Termed the photoelectric effect, it was first conceptualized by Albert Einstein, earning him the Nobel Prize in physics in 1921.

Electromagnetic energy occurs in many forms, including gamma rays, x-rays, ultraviolet radiation, visible light, infrared radiation, microwaves, and radio waves. It is characterized by three important variables: 1) speed, 2) wavelength, and 3) frequency. The speed of electromagnetic energy is a constant of 186,000 miles/second, or 3×10^8 meters/second, which is the speed of light. Wavelength is the distance between the same two points on consecutive waves and is commonly depicted as the distance from the peak of one wave to the peak of the next, as shown in figure 3.3. Frequency is the number of wavelengths per unit time.

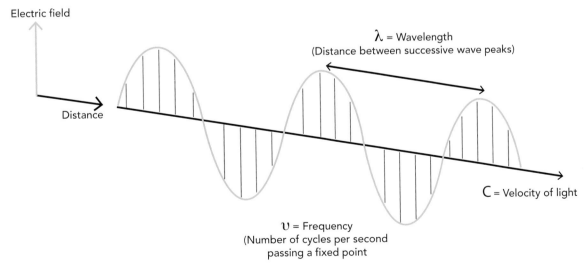

Figure 3.3. Diagram demonstrating the concepts of electromagnetic wavelength and frequency

The relationship between wavelength, wave speed, and frequency is expressed as

Wave Speed = Wavelength × Frequency.

Because electromagnetic energy travels at the constant speed of light, when wavelengths increase, frequencies decrease, and vice-versa (i.e., they are inversely proportional to each other). Photons with shorter wavelengths carry more energy than those with longer wavelengths. Remote sensing systems capture electromagnetic energy emitted or reflected from objects above 0 degrees Kelvin (absolute 0).

Electromagnetic energy is typically expressed as either wavelengths or frequencies. For most remote sensing applications, it is expressed in wavelengths. Some electrical engineering applications such as robotics and artificial intelligence express it in frequencies. The entire range of electromagnetic wavelengths or frequencies is called the electromagnetic spectrum and is shown in figure 3.4.

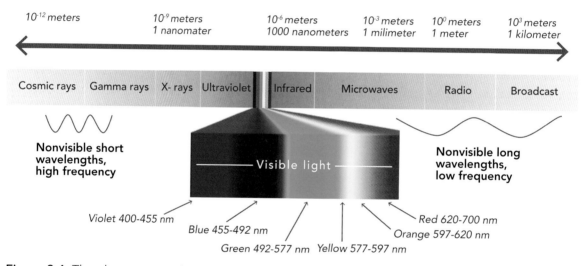

Figure 3.4. The electromagnetic spectrum

The most significant difference between our eyeballs and digital cameras is how the imaging surfaces react to the energy of photons. As shown in figure 3.4, the retinas in human eyes sense only the limited visible light portion of the electromagnetic spectrum. While able to capture more of the spectrum than human eyes, film is limited to wavelengths from 0.3 to 0.9 micrometers (i.e., the ultraviolet, visible, and near infrared). CCD or CMOS arrays in digital sensors are sensitive to electromagnetic wavelengths from 0.2 to 1400 micrometers. Because remote sensors extend our ability to measure more portions of the electromagnetic spectrum than our eyes can sense, remote sensors extend our ability to "see."

Film versus Digital Array Imaging Surfaces

The imaging surfaces of our eyes are our retinas. Cameras once used only film, but now primarily use digital (CCD or CMOS) arrays. From its beginnings in the late 1800s to the 1990s, most remote sensing sensors relied on film to sense the electromagnetic energy being reflected or emitted from an object. Classifying the resulting photographs into information required manual interpretation of the photos. In the 1960s, digital sensors were developed to record electromagnetic energy as a database of numbers rather than a film image. This enabled the development of sensors that can sense electromagnetic energy across the range from ultraviolet to radio wavelengths. Now, most remote sensing systems use digital arrays instead of film. Because the values of the reflected and emitted energy are stored as an array of numbers, computers can be trained to turn the imagery data into map information by discovering correlations between variations in the landscape and variations in electromagnetic energy. While manual interpretation is still very important, objects that are spectrally distinct from one another can be readily mapped using computer algorithms.

The imaging surface of a digital camera is an array of photosensitive cells that capture energy from incoming photons. Each of these cells corresponds to a *pixel* in the resulting formed image. The pixels are arranged in rectangular columns and rows. Each pixel contains one to three photovoltaic cells or photosites, which use the ability of silicon semiconductors to translate electromagnetic photons into electrons. The higher the intensity of the energy reaching the cells during exposure, the higher the number of electrons accumulated. The number of electrons accumulated in the cell is recorded and then converted into a digital signal.

The size of the array and the size of each cell in the array affect the resolving power of the sensor. The larger the array, the more pixels captured in each image. Larger cells accumulate more electrons than smaller cells, allowing them to capture imagery in low-energy situations. However, the larger cells also result in a corresponding loss of spatial resolution across the image surface because fewer cells can occupy the surface.

Source of Energy: Active versus Passive Sensors

Passive sensors collect electromagnetic energy generated by a source other than the sensor. Active sensors generate their own energy, and then measure the amount reflected back as well as the time lapse between energy generation and reception. Figure 3.5 illustrates the difference in how active and passive sensors operate.

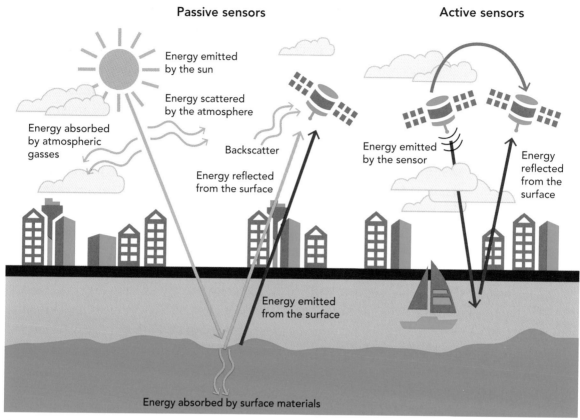

Figure 3.5. Comparison of how passive and active sensors operate

Most remote sensors are passive sensors, and the most pervasive source of passive electromagnetic energy is the sun, which radiates electromagnetic energy upon objects on the earth that either absorb/emit, transmit, or reflect the energy. Passive energy can also be directly emitted from the earth, as from the eruption of a volcano or a forest fire. Examples of passive remote sensors include film aerial cameras, multispectral digital cameras, and multispectral/hyperspectral scanners. Passive sensors are able to sense electromagnetic energy in wavelengths from ultraviolet through radio waves.

Passive sensors fall into three types: framing cameras, across-track scanners, and along-track scanners. Framing cameras either use film or matrixes of digital arrays (e.g., UltraCam airborne sensors, PlanetLabs satellite sensors). Each frame captures the portion of the earth visible in the sensor's field of view (FOV) during exposure. Often, the frames are captured with greater than 50 percent overlap, which enables stereo viewing. Each image of a stereo pair is taken from a slightly different perspective as the platform moves. When two overlapped images are viewed side by side, each eye automatically takes the perspective of each image, enabling us to now "see" the overlapped areas in three dimensions. With stereo frame imaging, not only can distances be measured from the aerial images, but so can elevations and the heights of vegetation and structures, discussed in detail in chapter 9.

Most across-track scanners (also called whisk broom scanners) move an oscillating mirror with a very small instantaneous field of view (IFOV) side to side as the platform moves. Each line of the image is built, pixel by pixel, as the mirror scans the landscape. Developed decades before the digital frame camera, across-track scanners were the first multispectral digital sensors and were used in multiple systems including the Landsats 1-7, GOES, AVHRR, and MODIS satellite sensors, and NASA's AVIRIS hyperspectral airborne system.

Along-track scanners (also called push broom scanners) rely on a linear array to sense entire lines of data simultaneously. Rather than mechanically building an image pixel by pixel or by groups of pixels, the along-track scanner builds an image line by line. Along-track scanners have higher spectral and radiometric resolution than across-track scanners because the sensor can spend more time (termed dwell time) over each area of ground being sensed. Like across-track scanners, along-track scanners often also use a dispersing element to split apart the incoming beam of electromagnetic energy into distinct portions of the electromagnetic spectrum to enable the collection of multispectral imagery. Developed 30 years ago, along-track scanners are a more recent development than across-track scanners. Many multispectral satellite systems (e.g., WorldView-3, Landsat 8) rely on along-track sensors, as do the Leica Airborne Digital Sensors.

Active sensors send out their own pulses of electromagnetic energy, and the sensor measures the echoes or returns of the energy as they are reflected by objects in the path of the pulse. For example, consumer cameras with flash attachments are active systems. Active remote sensors include lidar (light detection and ranging) systems, which generate laser pulses and sense electromagnetic energy in the ultraviolet to near-infrared regions of the spectrum, and radar (radio detection and ranging) systems, which generate and sense energy in the microwave range. An advantage of active systems is that they do not rely on the sun, so acquisitions can be made at times when the sun angle is low or at night. An additional advantage of radar systems is that the long wavelengths of microwaves can penetrate clouds, haze, and even light rain.

Wavelengths Sensed

Passive Sensors

Most images are collected by panchromatic or multispectral passive sensors that are able to sense electromagnetic energy in the visible through infrared portions of the electromagnetic spectrum. To separate different optical and midinfrared wavelengths from one another, passive remote sensors place filters or dispersing elements between the opening and the imaging surface to split different wavelengths or "bands" of the electromagnetic spectrum from one another. Filters are usually used with framing cameras and include the following:

- Employing a Bayer filter over the digital array, which restricts each pixel to one portion of the electromagnetic spectrum, but alternates pixels in the array to collect at different wavelengths. The computer then interpolates the values of the non-sensed wavelengths from the surrounding pixels to simulate their values for each frequency at each pixel. This is how consumer cameras and many of the high-resolution small satellite constellations (e.g., Planet Doves) collect multispectral imagery.
- Placing separate filters on multiple cameras, each filtered to accept energy from a distinct portion of the electromagnetic spectrum, allows each focal plane to be optimized for that portion of the spectrum. Many four-band (red, green, blue, and infrared) airborne image sensors (e.g., Microsoft Ultracam and Leica DMC sensors) use this approach, which requires that the images simultaneously captured with the separate cameras be coregistered to one another after capture.
- Placing a spinning filter wheel in front of one camera so that each exposure of the image surface is in one portion of the electromagnetic spectrum. This approach is very useful for fixed platforms, however it requires very complex postcollection registration for systems with moving platforms and is rarely used in remote sensing systems.

Alternatively, a dispersing/splitting element can be placed between the lens and a series of CCD arrays to split the incoming energy into its discrete portions of the electromagnetic spectrum. Many multispectral and most hyperspectral sensors employ dispersing/splitting elements (e.g., Leica Airborne Digital Sensors, NASA AVIRIS).

Figures 3.6 to 3.8 illustrate how Bayer filters, framing cameras, and dispersing elements are typically used to create multispectral images. In general, because pixel values are interpolated for two values out of every three, Bayer filters will always have lower spectral resolution than multiheaded frame cameras or systems using dispersing elements.

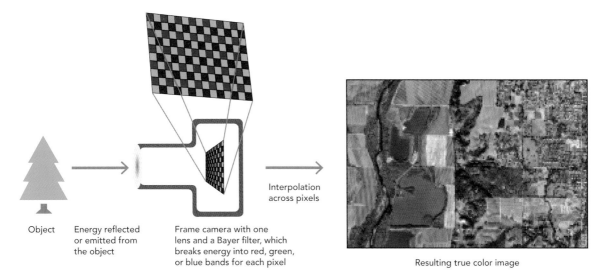

Figure 3.6. How a Bayer filter framing camera system works. While the figure shows a true color image, Bayer filters can also be used to collect in the near-infrared portions of the electromagnetic spectrum, resulting in infrared imagery.

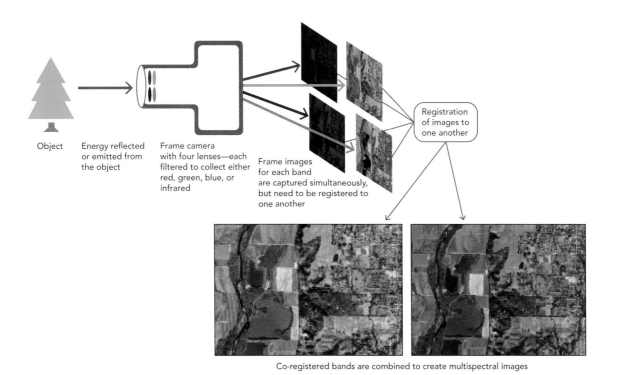

Figure 3.7. How a multilens multispectral framing camera system works

Chapter 3 : Imagery Fundamentals   37

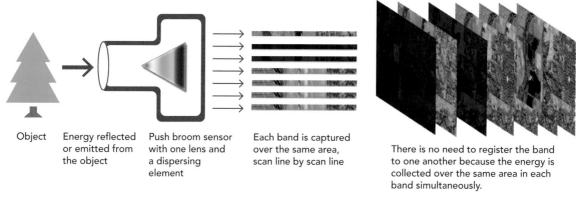

Figure 3.8. How a push broom multispectral scanner works with a dispersing element

Active Sensors

The most common active remote sensors are lidar and radar systems. As mentioned earlier, all active instruments work similarly by transmitting electromagnetic energy that is bounced back to the sensor from the surface of the earth. Because active sensors generate their own energy, they can capture imagery at any time of the day or night.

Radar imagery is often used to create digital surface and digital elevation models over large regions, and to map sea or land cover in perpetually cloudy areas where optical imagery can't be effectively collected. Figure 3.9 shows an example of a radar image of Los Angeles, California. Radar imagery is collected over a variety of microwave bands, which are denoted by letters and measured in centimeters as follows: Ka, 0.75 to 1.1 cm; K, 1.1 to 1.67 cm; Ku, 1.67 to 2.4 cm; X, 2.4 to 3.75 cm; C, 3.75 to 7.5 cm; S, 7.5 to 15 cm; L, 15 to 30 cm; and P, 30 to 100 cm. Usually, radar imagery is collected in just one band, resulting in a single band image. Bands X, C, and L are the most common ranges used in remote sensing. Some radar systems are able to collect imagery in several bands, resulting in multispectral radar imagery.

Varying antenna lengths are required to create the radar signal at these different wavelengths. Because it is often not viable to have a long antenna on a platform moving through the air or space, the length of the antenna is extended electronically through a process called synthetic aperture radar.

Radar signals can also be transmitted and received in either horizontal or vertical polarizations or a combination of both. HH imagery is both transmitted and received in a horizontal polarization, and VV imagery is both transmitted and received in a vertical polarization (i.e., like-polarized). HV imagery is transmitted horizontally and received vertically, and VH imagery is transmitted vertically and received horizontally (i.e., cross-polarized). The different polarizations can be combined to create a multipolarized image, which is similar to a multispectral image as each polarization collects different data about the ground.

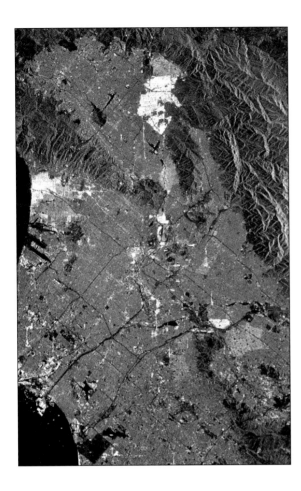

Figure 3.9. An example radar image captured over Los Angeles, California (esriurl.com/IG39). Source: NASA

Over the last 20 years in much of the world, airborne lidar has surpassed photogrammetric methods for measuring the 3-dimensional world. Lidar imagery is used to develop digital elevation models (DEMs), digital terrain models (DTMs), digital surface models (DSMs), digital height models (DHMs), elevation contours, and other derived datasets (chapter 8 provides more detail on the creation of DEMs). Additionally, NASA uses low-spatial-resolution satellite lidar to monitor ice sheet mass balance and aerosol heights and has recently initiated the Global Ecosystem Dynamics Investigation (GEDI) mission, which will result in the first global, moderate-spatial-resolution, spaceborne topographic lidar (**http://science.nasa.gov/missions/gedi/**).

Lidar sensors emit discrete pulses of electromagnetic energy that illuminate a given spot on the earth for an instant (less than 1/100,000 of a second). The energy emitted can be of ultraviolet through near-infrared wavelengths (250 nm to 10 μm), which are much shorter than those of radar pulses. The pulses of light then bounce back and are recaptured by the lidar instrument where the durations of their paths are recorded and analyzed to extract elevation information. The number of returns per unit area for discrete return lidar can be much higher than the number of pulses sent earthward, because each pulse can have multiple (typically three to five) returns.

There are two types of airborne lidar: topographic and bathymetric. Topographic lidar uses an infrared laser to measure elevations across the surface of the earth. Bathymetric lidar employs green laser light to penetrate water and measure the depth of water bodies. In topographic lidar, pulses of light encounter porous objects, such as vegetation, which will have multiple returns. For example, as shown in figure 3.10, a selected single pulse from this discrete return airborne lidar system has three returns from branches and a fourth return (the final return) from the ground. DTMs are generated from the last returns, DSMs from the first returns (buildings must be removed using specialized algorithms), and DHMs from the difference between the digital surface model and the digital terrain model. Lidar returns collectively form a lidar "point cloud" consisting of millions to billions of points that each contain the point's latitude, longitude, and elevation.

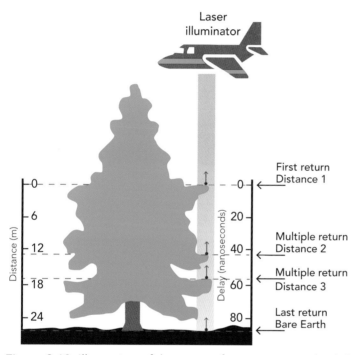

Figure 3.10. Illustration of the returns from a topographic lidar system. Source: Dr. Maggi Kelly

Lidar point density is measured by the average number of pulses sent downward from the aircraft per square meter of ground. As of this writing, "high density" airborne lidar is generally considered to have a point density of greater than eight points per square meter. In vegetated terrain, only a fraction of the pulses of light sent earthward by the lidar system penetrate all the way to the ground, and the number of ground returns decreases as the thickness of the vegetated canopy increases. The lack of ground returns in thickly vegetated areas can lead to inaccuracy in the digital terrain models derived from a lidar dataset. For this reason, the effective resolution of the digital terrain model and the digital height model depend on the point density of the lidar data. The higher the point density, the more ground returns and the higher the resolution of the derived DHM and DTM. It is recommended that lidar data be collected at a point density of at least eight pulses per square meter in project areas with dense forests. Eight pulses per square meter is the minimum point density that meets the US Geological Survey's (USGS) quality level 1 lidar data specification.[2] Figure 3.11 compares hillshades derived from a digital terrain models at USGS quality level 1 versus USGS quality level 2 lidar data, illustrating the enhanced detail and resolution gained by collecting lidar data at higher density.

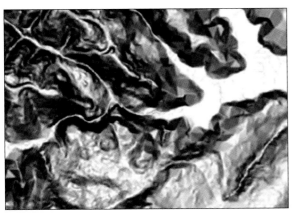 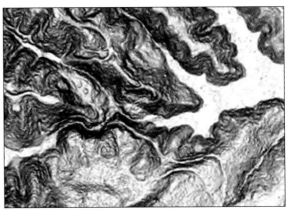

1.2 pulses/m² (0.91 meter post spacing)   8.0 pulses/m² (0.35 meter post spacing)

Figure 3.11. Comparison of a hillshade derived from 1.2 pulses/m² lidar to one derived from eight pulses/m² lidar. Source: Quantum Geospatial, Inc.

There are two common types of airborne topographic lidar: discrete return and waveform. Discrete return lidar provides elevation values at the peak intensity of each return. Typically, a maximum of between three and five returns is possible where there is vegetation, but only one return will occur in open areas. Each of the multiple returns is stored as a point in the point cloud, with its associated latitude, longitude, and elevation.

Full waveform lidar—which is mostly still in the R&D phase—provides the entire "waveform" graph associated with a lidar pulse. Because it records the entire waveform of a lidar pulse's returns and not just three to five discrete peaks, waveform lidar requires 30 to 50 times the amount of data storage as discrete return lidar.

Historically, lidar systems have been able to transmit energy in only one wavelength. However, recent advancements in lidar technology allow for transmitting energy in multiple wavelengths, making multispectral lidar images possible (Teledyne Optech Titan

---

[2] Hans Karl Heidemann, "Lidar Base Specification," ver. 1.2, November 2014, US Geological Survey Techniques and Methods, book 11, chap. B4, https://pubs.usgs.gov/tm/11b4/pdf/tm11-B4.pdf.

system). Additionally, new technologies such as Geiger-mode (Harris) and Single Photon (SigmaSpace/Hexagon) have been introduced that significantly improve the rate of data collection and resulting point density by increasing the sensitivity of the lidar sensors.

Lenses

Objects emit or reflect electromagnetic energy at all angles. The angles between an object and an imaging surface change as the imaging surface moves closer to or farther from the object. The purpose of a lens in a camera or in an eyeball is to focus the electromagnetic energy being emitted or reflected from the objects being imaged onto the imaging surface. By moving the lens back and forth relative to the imaging surface, we can affect the angle of electromagnetic energy entering and exiting the lens, and thereby bring the objects of interest into focus.

Most remote sensing systems capture electromagnetic energy emitted or reflected from objects at a great distance from the sensor (i.e., at an effectively infinite distance), from hundreds of feet for a sensor in an aircraft to hundreds of miles for a sensor in a satellite. Because these distances approach infinity relative to the focal length, the lenses have a fixed focus.

The combination of the sensor's lens and the resolution of the imaging surface will determine the amount of detail the sensor is able to capture in each image—its resolving power. The resolution of a digital image is determined by the format size of the digital array of the imaging surface.

Openings

The purpose of a sensor opening is to manage the photons of electromagnetic energy reaching the imaging surface. Too large an opening results in the imagery being saturated with photons, overexposing the imaging surface. Too small an opening results in not enough photons captured to create an image.

Our irises manage the amount of light reaching our retinas by expanding and shrinking to let more or less light onto our retinas. In a camera, the diameter of the opening that allows electromagnetic energy to reach the imaging surface is called the aperture, and the speed at which it opens and closes is called the shutter speed. Together, aperture and shutter speed control the exposure of the imaging surface to electromagnetic energy. In a digital camera, the CCD array is read and cleared after each exposure.

Bodies

Remotely sensed imagery can be used for visualization—to obtain a relative concept of the relationship of objects to one another—or to measure distances, areas, and volumes. For either visualization or measurement, the geometry of the lenses, opening, and imagery surface within the camera body must be known. In addition, for measurement the location and rotation of the imagery surface when the image is captured must also be known.

Sensor Summary

While remote sensor components share similarities with our eyes and consumer cameras, they differ in the following fundamental ways:
- Imaging surfaces must be absolutely flat to minimize any geometric distortion.
- The energy sensed may be passively received by the sensor from another source (commonly the sun) or actively created by the sensor and then received back by the sensor.
- Because most remotely sensed images are taken from high altitudes, their lenses are commonly designed for an infinite object distance; i.e., the lenses are fixed.
- Shutter speeds are usually extremely fast because most platforms are moving at high speeds.
- Remote sensor camera bodies must be able to withstand the extreme temperatures and vibrations encountered by the vehicle, boat, aircraft, or satellite platform. Additionally, for mapping purposes, the precise internal geometry of the sensor components within the body must be known as well as the location of the imaging surface when an image is collected so that the imagery can be accurately terrain corrected and georeferenced to the earth.

Platforms

This section reviews remote sensing platforms by examining platform features. Seven major features distinguish platforms from one another: whether they are manned or unmanned, and their altitude, speed, stability, agility, and power.

Different Types of Platforms

Syncom

Geosynchronous—22,236 miles
Satellites that match Earth's rotation appear stationary in the sky to ground observers. While most commonly used for communications, geosynchronous orbiting satellites like the hyperspectral GIFTS imager are also useful for monitoring changing phenomena such as weather conditions. NASA's Syncom, launched in the early 1960s, was the first successful "high flyer."

Landsat 8

Sun synchronous—375-500 miles
Satellites in this orbit keep the angle of sunlight on the surface of the earth as consistent as possible, which means that scientist can compare images from the same season over several years, as with Landsat imagery. This is the bread-and-butter zone for earth observing sensors.

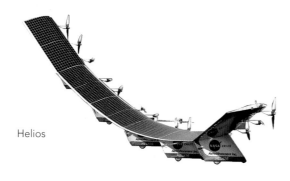
Helios

Atmospheric satellite—100,000 feet
Also known as pseudo-satellites, these unmanned vehicles skim the highest edges of detectable atmosphere. NASA's experimental Helios craft measured solar flares before crashing in the Pacific Ocean near Kauai.

SR71 Blackbird

Jet aircraft—90,000-30,000 feet
Jet aircraft flying at 30,000 feet and higher can be flown over disaster areas in a very short time, making them a good platform for certain types of optical and multispectral image applications.

44 Chapter 3 : Imagery Fundamentals

Cessna

Ultralight

US Navy Silver Fox

Helicopter

General aviation aircraft—100-10,000 feet
Small aircraft able to fly at low speed and low altitude have long been the sweet spot for high-quality aerial and orthophotography. From Cessnas to ultralights to helicopters, these are the workhorse of optical imagery.

3DR Solo private drone

Drones—100-500 feet
Drones are the new kid on the block. Their ability to fly low, hover, and be remotely controlled offer attractive advantages for aerial photography, with resolution down to sub-1 inch. Military UAVs can be either smaller drones or actual airplanes.

Smartphone   Handheld spectrometer

Ground based/handheld—ground level
Increasingly, imagery taken at ground level is finding its way into GIS workflows. Things like Google Street View, HERE street-level imagery, and Mapillary; handheld multispectral imagers; and other terrestrial sensors are finding applications in areas like pipelines, security, tourism, real estate, natural resources, and entertainment.

Street-level mapping car

Chapter 3 : Imagery Fundamentals   45

Piloted or Unpiloted

Until recently, most satellite platforms were unpiloted, and most airborne platforms were piloted. However, with the advent of unmanned aerial vehicles, most airborne platforms are now unpiloted, but piloted aircraft still capture much larger areas than unpiloted platforms. While less used, piloted satellite platforms have been very important in remote sensing. Starting with the Apollo space mission in the late 1960s and continuing with the International Space Station today, piloted satellites have completed many successful remote sensing missions including NASA's Shuttle Radar Topography Mission, which generated global digital elevation models of the earth from 56 degrees south to 60 degrees north. In 2019, the International Space Station will deploy the GEDI lidar to produce a 3D map of the earth's forests.

Most of the areas captured by airborne platforms used for mapping today are flown over by a pilot residing in the platform. UASs are either autonomous or have a pilot operating them from the ground. Originally developed and used by the military, the use of UASs in civilian markets is exploding because of their low cost, their ability to collect imagery over inaccessible or dangerous areas, and their ability to fly low and slow, enabling the capture of high-resolution imagery over small areas that would be too expensive to capture with piloted aerial systems. Hobbyist use in the United States has skyrocketed since 2010, but commercial use was stalled because of cumbersome FAA regulations. In 2015, the FAA streamlined the process for gaining authorization to commercially operate UASs in the US, resulting in a 500 percent increase in applications in the first six months of 2015 over all of 2014 (Andelin and Andelin, 2015). Outside the United States, UAS use is also rapidly increasing with successful deployments to map archeological sites, establish property rights, monitor illegal resource extraction, and support disaster response (Pajares, 2015).

While civilian drones do not currently have the capacity to capture imagery over large areas, the use of UASs is likely to continue to rapidly expand and evolve. As stated in the primer *Drones and Aerial Observations*:

> *Technology will change. Faster processors will stitch together and georectify images more quickly. The acuity of photographic sensors will improve, as will the endurance and range of drones. Increasing levels of autonomy in both flight software and post-processing software will allow for the creation of cheap maps with increasingly less direct human intervention (Kakaes et al., 2015).*

Altitude

Altitude is an object's height above sea level. The altitude of a remote sensing platform can vary between below sea level (in bathymetric projects) to more than 20,000 miles above sea level. Remote sensing platforms are classed into three types based on their range of distance from the earth:
1. Terrestrial and marine platforms, including elevated work platforms, mobile vehicles, buildings and towers, lampposts, buoys, boats, and humans.
2. Airborne platforms including UASs, fixed-wing aircraft, helicopters, and balloons.
3. Spaceborne platforms, which are either geostationary or orbit the earth.

Terrestrial platforms operate from beneath the ocean to the highest buildings on earth and may be fixed (e.g., ATM video cameras) or mobile (e.g., cars and boats). Airborne platforms fly within the earth's atmosphere up to an altitude of typically 9.5 miles (15.3 kilometers) and include fixed-wing aircraft, UASs, helicopters, and balloons. Fixed-wing aircraft are the most common type of remote sensing platform and are used by many private companies and governments for imaging purposes. High-altitude piloted aircraft platforms have pressurized cabins, enabling them to fly as high as 50,000 feet above sea level. Low-altitude piloted aircraft platforms operate at altitudes up to 30,000 feet (5.7 miles), but are generally used to collect data at lower elevations to gain higher spatial resolution. The hovering ability of helicopters (below 500 feet and up to 12,500 feet) allows them to collect imagery at lower speeds than fixed-wing aircraft. Balloons have a wide range of achievable altitudes, from as low as needed for a tethered balloon to around 20 km or more for a blimp. UASs can be fixed- or rotor-winged with altitudes ranging from very close to the ground to very high in the air.

At the highest altitudes, earth observation satellites carry remote sensors around the earth in orbit at altitudes ranging from 100 to over 22,000 miles above sea level. Maintaining orbital altitude is a constant requirement for satellites because of the earth's steady gravitational pull and atmospheric drag. Lower satellites must travel at higher velocities because they experience greater gravitational pull than satellites at higher altitudes. Thus, maintaining orbit requires a constant balance between gravity and the satellite's velocity. Satellites with fuel onboard maintain their orbital altitude by using the fuel to maintain their velocities. However, at some point all satellites fall back to earth and burn up in the atmosphere, usually in controlled descents.

Speed

Speed is the rate of motion of an object expressed as the distance covered per unit of time. It determines the level of detail and amount of area (extent) a remote sensing system can collect. The altitude and speed flown while collecting remotely sensed data are also determined by the desired resolution and coverage, as well as the sensor being used (e.g., digital or film camera, lidar). Remote sensing platform speeds can range from stationary (zero velocity) to over 17,000 miles per hour. Most terrestrial platforms are stationary. Mobile terrestrial platforms such as cars and boats tend to travel at low speeds to enable the collection of very-high-spatial-resolution imagery. Fixed-wing UASs and aircraft typically fly at 55 to 650 miles per hour. Helicopters and rotor UASs, with their ability to hover, typically fly at 0 to 150 miles per hour. The speed at which a satellite travels in orbit is determined by its altitude. The lower the altitude, the faster the satellite must travel to remain in orbit and not fall to earth. Satellites in near-circular orbits have near-constant speeds, while satellites in highly elliptical orbits will speed up and slow down depending on the distance from the earth and direction of motion.

Stability

Stability is the ability of an object to resist changes in position. Stability is an important feature of remote sensing platforms because platforms need to either maintain stability or precisely measure instability to ensure high-quality image capture and accurate registration of the image to the ground. The most stable platforms are fixed terrestrial platforms because they are structurally rigid and immobile, which also means that they have little or no agility. Satellite platforms are also relatively stable because they operate in the vacuum of space. Helicopters are less stable than fixed-wing aircraft because of the unequal lift and vibrations caused by the rotating blades. While balloons were an important platform in the early days of remote sensing, they are not widely used today because their flight is easily influenced by air currents and pressure changes resulting in minimal control of balloon flight path or position. Fixed-wing platforms are relatively stable airborne platforms. Because of this and their large range and speed, they remain the workhorse of airborne image collection.

Operating in the earth's atmosphere subjects aircraft to air pressure and wind variations that can result in changes in pitch, roll, and yaw (figure 3.11), causing a variety of displacements in the collected imagery. Pitch is rotation of the aircraft about the axis of the wings. Yaw is rotation about the axis that is perpendicular to the wings and directed at the nose and tail of the aircraft. Roll is rotation of the aircraft about the axis of the fuselage.

Figure 3.12. The effects of pitch, yaw, and roll on aircraft stability

Traditionally, aerial photography missions required the precise measurement of many ground control points in each photograph to establish the exact spatial position and orientation of the photograph relative to the ground at the moment the image was taken. In the late 1950s, a technique called bundle block adjustment was developed to reduce the number of expensive control points required. This was based on finding tie points between photographs and then solving least squares adjustment formulas. In the 1990s, the number of control points required was again reduced by the advent of accurate GPS positioning of the aircraft that effectively added control points in the air, further reducing the control required. The advent of lower-cost precise IMUs (inertial measurement units) has further reduced the number of control points required, so that for many applications sufficient accuracy can be achieved using only highly accurate GPS and IMU systems, which is referred to as direct georeferencing. These orientation parameters are used in image orthorectification (see chapter 6) to geometrically correct the images so that coordinates in the imagery accurately represent coordinates on the ground.

Agility

Agility refers to the ability of the platform to change position and can be characterized by 1) reach or the ability of a platform to position itself over a target, which is sometimes referred to as field of regard; 2) dwell time, which is how long the platform can remain in the target area working; and 3) the ability to slew across the target area.

Fixed platforms such as a traffic-light pole above a street intersection have no agility. Satellites are tied to their orbits, which restricts their agility. However, some satellites are pointable (e.g., able to slew off nadir), which makes them much more agile than nonpointable satellites. This, coupled with their ability to quickly orbit the earth, provides them with a long-range reach around the globe, which is not available to aircraft.

Within their range, aircraft and fixed-wing UASs are more agile than satellites, and helicopters are more agile than fixed-wing aircraft. The hovering abilities of helicopters and rotor-winged UASs allow them to obtain more target specific data than fixed-wing aircraft can collect, and they can more easily reach targets in a congested airspace. Blimps and remote-controlled balloons have greater mobility than hot-air balloons because they have engines and are more maneuverable.

Power

Power refers to the power source that runs the platform. The more powerful the engine or engines, the faster and higher the platform can travel and the greater payload it can carry. Satellites are propelled into space by launch vehicles to escape the earth's gravity. Afterward, they use electric power derived from solar panels for operation, and stored fuel for orbital maneuvering. Of critical importance is the amount of power remaining after launch for the sensor to operate. Size, weight, and power, coupled with communication bandwidth (the ability to offload the image from the focal plane) are the biggest drivers in satellite sensor design.

Fixed-wing aircraft are powered by piston engines, turbocharged piston engines, turboprops, or jet engines in single- or twin-engine configurations. High-altitude piloted aircraft platforms are usually powered by twin jet engines or turboprops. The high power of these aircraft and their ability to fly at high altitudes with large payloads results in large operational costs, but this can be offset by their broad spatial coverage abilities and fast data collection (Abdullah et al., 2004). Single-engine platforms are lighter and have fewer logistical concerns and lower operational costs, while twin-engine platforms offer more power and weight for larger payloads (Abdullah et al., 2004). Many low-altitude platforms employ a dual sensor configuration for collecting multiple types of data (e.g., lidar and optical), but aircraft with less powerful engines are less likely to be able to carry multiple sensors because the power requirements are too high and the combined payload becomes too heavy for the plane. However, over the last 10 years the weight, size, and power requirements of many sensors have rapidly decreased, making multiple sensor configurations more feasible.

Collection Characteristics

The components of sensors and the features of platforms combine to determine the collection characteristics of an image: its spectral resolution, radiometric resolution, spatial resolution, viewing angle, temporal resolution, and extent. Table 3.1 provides definitions of commonly used categories of the three most important collection characteristics: spatial, spectral, and temporal resolution.

Table 3.1. Commonly used categories of imagery collection characteristics

	Very high	High	Moderate	Low
Spatial resolution (ground length of one side of a pixel)	less than 1 meter	1.1 to 5 meters	5.1 to 30 meters	greater than 30 meters
Spectral resolution (number of bands)	greater than 50 bands (hyperspectral)	greater than 7 to 49 bands (superspectral)	2 to 7 bands (multispectral)	1 band (panchromatic)
Temporal resolution (minimum revisit period)	Once a day	Once a week	Once a month	Once a year

Spectral Resolution

The spectral resolution of an image is determined by the sensor and refers to the following:
- The number of bands of the electromagnetic spectrum sensed by the sensor
- The wavelengths of the bands
- The widths of the bands

Panchromatic sensors capture only one spectrally wide band of data, and the resulting images are shades of gray, regardless of the portion of the spectrum sensed or the width of that portion. Panchromatic bands always cover more than one color of the electromagnetic spectrum. Multispectral sensors capture multiple bands across the electromagnetic spectrum. Hyperspectral sensors collect 50 or more narrow bands. Traditionally, multispectral bandwidths have been quite large (usually 50 to 400 micrometers), often covering an entire color (e.g., the red portion). Conversely, hyperspectral sensors measure the radiance or reflectance of an object in many narrow bands (usually 5 to 10 micrometers) across large portions of the spectrum, similar to imaging spectroscopy in a chemistry laboratory.

Film images are stored as negative or positive film or paper prints. Remotely sensed digital data files are stored in a raster or rectangular grid format. When imaging, each picture element, or pixel, collects a digital number (DN) corresponding to the intensity of the energy sensed at that pixel for each specific band of the electromagnetic spectrum. Panchromatic data is stored in a single raster file. Figure 3.13 shows example infrared DNs for a small area.

6	7	10	109	98	107
11	5	9	97	100	99
112	4	110	112	95	114
114	107	86	113	174	180
117	116	96	114	147	169
110	95	118	99	177	183

Figure 3.13. Example infrared digital number (DN) values

Multispectral images store each band as a separate raster. Each band is monochromatic, but when they are combined they can be displayed in color. Figure 3.14 shows four separate bands of airborne digital imagery collected over a portion of Sonoma County, California. Each band is monochromatic. Figure 3.15 combines the bands to create true color and color infrared displays.

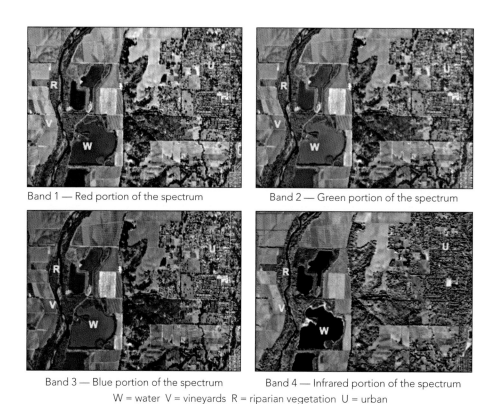

Band 1 — Red portion of the spectrum   Band 2 — Green portion of the spectrum

Band 3 — Blue portion of the spectrum   Band 4 — Infrared portion of the spectrum

W = water   V = vineyards   R = riparian vegetation   U = urban

Figure 3.14. Red, green, blue, and near infrared bands of airborne multispectral imagery captured over Sonoma County, California (esriurl.com/IG314)

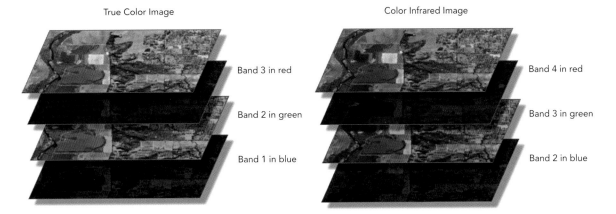

Figure 3.15. True color and infrared combination of bands of airborne multispectral imagery collected over Sonoma County, California (esriurl.com/IG315)

The bands shown in figures 3.14 and 3.15 are in the red, green, blue, and near-infrared portions of the electromagnetic spectrum. Each pixel of the imagery contains four numbers, one for the DN recorded in each of the four bands. Table 3.2 presents the range of DN values for each band of the different land-cover types depicted in figure 3.15.

Table 3.2. Range of sample DN values

Land-cover class	Water DNs	Riparian vegetation DNs	Vineyard DNs	Urban DNs
Band 1—blue	12–98	17–77	95–133	148–183
Band 2—green	32–120	23–115	111–152	142–194
Band 3—red	29–65	23–56	65–91	93–198
Band 4—infrared	0–5	144–204	165–203	111–200

Note: The range of values are in each band of airborne multispectral imagery for different types of land cover in Sonoma County, California. This imagery's DN range is from 0 to 254.

Notice how water is significantly lower in the infrared band than are the other land-cover types. Also, urban has high values in all bands relative to the other classes. Riparian vegetation and water are similar in the red, green, and blue bands, but significantly different in the infrared band, indicating that without the infrared band it might be difficult to distinguish the greenish water from the green vegetation.

At this point, we can begin to see how variations in land-cover types can be related to variations in spectral responses, and it becomes straightforward to group the similar pixels of the image sample in figure 3.14 together into land-cover classes, as depicted in figure 3.16. Of course, it is never quite this straightforward to turn image data into map information, which is why chapters 7 to 9 thoroughly examine the methods and tools for image interpretation and classification.

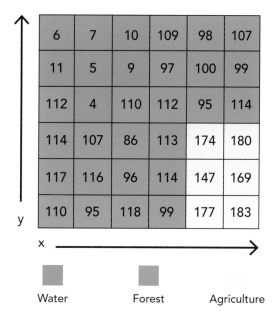

Figure 3.16. Infrared DN values from figure 3.13 combined into land-cover classes

Radiometric Resolution

Radiometric resolution is the minimum variation in electromagnetic energy that a sensor can detect, and therefore determines the information content of an image. Like spectral resolution, radiometric resolution is determined by the sensor.

In film systems, radiometric resolution is determined by the contrast of the film. Higher-contrast films will have higher radiometric resolutions than low-contrast films. In digital sensors, the potential range of DN values that can be recorded for each band determines the sensor's radiometric resolution. The larger the number of bits or intensities discernible by the sensor, the higher its radiometric resolution and the better the sensor can detect small differences in energy. In general, higher radiometric resolution increases the ability to more finely distinguish features on the imagery. Discerning objects within shadowed areas or extremely bright areas is particularly enhanced by higher radiometric resolution.

Digital data is built with binary machine code, therefore each bit location has only two possible values (one or zero, on or off), and radiometric resolution is measured as a power of 2. One-bit data would result in image pixels being either black or white, so no shades of gray would be possible. The first digital sensors were 6 bit, allowing 64 levels of intensity. More recent sensors such as Landsat 8, Sentinel-2, and WorldView-3 have 11- to 14-bit radiometric resolutions (for a range of from 2,048 to 16,384 levels of intensity).

The range of electromagnetic energy intensities that a sensor actually detects is termed its *dynamic range*. Specifically, dynamic range is defined as the ratio of the maximum intensity that can be measured by a device divided by the lowest intensity level discernible. It is important to note the difference between radiometric resolution and dynamic range. The radiometric resolution defines the *potential* range of values a digital remote sensing device can record. Dynamic range is calculated from the *actual* values of a particular image. Dynamic range is defined by the difference between the lowest detectable level and the brightest capturable level within one image. It is governed by the noise floor/minimal signal and the overflow level of the sensor cell.

The sensor used to originally capture an image determines the radiometric resolution of the image. Thus, scanning a film image to create a digital version results in a digital image with the radiometric resolution of the film sensor, not of the digital scanner, even though the radiometric resolution of the scanner may be better than that of the film image.

Spatial Resolution

An image's spatial resolution is determined by the altitude of the platform, and the viewing angle, lens focal length, and resolving power of the sensor. Spatial resolution has two different definitions:

- The smallest spatial element on the ground that is discernible on the image captured by the remote sensing system. The definition of "discernible" can refer to the ability to detect an element as separate from another, or to both detect and label the different elements. This definition was commonly used when remotely sensed images were collected primarily on film.
- The smallest spatial unit on the ground that the sensor is able to image. This is the more common meaning and is the one relied upon by makers and users of digital remote sensing systems. Usually, it is expressed as the ground sample distance (GSD), which is the length on the ground of one side of a pixel.

GSD is a function of sensor pixel size, height above terrain, and focal length, as expressed in the following equation:

GSD = (sensor pixel size × height above terrain) / (focal length) .

The distance to ground is a function of platform altitude and sensor viewing angle. If focal length and sensor resolving power are held constant (as they are in most airborne systems), then the lower the altitude of the system, the smaller the GSD and the higher the spatial resolution of the resulting imagery. If focal length and distance to ground are held constant (as they are in satellite systems), then the higher the sensor resolving power, the higher the spatial resolution. If sensor resolving power and distance to ground are held constant, then the longer the focal length, the higher the spatial resolution of the sensor. Because the sensor and the altitude of satellite remote sensing systems are constant over the usable life of the system, their spatial resolutions are also fairly constant for each satellite system and change only when the viewing angle is changed.

Airborne systems have varying spatial resolutions depending on the sensor flown and the altitude of the aircraft platform. Spatial resolution is also affected by whether the sensor has a stabilized mount, a forward motion compensation unit, or both, which compensate for the forward motion of the aircraft and minimize the blur caused by the motion of the platform relative to the ground by moving the sensor in the reverse direction of that of the platform (and at the ground speed of the platform) during sensor exposure. Figure 3.17 compares the spatial resolution of 15-meter pan-sharpened Landsat imagery to that of airborne 1-meter National Agriculture Imagery Program (NAIP) imagery over a portion of Sonoma County, California. Figure 3.18 compares the NAIP imagery to 6-inch multispectral imagery over a subset of the same area.

True color pan-sharpened Landsat imagery. Pixel size = 15 meter.

True color NAIP airborne imagery. Pixel size = 1 meter.

Figure 3.17. Comparison of Landsat 15-meter pan-sharpened satellite imagery to 1-meter National Agriculture Imagery Program (NAIP) airborne imagery over a portion of Sonoma County, California. Color differences are due to sensor differences and the imagery being collected in different seasons. (esriurl.com/IG317)

True color NAIP airborne imagery. Pixel size = 1 meter. Spring 2015.

True color airborne imagery flown for the County of Sonoma. Pixel size = 6 inches. Fall 2013.

Figure 3.18. Comparison of 1-meter National Agriculture Imagery Program (NAIP) imagery to 6-inch airborne imagery over a subset of the area of figure 3.17. Color and shadow differences are due to sensor differences and the imagery being collected in different seasons. (esriurl.com/IG318)

The highest spatial resolution obtainable from a civilian satellite is WorldView-4's 30 centimeters (11.8 inches). High-resolution airborne multispectral sensors have spatial resolutions of 2 to 3 centimeters at an altitude of 500 feet (e.g., UltracamEagle). Because they can fly lower than piloted aircraft, UASs can obtain higher spatial resolutions than manned aircraft.

Viewing Angle

Viewing angle is often used to refer to one or both of the following angles:
- *The maximum angle of the IFOV of the sensor*, from one edge of the sensor view to the other, as shown in figure 3.19. Traditional film-based aerial survey cameras often used wide-angle cameras with a 90-degree IFOV. When they took photographs vertically, the features at the edges of the frames were captured at an angle of about 45 degrees to vertical. With the advent of digital photography, many digital aerial survey cameras have a narrowed IFOV, and coverage is achieved by taking more images. Most satellite imagery is collected with an even narrower IFOV. For example, a vertical WorldView-3 scene captures a strip about 13.1 km wide from an altitude of 617 km, with an IFOV of about 1 degree.
- *The pointing angle of the sensor* as measured from directly beneath the sensor (0°, or nadir) to the center of the area on the ground being imaged. This angle is also

Chapter 3 : Imagery Fundamentals

referred to as the elevation angle. Sensor viewing angles are categorized as vertical or oblique, with oblique being further divided into high oblique (images that include the horizon) and low oblique (images that do not include the horizon), as shown in figure 3.20.

Traditionally, with aircraft imagery, images captured with the sensor pointed at less than ± 0 to 3 degrees off nadir are considered vertical, and images collected at greater than ±3 degrees are considered oblique (Paine and Kiser, 2012). However, with the plethora of pointable high-resolution satellites, satellite companies tend to define images captured with a sensor viewing angle of ± 0 to 20 degrees as vertical images, and images collected with sensor angles greater than ±20 degrees as oblique.

Viewing angle is important because it affects the amount of area captured in an image, whether only the top of an object or its sides are visible, and the spatial resolution of the imagery. The larger the viewing angle from the sensor to the object, the longer the distance to the ground and the lower the spatial resolution of the pixels. For example, DigitalGlobe's WorldView-3's nadir spatial resolution of its panchromatic band is 0.31 meter on-nadir, and 0.34 meter at 20 degrees off nadir. The spatial resolution and scale within an oblique image change more rapidly than across a vertical image.

The primary advantage of a vertical image is that its scale and illumination are more constant throughout the image than those of an oblique image. While a vertical image's scale will be affected by terrain and the slightly off-nadir pixels at the edge of the frame or scan line, a vertical image will always have more uniform scale than an oblique image. As a result, measurements are easier and directions can be more easily determined, allowing the image to approximate a map and be used for navigation (as long as the impacts of topography are considered).

On the other hand, an oblique image will show the sides of an object instead of just the top, allowing for realistic 3D rendering. Because humans spend much of their time on the ground, an oblique view is more intuitive to us and we are easily able to judge distances to objects seen in an oblique view (Paine and Kiser, 2012). Much imagery for military surveillance applications was captured as oblique or nonvertical, providing the advantage of showing objects farther away and showing more of the sides of the features, which often provide significant details for interpretation.

The very first aerial photographs were mostly oblique. However, for 70 years vertical photographs became the basis for most maps because the geometrical relationship between the sensor and the ground is fairly straightforward to determine with vertical images. In addition, the scale and illumination of vertical images are relatively constant within the image, and stereo models can be easily created by overlapping vertical images. Usually, the photographs were collected with at least a 50-percent overlap to enable stereo viewing and photogrammetric measurements (see chapter 6 for more detail on photogrammetry). Similarly, since the first launch in 1972, all nine Landsat satellites were designed to collect vertical images, but the systems are incapable of stereo except over higher latitudes, where there is enough overlap to allow some stereo collection.

When a vertical object such as a building is viewed at nadir, the sides of the building are not visible; only the top of the building is visible. If that vertical object is not located directly below the sensor at nadir, then one or more sides of the object will be visible in the image; this is termed an off-nadir view, and the effect is called relief displacement. We can refer to the angle between the nadir and the ray of light between the sensor and the vertical object as the off-nadir angle. This angle can be the result of a ray being off the center of a vertical image, meaning that it is not the principal axis of the image, or it can be the result of a ray from an oblique image. In either case, you can see the side of the vertical object, and this view allows for height measurements as well as being the basis for parallax between two images, which provides stereo imagery. Parallax is the apparent displacement of the position of an object relative to a reference point due to a change in the point of observation. This off-nadir angle may be small across the image when the imagery is vertical and the IFOV is small. Larger off-nadir angles are seen when the imagery is captured as oblique imagery or if the camera has a large IFOV.

These geometric shifts due to sensor perspective and collection geometry enable some good things like stereo imagery, but they also lead to occlusion of objects and variation from image to image that adversely affect image classification and other automated processes if elevation is not modeled at a high fidelity.

The 1986 introduction of the French SPOT systems brought off-nadir pointability to civilian satellite image collection, allowing for the collection of off-nadir stereo pairs of imagery to support the creation of DEMs. Now, most very-high-spatial-resolution satellites and airborne systems are able to collect both nadir and off nadir to oblique imagery either through pointing the system as shown in figure 3.20 or through the use of multiple sensors on the platform, some collecting at nadir and others collecting off nadir, as shown in figure 3.21. Recently, with advances in photogrammetry and computing power, airborne and terrestrial oblique images have been used to created detailed and accurate 3D representations of the landscape.

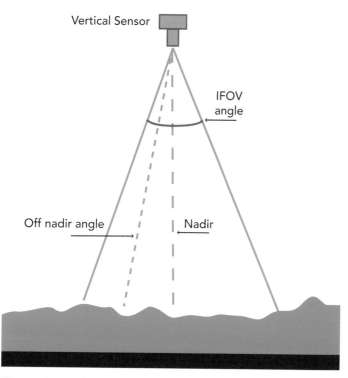

Figure 3.19. The concepts of the instantaneous field of view (IFOV), nadir, and off-nadir angles of a vertically pointed sensor

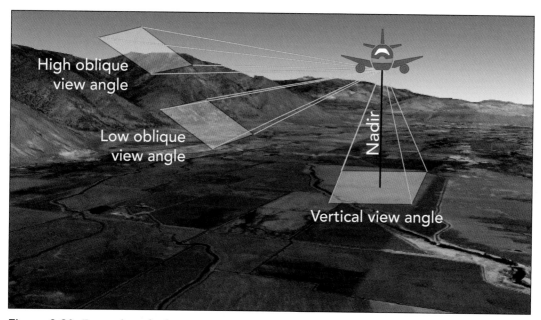

Figure 3.20. Examples of a framing camera's vertical, low-oblique, and high-oblique viewing angles

60   Chapter 3 : Imagery Fundamentals

Figure 3.21. Diagram showing how pointable satellites can pitch from side to side during orbit to collect off-nadir stereo images. Source: DigitalGlobe

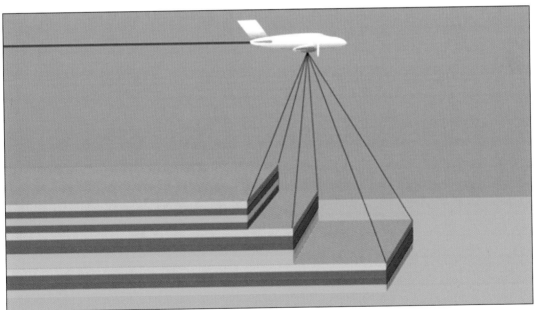

Figure 3.22. Conceptual diagram of an aircraft with a Leica ADS100 with three beam splitters: two tetrachroid beam splitters in forward and backward directions with multispectral red, green, blue, and near-infrared (RGBN) bands and one bi-tetrachroid beam splitter in nadir with multispectral red, green, blue, and near infrared bands with staggered green bands. Source: Hexagon

Temporal Resolution

Temporal resolution is determined primarily by the platform and refers to the flexibility regarding the day, time of day, and time between capture of remote sensing images of the same feature by a sensor on the platform.

The highest temporal resolution will always be obtained with a geostationary system. Because weather is constantly changing and must be constantly monitored, many of our weather satellites are geostationary—that is, they rotate above the earth at the same speed as the earth allowing them to remain stationary over a particular region. To remain geostationary, they require an orbit of about 35,786 kilometers above ground, and as a result their spatial resolution is very low. Other examples of high temporal resolution geostationary systems are video cameras placed in banks, at many road intersections, and in high-risk areas such as subway stations and airports. Other applications for fixed platforms include continuous weather observation using a Doppler radar system or continuous spectral monitoring of a specific ground point (e.g., monitoring spectral characteristics of a specific farm crop). However, terrestrial platforms are not practical for most large mapping missions.

Airborne platforms usually offer higher temporal resolution than orbiting spaceborne platforms because orbiting platforms are restricted by their orbits, which determine how often and when a spaceborne platform will pass over a specific location on the earth. Conversely, airborne platforms can be flown at any time of day or night but may be restricted from flying over a specific area such as a war zone.

Because passive systems rely on the energy of the sun, they will always have lower temporal resolution than active systems, which can be flown at any time of the day or night. Because satellites are tied to their orbits, they will never have the temporal resolution of aircraft systems in countries where airspace is relatively unrestricted (as opposed to severely restricted airspace over countries such as Iraq or North Korea). Sun-synchronous satellites capture imagery during the same period every day, which reduces their versatility. Aircraft can often fly under clouds, and image collections can be specifically timed to tidal stages, crop calendars, or deciduous leaf conditions.

Extent

The term extent is used to refer to the area on the ground that can be captured with each exposure of the sensor. It is often used relative to one mission or collection and is determined by the sensor size, its focal length, and its distance to ground. Many satellite systems collect imagery continually along their orbital paths. The area collected is, therefore, constrained by the width of the sensor's swath, but not by the strip length. The length of individual satellite scenes is arbitrary and is determined by the operator of the system. Most scenes are approximately square. For example, a Landsat scene is 170 by 183 kilometers.

The size and shape of a project area will affect what remote sensing systems are most suitable for imaging it. Because of their altitude, satellites can capture large areas in individual satellite scenes (e.g., a Landsat scene covers just over 12,000 square miles). However, satellite images are restricted to the satellite's orbital paths, making aircraft systems more effective in collecting linear or sinewy project areas (such as riparian areas, transmission lines, and coastlines) because aircraft are not tied to an orbit.

Helicopters are ideally suited to collect data over multiple distributed points because they are more agile than airplanes. They are also suitable for collecting data along corridors where frequent turning may be necessary. Fixed-wing aircraft are more suitable for collecting imagery over large areas because they can quickly collect large swaths of data. For extremely large areas, a high-altitude aircraft or satellites might be employed to maximize ground coverage.

For example, a riparian mapping project, following a long sinewy river with a required spatial resolution of 1 meter will probably be better accomplished with an airborne system than with a satellite system. The airborne platform can follow the path of the river and constrain its data collection to only the river area. Using data from a satellite system would require collecting multiple scenes, and then extracting the river areas from the larger scenes. More area and thus more data than required would be collected, which would increase the cost of the project.

An advantage that moderate-spatial-resolution, large extent satellite systems (e.g., Landsat and Sentinel) have over airborne collections is that the entire scene is captured at once with instantaneous sun illumination, vegetative condition, and atmospheric conditions fixed across the scene. Capturing the same area as a Landsat scene with an aircraft system would take several days, with the sun illumination, vegetation, and weather conditions changing throughout each day and from day to day. These variations can introduce confusion when the images are manually interpreted or classified to create maps.

High-spatial-resolution systems (both airborne and satellite) have a smaller extent than moderate- and low-spatial-resolution systems, but individual scenes can be mosaicked together to represent a larger area. Mosaicking is discussed in more detail in chapter 5.

Organizational Characteristics

Introduction

The choice of what imagery best meets a project's requirements will be determined not only by the imagery collection characteristics but also by the imagery's organizational characteristics. Organizational characteristics are determined by the organization(s) funding the imagery acquisition and distribution. Types of organizations include public agencies, private companies, and organizations with a combination of public and private funding. Organizational characteristics affect the imagery's accessibility and price to users. This section introduces and reviews imagery organizational characteristics—its pricing and licensing and its accessibility.

Pricing and Licensing

Price is the amount a user pays to gain access to imagery. Licensing refers to the restrictions placed on the use of the imagery. Licensing and pricing are often linked. Much of the medium- and low-spatial-resolution remote sensing data collected is free and its use is unrestricted (i.e., the data is in the public domain). Other imagery, especially high-resolution satellite imagery, is either severely restricted by government policy or accessible only through the purchase of a license with associated restrictions on the user's ability to share the imagery with other users.

Because the primary demand for low- and moderate-spatial-resolution imagery is from the public sector, acquisition and distribution of much of the low- and moderate-spatial-resolution civilian satellite imagery acquired by the US government and the European Space Agency (ESA) is funded by taxpayers and available to most users at no charge with few or no use restrictions. As a result, NASA earth observation data, Landsat imagery, National Oceanic and Atmospheric Administration (NOAA) weather imagery, and the ESA's Sentinel imagery are all freely available to most users worldwide.

The collection of high-resolution airborne imagery in the United States is usually purchased as a service by government agencies from commercial providers. The agencies pay the provider to collect and process the imagery, with the agency retaining all or most rights to the imagery. Similar to the availability of low- and moderate-resolution satellite imagery, most high-resolution airborne imagery is made available by agencies to the public at no cost, although some charge user fees. The USDA NAIP collects high-resolution, 4-band multispectral imagery over one-third of the continental United States every year at 1-meter spatial resolution. The imagery is available to the public at no cost and with no user restrictions. Most of the NAIP commercial providers also offer the ability to "upgrade" the imagery to 30- centimeter spatial resolution on a paid subscription basis. The upgrades are available because the providers capture the imagery at the higher resolution and resample it to the lower resolution for the public domain product. The provider then makes the higher resolution product available through a licensing agreement that restricts the use of the imagery (i.e., the purchaser is restricted in some way from copying or sharing the imagery with other organizations). Besides NAIP, many states and local governments retain commercial firms to collect airborne high-resolution multispectral and lidar data over their jurisdictions. Usually, the imagery is made available to the public at low or no cost, and with few, if any, use restrictions. Private companies such as utility and forestry firms also contract with airborne providers to produce high-resolution imagery of their properties.

Until recently, high-spatial-resolution satellite imagery was either completely government funded with use severely restricted, or partly government funded, with use restricted by licensing. For example, the United States, Russia, China, India, and Israel all have constellations of satellites that are fully funded by their government agencies but whose imagery use is strictly restricted to security agencies.

The passage of the 1992 Land Remote Sensing Act made it possible for US commercial companies to build, launch, and operate satellite sensors able to collect high-resolution imagery globally. Although fully commercial, the first companies to launch high-resolution systems received large contracts from the National Geospatial Agency of the Department of Defense for imagery. As a result, the funding for the imagery is part government and part commercial. The commercial companies distribute the imagery through licensing agreements that restrict either the amount of time the imagery is available for use or the sharing of the imagery with other organizations. This quasi-public/private funding model for high-resolution satellite imagery with licensing restrictions has since been replicated by several companies (e.g., DigitalGlobe, Airbus, Planet, and DMC constellations).

Access

Organizations make imagery available in a variety of ways. It can be delivered on a hard drive, downloaded from the web, or served as image services. Because imagery files are very large, access can be problematic and can affect the cost of working with imagery. Free imagery with no license restrictions can still be difficult to use if its access is cumbersome.

Before digital sensors, imagery was accessed as hard copy negatives and photographs. Reproduction of the negatives and photographs was very expensive and, as a result, access to them was limited. With the adoption of digital sensors, digital imagery was initially accessed from tape, and then from hard drives and CDs, and processed first on mainframe computers and then on desktop computers.

Until recently, the most efficient way to deliver and gain access to high-spatial-resolution imagery for analysis was still by shipping hard drives and then using on desktop machines or serving the imagery locally. With increases in Internet bandwidth, imagery is increasingly accessible by FTP download or direct access from cloud storage. In this way, imagery can be downloaded to desktop machines or directly used in the cloud infrastructure.

Over the last five years, several imagery providers and software companies have begun to host imagery in the cloud and offer direct visualization, analysis, and processing of the imagery. Most notable is Esri's Landsat services, which obtain Landsat imagery hosted on Amazon Web Services and provide access and on-the-fly processing of large collections of multitemporal multispectral Landsat imagery that is updated daily as imagery is acquired by the USGS. Google also hosts archives of Landsat imagery and provides processing to educational and research organizations.

Case Study—the Effects of Price and Licensing on the Use of Landsat Imagery

The history of Landsat imagery is a good example of how organizational characteristics affect imagery use. Landsat satellite imagery is moderate resolution, multispectral, and funded by US taxpayers. NASA launched the first Landsat satellite in 1972. The spatial resolution was coarse (80 meters) and included only four bands (green, red, and two infrared bands). Technological barriers slowed the use of the imagery because the knowledge base was small, little image processing software existed, and the files were huge for that time, requiring mainframe computers. Most users were NASA or academic scientists and government agencies. Landsats 2 and 3 were similar to Landsat 1.

In 1979, the Landsat program was moved from NASA to NOAA. In 1982, Landsat 4 was launched and included a 30-meter resolution instrument that collected seven bands of imagery, adding two middle-infrared and one thermal band. A similar system, Landsat 5, soon followed in 1984. However, Congress passed the Land Remote Sensing Commercialization Act of 1984, which directed NOAA to migrate Landsat imagery distribution from the federal government to the private sector with the hope that revenue from imagery sales would support the continuation of the Landsat program. As a result, the cost of Landsat imagery increased from $2,800 per scene from NOAA to $6,000 per scene from the commercial company

EOSAT, and use of the imagery was license restricted. The demand for imagery sharply declined, as did Landsat research and innovation (Draeger et al., 1997).

In 1992, Congress passed the Land Remote Sensing Policy Act (Public Law 102-555), which ended Landsat commercialization by designating the USGS to take over distribution of Landsat 7 imagery when it was launched (Landsat 6 failed to reach orbit). The act required that imagery be priced at the cost of fulfilling user requests and have no licensing restrictions. Landsat 7 was successfully launched in April 1999, and the USGS initially set the price of a scene at $600. The lower price of Landsat 7 imagery forced the company distributing Landsat 4 and 5 data to match the price of Landsat 7 imagery. Unable to run Landsats 4 and 5 profitably, the company returned its rights to distribute Landsat 4 and 5 imagery to the federal government in 2002. The lower price and unrestricted licensing for all Landsat imagery resulted in a dramatic increase in the operational use of Landsat imagery, with government revenue from image sales growing from $4 million in 1999 to $11 million in 2002. However, access to the imagery was still cumbersome and slow, requiring the manual ordering and writing of CDs.

With improvements in the web and automation of the USGS distribution processes, the agency made Landsat imagery free and downloadable from the web in 2009. As a result, the use of Landsat imagery skyrocketed from 20,000 scenes to 2,000,000 scenes a year, and commercial companies such as Esri are hosting Landsat imagery and processing services, which further increases global access to the imagery.

Summary—Practical Considerations

In this chapter, we have learned how imagery is differentiated by a combination of technical and organizational characteristics. An image's sensor and platform determine its technical characteristics—its spectral, radiometric, spatial, and temporal resolutions, as well as its viewing angle, and extent. In summary:
- Spectral resolution—Terrestrial, airborne, and satellite platforms can and do carry all types of sensors. Currently, panchromatic, multispectral, and hyperspectral sensors can be found on terrestrial, airborne, and satellite platforms, as are active and passive sensors.

- Radiometric resolution—Older sensors will often have lower radiometric resolution than newer sensors because newer sensors can take advantage of continual improvements in digital arrays, memory, and storage.
- Spatial resolution—Airborne systems are more commonly used to collect high-spatial-resolution imagery than spaceborne systems if the infrastructure to support aircraft is available and if the aircraft have access to airspace. If access to the air is limited, satellite systems or drones can be used to collect high-resolution imagery. Moderate-and low-spatial-resolution imagery is best captured from satellites.
- Temporal resolution—Geostationary systems offer the highest temporal resolution, but at either a lower spatial resolution (e.g., weather satellites) or a smaller extent (e.g., video cameras at ATM machines) than airborne or satellite systems. Airborne systems are more flexible than satellite systems and are limited only by aircraft access and fuel capacity. Additionally, cloud interference can be avoided by positioning airborne systems below the cloud ceiling or by timing flights to avoid cloud cover (e.g., flying after fog has burned off in a coastal area). However, the marginal cost of mobilization for each image is higher for airborne systems than for satellite systems.
- Extent—Depending on the resolving power of the sensor, high-altitude platforms will generally result in greater area imaged per exposure (i.e., larger extent), but at coarser spatial resolution than platforms operating at low altitudes. Airborne systems are usually more effective than satellite systems in collecting long and sinewy project areas.

Technical characteristics are not the only factors differentiating imagery types from one another. Often more important are the organizational characteristics, which will determine an image's price, licensing, and accessibility. Choosing what imagery to use in a project requires making trade-offs between technical and organizational characteristics. In the next chapter, we will learn how to match imagery characteristics with user requirements to decide what type of imagery will best meet user needs.

Chapter 4
Choosing and Accessing the Right Imagery

Chapter 3 introduced the characteristics that differentiate the types of imagery available to civilians. This chapter examines those characteristics through the eyes of the image user, first by presenting a framework for matching imagery characteristics to user requirements so as to ensure the best imagery is chosen for each project, and then by describing and cataloging the wide variety of imagery datasets available at the time of the printing of this book. However, please be aware that the supply of both public and commercial imagery is very dynamic and constantly changing, especially with the advent of unmanned aerial systems (UASs) and small-payload earth observing satellites (often referred to as cubesats, microsats, or smallsats). There is a high probability that new imagery sources have arisen and others have failed since this book has been published.

Selection Framework—What's Required versus What's Available

The usefulness of a particular imagery product will depend on its technical and organizational characteristics as well as how those characteristics meet the needs of users. Sometimes an analyst has no choice regarding the type of imagery to be used and has to make do with what has been provided by their organization. However, in today's ever-expanding imagery marketplace, multiple datasets are often available and accessible. Because so much imagery is available, it is also now common for a mapping project to use multiple datasets. For example in the Sonoma Vegetation Mapping project, multiple imagery datasets were used including two years of National Agriculture Imagery Program (NAIP) imagery (4 bands, 1-m spatial resolution), multiple years of Landsat 8 imagery (11 bands, 30-m spatial resolution), Airborne Visible/Infrared Imaging Spectrometer (AVIRIS) hyperspectral imagery, 4-band multispectral, 1-foot imagery collected in 2011 by the county, and quality level 1 lidar and 4-band optical imagery (6-inch spatial resolution) flown specifically for the project.

No matter how large or small, each image acquisition will require trade-offs between imagery characteristics. It's not uncommon for imagery analysts to want to use the best spatial, temporal, and spectral resolutions available. However, not every project requires or can afford the highest resolutions possible. Higher resolutions usually translate into higher costs and limited accessibility or both, due to licensing, increased storage, and processing times. Rather, imagery should be chosen to fulfill the project requirements. Products should be substituted for one another and trade-offs made, especially between price, spatial resolution, spectral resolution, temporal resolution, and licensing restrictions.

To determine what is required by a project instead of what is desired or available, the analyst can ask several simple questions that will narrow the choice of imagery. They are:
1. Will the imagery be used for visualization or to make a map?
2. What is the smallest item to be identified on the ground?
3. What is the time frame of the project and its results?
4. What types of features need to be mapped?
5. What is the size, shape, and accessibility of the project area?
6. What are the requirements for spatial and spectral accuracy?
7. Will the imagery be shared with other organizations?
8. Is the imagery accessible?
9. What is the project budget?

It is the combined answers to all these questions that will determine which type(s) of imagery will best meet the needs of the project. Often some project requirements must be relaxed, and the questions asked and reasked multiple times before the answers converge on a particular set of imagery products.

Will the imagery be used for visualization, or to make a map?

Imagery is most often used as an aid in the visualization of map information. Imagery used in map visualization allows the user to understand the context of a location—to be able to visualize its surroundings. The most readily available imagery is served for visualization to mapping websites such as ArcGIS Online, Bing, Apple, and Google at no charge to the user. Usually, imagery for visualization is optical and served in true color. For example, a common base image, Esri's World Imagery, is served at no charge in true color and is created from a variety of sources at multiple scales including Landsat, NAIP, WorldView, and Pleiades imagery. Imagery used for visualization is always dated, because it is derived from archive imagery. However, the currency of visualization imagery is rapidly increasing with the increasing supply of worldwide high- and very-high (HVH) spatial-resolution imagery.

However, the imagery in visualization applications is available only for viewing and not for analysis. It is made available as cached, tiled services, which offer imagery in highly compressed JPEG or PNG format (please see chapter 5 for more detail on image compression). As a result, the band combination that is displayed (usually natural color or color infrared composite) cannot be changed by the user, and the pixels values are not available for analysis. These constraints limit how much information can be derived from the imagery.

While suitable for heads-up digitizing of objects easily discernible (e.g., streets, buildings), the cached imagery does not provide enough data for deciphering subtle changes such as vegetation types or camouflaged items. The creation of many maps requires that the values of the imagery pixels be available for digital analysis. Most cached imagery is also not current enough to support disaster response activities.

What is the smallest item to be identified on the ground?

The imagery spatial resolution requirements of a project are determined by the smallest item to be identified on the ground—the project's minimum mapping unit (MMU). Smaller MMUs require higher spatial resolutions. For example, mapping forest types with an MMU of 10 acres can easily be accomplished using 30-m Landsat imagery. However, mapping individual trees will require 1-meter or higher spatial resolution imagery. Similarly, mapping four-lane freeways can just barely be accomplished with 30-m imagery, but mapping two-lane residential streets or small unpaved secondary roads will require a higher spatial resolution.

What is the time frame of the project and its results?

The time frame of the project and its results will affect the temporal resolution required and can also affect the choice of using an active or a passive sensor. Land-use and land-cover information required immediately for decision-making will often be based upon readily available archived imagery, so that the user does not have to wait for new imagery to be collected and processed. Using archived imagery is viable as long as the date of imagery capture is not so distant as to make the imagery obsolete. Conversely, disaster response requires immediate postevent imagery that shows the extent and impact of the disaster. Similarly, imagery used for weather prediction must be up to date so that weather models can be run from current weather conditions. Obviously, mapping troop and military equipment movements also requires high-temporal-resolution imagery. Mapping perpetually clouded areas such as Central America can force the use of radar imagery, which can penetrate through the clouds, rather than waiting for a cloud-free period to capture optical imagery. Figure 4.1 shows different applications plotted against their general temporal and spatial-resolution requirements—the requirements that tend to most influence imagery choices.

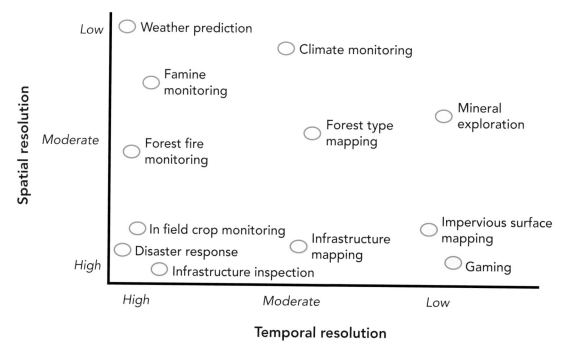

Figure 4.1. Comparison of required spatial and temporal resolutions of different mapping applications

What types of features need to be mapped?

The types of features to be mapped will affect the spectral and temporal resolutions required. Mapping general land-use land-cover classes (e.g., urban versus agricultural versus water versus forests) can be accomplished with one date of panchromatic imagery. Identifying tree species usually requires multispectral imagery and is greatly enhanced if lidar is also available to measure tree height. Adding in multitemporal imagery will help distinguish deciduous from evergreen tree species, or to map different deciduous species from one another if they change colors differently during the fall or spring. Mapping crop types also requires multispectral imagery taken when the crops are established and growing. Mapping crop yield requires multispectral and multitemporal imagery. Coastal wetland mapping often requires careful coordination of the imagery collection with tidal and weather conditions, often necessitating collection of imagery at low tide on calm days so that a maximum amount of wetland vegetation is exposed, and wave action does not interfere with the vegetation's spectral response. Mapping evapotranspiration (the transfer of water from land and plants to the atmosphere) requires thermal imagery.

Change detection requires multitemporal imagery. Change detection of obvious changes such as flooding, forest harvesting, or urban expansion can often be accomplished with multiple dates of panchromatic imagery because the spectral differences between to/from classes are very distinct. For example, clear-cuts do not look like forests, wildland and crop lands do not look like subdivisions, and the spectral response of water versus other land-cover types makes flood mapping fairly straightforward. Mapping subtle changes such as tree growth or crop production requires multitemporal, multispectral imagery and can be greatly aided by lidar.

Also important is the quality of calibration of the imagery. Calibration compensates for radiometric variation caused by sensor defects, system noise, and scan angle. Landsat sensors are methodically calibrated using preflight, postlaunch-onboard, and ground reference data. As a result, Landsat data is considered the gold standard of radiometric quality, and many other satellite systems calibrate their imagery to Landsat.

What is the size, shape, and accessibility of the project area?

The size, shape, and accessibility of a project area often determine the platform used to collect the imagery. Utility corridors, coastlines, and sinewy river corridors are often mapped best with airborne platforms instead of satellite imagery because, unlike satellites, aircraft can closely follow the shape of the project area. Large statewide, regional, or country-sized areas can often best be mapped from satellites with large image footprints. Areas inaccessible to aircraft because of government restrictions are best imaged with satellites. Small areas inaccessible to aircraft because the infrastructure does not exist to support aircraft operations might be best imaged by a UAS.

What are the requirements for spatial and spectral accuracy?

All imagery used to make a map or measure distances on the landscape needs to be registered to the ground and have the effect of terrain displacement removed (see chapter 6 for more detail on these topics). While once a cumbersome and difficult task, georeferencing and terrain correction have become much easier with the development of worldwide digital elevation models, control points, and image matching algorithms. However, not all remote sensing systems have the same quality of instrumentation, and not all remote sensing companies have the same quality of processing systems, access to high-quality digital elevation models, existing accurate orthoimagery, or ground control points. Additionally, many remote sensing companies sell products with different levels of spatial accuracy.

When acquiring imagery, the analyst should always understand its stated spatial accuracy, which is usually expressed as the maximum circular error in meters at a 90 percent confidence level—termed "CE90" (see chapter 12 for a detailed discussion of how spatial accuracy is determined). Additionally, the accuracy of the imagery should be checked against ground control points or a GIS dataset known to be more accurate than the imagery. It is not unusual for spatial accuracies to be less than stated, especially in areas with little ground control or of high terrain relief. Additionally, spatial accuracy is also affected by the viewing angle of the collection. Usually, the more off-nadir the collection, the lower the spatial accuracy.

Equally important but less frequently discussed is the spectral accuracy of the sensor—do the sensor's measured data recordings match its expected data recordings? Atmospheric interference, sensor defects, system noise, and variations in scan angle can result in a device recording data for an object that is different from the true spectral reflectance or emission of the object. There are two important questions to ask regarding a sensor's spectral accuracy. First, is the sensor regularly tested to determine whether it records data precisely and accurately? Second, does the sensor operator either calibrate their imagery to correct for sensor errors or provide calibration statistics and algorithms so that the user can correct for errors?

For example, the spectral accuracy of Landsat data is continually tested using preflight, postlaunch-onboard, and ground reference data. USGS also provides calibration parameter files of geometric and radiometric coefficients needed for correcting raw Landsat image data. The calibration of Landsat imagery is considered the gold standard and is so good that other satellite image providers often calibrate their imagery against Landsat imagery rather than collecting their own ground reference data. Most, but not all, sensor operators include some metadata with their image data, which can support calibration. Calibrating imagery is a common preprocessing step that is discussed in more detail in chapter 6.

Will the imagery be shared with other organizations?

The need to share imagery with others will affect the type of imagery license chosen. High- and very-high-spatial-resolution satellite imagery often has some sort of license restriction, although licenses that allow some sharing (e.g., within an agency, or across federal agencies) are common. If the imagery is to be shared with multiple users both inside and outside of your organization, it might be best to focus on nonlicensed, unrestricted imagery available in the public domain, or to acquire new imagery from an organization that does not license restrict their products. For example, a great deal of the high-spatial-resolution imagery captured over the United States is in the public domain and is not license restricted. This includes NAIP 1-m multispectral imagery and even higher-spatial-resolution-imagery funded and collected for many local and regional government agencies. Moderate-resolution imagery is also available in the public domain worldwide from either the Sentinel (10 to 20 m) (https://sentinel.esa.int/web/sentinel/sentinel-data-access) or Landsat (30 m) (http://landsat.usgs.gov/Landsat_Search_and_Download.php) programs, including Landsat's archive of more than 40 years. NOAA weather data and NASA earth science data are also freely accessible and shareable. Aside from military systems or imagery captured by UASs, access to high- or very-high-resolution imagery outside of the United States is usually available only from commercial satellite companies who restrict sharing of their imagery through licensing.

Is the imagery accessible?

Millions of images have been taken of the earth, but not all of them are accessible. Current reconnaissance imagery is not shared broadly, and many archives of imagery exist but are not easily searchable. Actually obtaining imagery can be problematic because there currently is no coordinated repository for imagery metadata. Some datasets such as NAIP and Landsat are easily searchable on the web, downloadable, and dynamically served. Others, such as imagery in private photogrammetry company archives, are usually searchable and accessible only by contacting personnel at the firm.

What is the project budget?

Ultimately, the project budget will limit the maximum expenditure on imagery and budgets often force trade-offs in project requirements. In the United States, multiple spatial and temporal resolutions are abundant and accessible at no charge (e.g., NAIP, Landsat, or imagery acquisitions funded by local agencies). More costly new or higher-spatial-resolution imagery can be acquired from commercial airborne and satellite operators. Outside of the United States, moderate-spatial-resolution imagery (e.g., Landsat and Sentinel) is freely available, but high-resolution imagery usually is either restricted by government programs or is sold by commercial companies under a license agreement.

Summary

We cannot emphasize enough how your choice of imagery will be dependent upon the requirements of your projects and not on the newest technology available. Different applications and organizations will have different requirements. A project to map impervious surfaces will have requirements different from one for property assessments, environmental monitoring, parcel mapping, weather prediction, road design, pipeline monitoring, forest inventory, change detection, soils, or geology. The imagery needs of a federal agency will likely be very different from those of a local or state agency, nongovernment organization (NGO), or private landowner. However, often one dataset can meet multiple organizations' needs. Such is the case with the lidar and optical imagery acquired for Sonoma County, California, which is continually used by multiple organizations including NASA and USGS researchers, the Sonoma County Open Space and Agricultural Preservation District, Sonoma County Permit and Resource Management Department, Sonoma Ecology Center, Sonoma County Water Agency, San Francisco Estuary Institute, and many private companies (Green, 2017). It is essential that you carefully evaluate your and your partner organization's proposed imagery uses before you purchase or acquire imagery. Considering the questions posed in this chapter forces imagery users to fully analyze and understand the benefits and costs of their imagery requirements, allowing them to make fully informed trade-offs when necessary.

Imagery Sources

Overview

Sources of imagery are globally distributed, highly varied, and often confusing and challenging to navigate or understand. This section reviews the major sources of both archival imagery and new imagery collects. The following sections provide more detailed information about imagery sources, organized by spectral resolution.

In general, the major civilian sources of imagery for analysis or visualization are either public agencies or private companies from whom imagery is both served and can be downloaded. There are many sources of imagery worldwide. The following list of airborne and satellite sources is not exhaustive but includes the most prominent sources. Table 4.1 summarizes and compares the sources by spatial resolution, spectral resolution, and availability. For comprehensive information about earth observing satellites, three websites offer up-to-date and detailed information:

- http://database.eohandbook.com/. The Committee on Earth Observation Satellites Earth Observation (EO) Handbook and Database provides a detailed database on all civilian government earth observing satellites, past and present, which is searchable by agency, missions, and instruments.
- https://directory.eoportal.org/web/eoportal/home. The European Space Agency's (ESA) EO Portal provides a directory of past and planned satellite nonclassified missions from 1959 to 2020, categorized by space agency and from A to Z. It also includes a directory of 40 government scientific airborne flight campaigns.
- http://space.skyrocket.de/. Known as Gunter's Space Page, the website has information about all civilian satellites in orbit, not just earth observing satellites, and is searchable by nation and type.

ArcGIS Online

ArcGIS Online (https://www.arcgis.com/home/index.html) serves a huge array of imagery datasets for visualization with some imagery also dynamically served across the web and available for analysis. ArcGIS Online is perhaps the most comprehensive and best-organized source for cached worldwide imagery served online. Hundreds of datasets are available for most of the world that include a rich variety of imagery types from high-spatial-resolution, true color world imagery collected by commercial companies, to low-spatial-resolution weather data from NOAA, to images of global ozone and precipitation from NASA. Esri also dynamically serves the pixel values of several sources of imagery including NAIP and Landsat 8.

Table 4.1. Comparison of the major sources of imagery

Source	High and very-high spatial resolution panchromatic & multispectral	Moderate spatial resolution panchromatic & multispectral
ArcGIS Living Atlas	A, WS	A, WS
Commercial Photogrammetry and Remote Sensing Firms	A, NC, DL, WS	
Commercial Satellite Firms	A, NC, DL, WS	A, NC, DL, WS
USGS EROS	A, DL	A, DL
USDA APFO	A, O	
BLM Aerial Photo Archive	A, O	
NASA DAACs		
NASA GIBS		A, DL, WS
NOAA Digital Coast	A, DL	
NOAA Climate Data Online		

A=Archive NC=New Collect DL=Download WS=Web Services O=Order/Process

Commercial Photogrammetry and Remote Sensing Firms

Commercial Photogrammetry and remote sensing firms operate aircraft in countries with open access to airspace such as Canada, the United States, Australia, South Africa, and many of the countries of Europe and South America. In fact, most of the imagery collected over those areas is acquired by private commercial firms who primarily collect mono and stereo, oblique and nadir, and passive and active very-high and high-spatial-resolution imagery from airplanes, helicopters, vehicles, and UASs. The firms primarily acquire imagery on an as-needed basis with the specifications of the collect determined by the image purchasers who are either private firms, public agencies, or NGOs. However, some companies collect imagery speculatively and then license access to the imagery to their customers. Links to information about and websites for many commercial photogrammetry and remote sensing firms can be found on the following websites:
- Management Association of Private Photogrammetric Surveyors (http://www.mapps.org/search/custom.asp?id=196)
- American Society of Photogrammetry Remote Sensing (ASPRS) (http://www.asprs.org/)
- International Society for Photogrammetry and Remote Sensing (http://www.isprs.org/)

Table 4.1 cont. Comparison of the major sources of imagery

Low spatial resolution panchromatic & multispectral	Hyperspectral	Thermal	Radar & radar derived products	Lidar & lidar derived products
A, WS	A, WS	A, WS	A, WS	A, WS
	A, NC, DL, WS	A, NC, DL, WS	A, NC, DL, WS	A, NC, DL, WS
			A, NC	
A, DL	A, DL	A, DL	A, DL	A, DL
A, DL	A, DL	A, DL	A, DL	
A, DL, WS		A, DL	A, DL	
				A, DL
A, DL		A, DL	A, DL	

Commercial Satellite Companies

Commercial satellite companies offer imagery collected worldwide. These firms collect passive, multispectral, very-high, and high-resolution imagery. The Satellite Imaging Corporation's website has a listing and summary of many, but not all, of the types of commercial satellite imagery available (http://www.satimagingcorp.com/satellite-sensors).

Government Agencies

Government agencies including local, state, federal, and international agencies make imagery available. The primary United States government sources of imagery both over the United States and worldwide are
- the USGS Earth Resources and Science Center (EROS) in Sioux Falls, South Dakota, which has an enormous archive of global satellite and airborne passive and active remote sensing data that is well organized, easily accessible, and mostly downloadable (http://eros.usgs.gov/find-data);
- the USDA Aerial Photography Field Office (APFO), which has an archive of aerial high-resolution photography and imagery collected over the United States by

a variety of USDA and USGS government agencies (http://www.fsa.usda.gov/programs-and-services/aerial-photography/index);

- the NASA Distributed Active Archive Centers, which act as custodians of NASA earth science data and make it available to users (https://earthdata.nasa.gov/about/daacs);
- NASA, which also serves NASA data through its Global Imagery Browse Services (https://earthdata.nasa.gov/);
- NOAA's Digital Coast, which archives and provides access to multiple lidar (both topographic and bathymetric) and multispectral airborne and satellite imagery datasets collected primarily in the coastal areas of the United States (https://coast.noaa.gov/dataviewer/#/imagery/search/);
- NOAA Climate Data Online, which provides access to NOAA's archive of worldwide climate and weather data (https://www.ncdc.noaa.gov/cdo-web/); and
- the Bureau of Land Management Aerial Photo Archive, a collection of aerial film at the National Operations Center in Denver, Colorado https://www.blm.gov/nstc/library/aerial/. Additionally, many states and local governments manage and serve image collections and make available their archives of both active and passive HVH-resolution imagery. Many of these image datasets can be found in ArcGIS Online.

Many national governments also collect and archive imagery. International sources of imagery are referenced in the relevant sections below.

Unmanned Aerial Systems

For the first time since the advent of remote sensing, technologies are available that allow anyone to collect imagery. UASs (sometimes called drones) have long been recognized for their potential as a means of accomplishing tasks that are too repetitive, inaccessible, or dangerous for manned aircraft. The military implications of such a device are obvious, and most development has occurred in the military as a result. However, in the last decade or so, the use of UASs for civilian purposes has grown tremendously. The proliferation of better and more effective software for processing imagery coupled with the miniaturization of sensors has allowed UASs to successfully collect remotely sensed data for a very large number of applications.

Today, virtually any sensor from cameras, video, multispectral and hyperspectral sensors, thermal imagers, radar, lidar, and others can be flown on a UAS. Some of these platforms are fixed-wing aircraft but many more are some type of helicopter, often with four, six, or eight propellers. These platforms can be small enough to fit in your hand or large enough to carry a substantial payload of sensors and equipment. Software for processing this data is available both commercially and in the public domain to create mosaicked images, thematic maps, topographic data, and other cartographic output. Esri has developed

Drone2Map for the creation of professional imagery products from UAS-captured still imagery for visualization and analysis in ArcGIS.

UASs are making the collection of ultrahigh-resolution imagery a reality for small geographic study areas. The software allows users to process the data into custom products, often in the same day. For the first time, UASs are offering the promise of making remote sensing technology personal, in much the same way that PCs made computers personal.

Many companies now sell small UASs (sUASs) at reasonable costs that allow farmers, ranchers, environmentalists, researchers, utility companies, academics, and others to take advantage of this rapidly growing technology. Federal agencies from the United States and many other countries fly large UASs with heavy payloads for extended missions of both military importance and civilian usefulness. However, the growth in this technology is primarily in smaller UASs that fly short missions with small payloads. The list of applications for such remotely sensed data is endless and includes archaeology, engineering, wildlife habitat analysis, agricultural mapping, forest inventory, disaster monitoring, road and bridge inspection, and many, many more.

Perhaps the biggest stumbling block for the use of UASs today is the regulations surrounding their use. Many countries have modernized their regulations, clearly separating the use of manned versus unmanned systems, especially for sUASs. The United States has lagged behind in this adjustment and therefore is behind many countries in the use and development of this technology. New regulations for sUASs were recently published by the FAA (https://www.faa.gov/uas/media/Part_107_Summary.pdf).

Mapping Woody Debris in the Great Brook

Flows of the Great Brook in Vermont long caused problems for the residents of Plainfield. Over time, bank erosion resulting from natural and anthropogenic forces increased the amount of large woody debris in the stream. During extreme precipitation events, the debris moved downstream, collecting at the first bridge and forming an artificial dam, which diverted water out and over the stream bank causing tens to hundreds of thousands of dollars of damage to the bridge, roads, and surrounding homes. The damage occurred so often that it became economically unviable for the town to continue to make regular repairs. As a result, the town retained a consulting engineering team to evaluate bridge alternatives. Key to the development of alternatives was an estimate of the amount of woody debris predicted to move through the bridge during a storm. While only a dozen or so logs often caused the jam at the bridge it was unclear how much more woody debris there was moving downstream.

The town tried for several years to carry out woody debris inventories of the Great Brook, but the process was slow, cumbersome, costly, and dangerous. Remote sensing approaches, while compelling, were also not feasible because even the best commercial satellite imagery lacked the spatial and temporal resolution required, and imagery acquired through manned flights was far too costly. With funding from the US Department of Transportation and the Vermont Agency of Transportation, the University of Vermont's (UVM) UAS Team began long-term monitoring of Great Brook starting in December 2014 with a goal of mapping and tracking the movement of woody debris through the 2015 spring flood season. The UVM UAS Team employed the senseFly eBee, a small, lightweight UAS specifically designed for mapping. The workflow for the eBee essentially consists of the operator using flight planning software to specify the flight area and flight parameters (e.g., desired ground resolution and maximum altitude), flight operations, and postprocessing. The eBee flies autonomously, following its preprogrammed flight path, acquiring imagery with the requisite overlap and at the appropriate angle for generating orthorectified imagery. Once the eBee is recovered the imagery is fed into photogrammetric software where it is orthorectified, making it suitable for using in GIS software.

Throughout spring 2015, the UVM UAS Team conducted multiple flights of a three-mile stretch of the Great Brook. Each time, the orthorectified imagery was brought into ArcGIS where technicians digitized the location of all of the woody debris, populated the attribute table with size information, and noted if it had moved or if the size had changed. A dry winter, combined with little in the way of spring participation, resulted in minimal changes to the woody debris conditions in the stream. Then, in July 2015, a highly localized storm dumped nearly half a foot of rain on the area during a Sunday evening. Floodwater moved rapidly down the Great Brook, trees piled up at the bridge, and the residents of Plainfield awoke to find a bridge that was in need of major repairs. The UVM UAS Team responded, collecting imagery of the damaged bridge along with the upstream area of interest. The day after the flight, technicians once again combed through the data and noticed that most of the previously mapped debris was gone, replaced by new debris from the surrounding slopes and upstream. A series of analyses was performed within ArcGIS to summarize the amount of woody debris by stream segment (figure 4.A). The data showed how dynamic the woody debris situation was in the Great Brook: hundreds of pieces of large woody debris were in motion during the flooding, but only a small percentage of them caused the problems at the bridge. Any new bridge would have to be designed to accommodate this reality.

The UAS proved to be a cost-effective, safe, efficient, and rapid way of mapping woody debris in the Great Brook. Without UAS technology, it is highly unlikely that the engineering team would have had the information they needed to complete their study. Fundamentally, the data stemming from the UAS was like any other remotely sensed data. The information gleaned from it, through manual interpretation, was done using methods employed by humans for decades. What was unique was the low-cost, flexible, and rapid response that the UAS offered.

Pre-Storm and Post Storm UAV Imagery of Great Brook

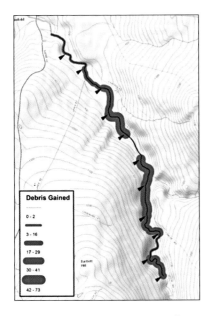
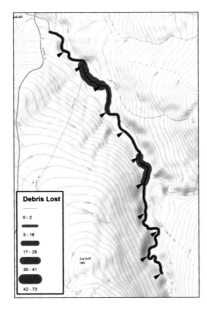

Maps of Areas of Debris 6/25/2015 through 7/21/2015

Figure 4.A. UAV images and maps of debris along the Great Brook pre- and post-storm. Source: University of Vermont Spatial Analysis Lab

Chapter 4 : Choosing and Accessing the Right Imagery   83

Sources of Passive Panchromatic and Multispectral Imagery

High- and Very-High Spatial Resolution

By far the largest amount of imagery is collected from HVH-spatial-resolution multispectral and panchromatic passive sensors on all platform types including UASs. These sensors provide the imagery that is the source of most maps of streets, buildings, soil types, hydrologic features, airports, crops, forests, wetlands, military facilities, and topography.

All early passive sensors were flown on aircraft (initially balloons) and relied upon panchromatic film imaging surfaces. In the 1930s, Kodak introduced color aerial camera film. Satellite civilian digital satellite scanners were introduced with the launch of Landsat 1 in 1972, and civilian digital airborne systems were introduced in the 1990s. While it is still possible to collect airborne imagery with film cameras, they are in little use. However, much of the long-term archive of remotely sensed imagery exists in film archives, and any change detection for the years before the 1970s will need to rely on scanned images captured from film positives or negatives.

Most HVH-resolution imagery is collected in either panchromatic or blue, green, red, and near-infra wavelengths. Some systems (e.g., WorldView-2 and -3) collect additional bands. Table 4.2 summarizes the current satellite sources of passive HVH-spatial-resolution panchromatic and multispectral imagery. Within five years, the supply of this type of imagery is anticipated to grow exponentially with the continued adoption of UAS use and the launches of several constellations of satellites by private companies, as shown in figure 4.2 and detailed in table 4.3.

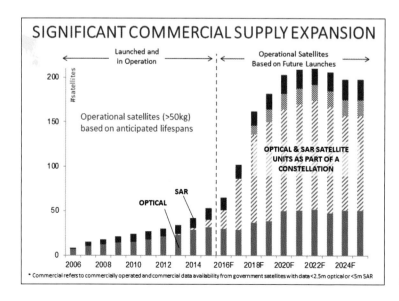

Figure 4.2. The growing supply of high- and very-high-resolution satellite imagery. Source: Euroconsult

Sources of HVH-resolution panchromatic and multispectral imagery include the following:

- *ArcGIS Online*, which offers multiple cached high- and very-high-resolution imagery datasets. Its World Imagery comprises imagery at multiple scales and from multiple sources. ArcGIS Online also dynamically serves four-band NAIP imagery over the lower 48 states (at no charge), and four bands of very-high-resolution Hexagon imagery over parts of the United States, Canada, and Western Europe (for a subscription charge).
- *Private remote sensing and photogrammetry firms*. Many remote sensing and photogrammetry firms operate in countries with open airspace. The companies offer new flights to collect imagery for a charge, with the rights of the imagery usually passing to the purchaser. Most remote sensing firms also archive images captured over the life of the firm. Unfortunately, these archives are often dispersed, held by either the company that collected the photos or the organization that funded the collection. Recently the American Society of Photogrammetry and Remote Sensing established the ASPRS Aerial Data Catalog, which is a tool for locating aerial photography throughout the world (**http://www.asprs.org/DPAC/index.php/?view=listmanagerfront**). The site is fairly new and has records on only a limited number of archives, but several firms have committed to including their information.
- *UASs*. With relatively little investment (compared to buying an airplane), analysts can purchase their own UASs with panchromatic and/or multispectral sensors. However, the capacities of these systems are still limited, making only relatively small collects possible. A good primer on using UASs can be found and accessed at **http://drones.newamerica.org/primer/**.
- *NAIP* imagery is collected at a 1-meter resolution over the lower 48 states of the United States over three-year cycles by the USDA Farm Services Agency. The program initially relied on true color film sensors when it started in 2003, but transitioned to four-band (R, G, B, and near-infrared [NIR]) digital sensors in 2009. The imagery is available for free download and without license restriction from APFO as compressed .sid (2003–2007) or JPEG 2000 (2008 to the present) files of digital ortho quarter quad mosaics. NAIP imagery can also be downloaded from USGS EROS as JPEG 2000 digital quarter quads (**https://lta.cr.usgs.gov/NAIP**). Additionally, some states (e.g., California) both serve and provide access for downloading uncompressed NAIP imagery for their state and, as mentioned before, ArcGIS Online serves all four bands of the imagery, providing full access to the raw pixel data. Figure 4.3 shows the current cycle of NAIP imagery by state.

Table 4.2. Comparison of current commercial high- and very-high-spatial-resolution satellites
(esri.url.com/IGT42 for current version.)

Company/Agency	Digital Globe				
System/Constellation Name	Worldview1	WorldView2	WorldView3[1]	Worldview4	GeoEye-1
Status	Launched in 2007, projected EOL 2020	Launched in 2009, projected EOL 2021	Launched in 2014 with a 7.25 yr lifespan	Launched in 2016, projected EOL 2026	Launched in 2008 with 7+yr. planned lifespan
Orbit	Sun synchronous @ 1:30PM	Sun synchronous @ 10:30AM	Sun synchronous @ 10:30AM	Sun synchronous @ 10:30AM	Sun synchronous @ 10:30AM
Spatial Resolution at nadir in meters					
Panchromatic	0.5m	0.46	0.31	0.31	0.41
multispectral	NA	1.8	1.24	1.24	1.665
SWIR	NA	NA	3.7	NA	NA
SAR	NA	NA	NA	NA	NA
Thermal	NA	NA	NA	NA	NA
Video	NA	NA	NA	NA	NA
Spatial Accuracy (RMSE) excluding terrain effects	< 4m CE90	<3.5m CE90	<3.5m CE90	3m CE90	3m CE90
Bands	1	8	16	4	4
Spectral Resolution	panchromatic only	B, B, G, Y, R, and 3 NIR bands	B,B,G,Y,R, & 3 NIR, and 8 SWIR bands	R,G,B,NIR bands	R,G,B, NIR
Radiometric Resolution	11 bits	11 bits	11bits pan and MS, 14 bits SWIR	11 bits	11 bits
Temporal Resolution (nadir to 25 degrees)	4.6 days	3.7 days	4.5 days	4.5 days	8 days
Swath Width at nadir	17.6km	16.4km	13.1km	13.1km	15.2km
Stereo Capability	yes	yes	yes	yes	yes
Number of Satellites	1	1	1	1	1

[1] WorldView3 also carries the CAVIS sensor with 12 bands at 30m resolution in the range from 405 to 2245nm.

Table 4.2 cont. Comparison of current commercial high- and very-high-spatial-resolution satellites

Airbus Defense & Space			Urthecast		
Pleiades 1A&1B	SPOT 6 & 7	TerraSar-X & TanDEM-X	Deimos 2	Iris	Theia
1A launched in 2011 with 5yr planned lifespan; 1B launched in 2012 with a 5yr lifespan	Launched in 2012 and 2014 with 10 yr planned lifespans. Co-orbital with Pleiades very high resolution satellites.	TerraSar-X launched in 2007 with expected life to 2020, TanDEM-X launched in 2010 with expected life to 2022	Launched in 2014 with a planned lifespan of 10 years	Launched in 2014 and carried on the International Space Station	Launched in 2013 and carried on the International Space Station
Sun synchronous @ 10AM	Sun synchronous @ 10AM	Sun synchronous	Sun synchronous @10:30AM	International Space Station	International Space Station
0.7	2	NA	0.75	NA	NA
2.8	8	NA	4	1	5
NA	NA	NA	NA	NA	NA
NA	NA	5 modes: 0.25m, 1m, 3m, 18.5m, 40m	NA	NA	NA
NA	NA	NA	NA	NA	NA
NA	NA	NA	NA	NA	5
3m CE90	35m CE90		less than 20m	not stated	not stated
5	5	1	5	3	5
B,G,G,R, NIR	Pan, R,G,B,NIR	X-band	Pan, R,G,B,NIR	R,G,B	Pan, R,G,B,NIR
12 bits	12bit		10 bits		16 bit
4 days	Not stated for nadir. Otherwise daily.	11 days	2 days		3 days
20km	60km	5 modes: 4km-270km	12km		50km
yes	yes	yes	yes		no
2	2	2	1		1

Chapter 4 : Choosing and Accessing the Right Imagery

Table 4.2 cont. Comparison of current commercial high- and very-high-spatial-resolution satellites

Company/Agency	Planet				
System/Constellation Name	SkySat 1&2	SkySat-3	SkySat 4-7	Rapid Eye	Doves and Flock
Status	2 satellites launched in 2013 & 2014	Launched in 2016 with a lifespan of 6+ years	Launched in 2016 with a lifespan of 6+ years	Launched in 2008. Planned lifetime of 7 years, but currently projected to 2021	Currently 130 satellites with 178 satellites expected by the end of 2017, expected lifetime of 2-5 years for each satellite, constellation replenished over time
Orbit	Sun synchronous	Sun synchronous @ 9:30AM	Sun synchronous @ 9:30AM	Sun synchronous @ 11AM	Sun synchronous & ISS
Spatial Resolution at nadir in meters					
Panchromatic	0.9	0.75	0.75	NA	NA
multispectral	2	2	2	6.5	2.7-4.9
SWIR	NA	NA	NA	NA	NA
SAR	NA	NA	NA	NA	NA
Thermal	NA	NA	NA	NA	NA
Video	1.1	0.75	0.75	NA	NA
Spatial Accuracy (RMSE) excluding terrain effects	not stated	not stated	not stated	less than 10m CE90	not stated
Bands	5	5	5	5	3
Spectral Resolution	pan, R,G,B,NIR	pan, R,G,B,NIR	pan, R,G,B,NIR	B, G,R, R-edge, NIR	R,G,B or R,G,NIR
Radiometric Resolution	11 bit	11 bit	11 bit	12 bit	12 bit
Temporal Resolution (nadir to 25 degrees)	not stated	not stated	daily access	5.5 days	approx. daily
Swath Width at nadir	8km	6.6km	6.6km	77km	26km
Stereo Capability	no	no	no	no	no
Number of Satellites	2	1	4	5	178

Table 4.2 cont. Comparison of current commercial high- and very-high-spatial-resolution satellites

Korean Aerospace Research Institute		Earthi	Government of Singapore / AgilSpace	China Academy of Space Technology	BlackSky
KOMPSAT	KOMPSAT 3A	TripleSat	TeLEOS-1	Gaofen-2	Pathfinder-1
Korean satellite launched in 2012 with a 4 year lifespan	Korean sister satellite to KOMSAT 3A launched in 2015 with a 4 year lifespan	Launched in 2015 with 7 year lifespan	Launched in late 2015 with a 5 year lifespan	Launched in 2014 with a 5-8 year lifespan	Launched in 2016 with a 3 year lifespan, Pathfinder 2 to follow in 2017
Sun synchronous @ 1:30PM	Sun synchronous @ 1:30PM	Sun synchronous @ 10:30AM	Equatorial	Sun synchronous @ 10:30AM	Sun synchronous @ 9:20AM
0.7	0.55	0.8	1	0.8	NA
2.8	2.2-5.5	3.2	NA	3.2	2
NA	NA	NA	NA	NA	NA
NA	NA	NA	NA	NA	NA
NA	NA	NA	NA	NA	NA
NA	NA	NA	NA	NA	NA
48.5m CE90	not stated	23m CE90	not stated	not stated	not stated
5	6	5	1	5	4
pan, R,G,B,NIR	pan, R,G,B,NIR,MWIR	pan, R,G,B,NIR	pan	pan,B,G,R,NIR	pan, R,G,B
14 bit	14 bit	10 bits	10 bits	14 bits	not stated
28 days	28 days	daily	12-16 hours	60 days	not stated
15km	12km	23km	12km	45km	30km
yes	yes	yes	yes	yes	not stated
1	1	3	1	1	1

Chapter 4 : Choosing and Accessing the Right Imagery

Table 4.3. Planned near-term launches of high- and very-high-spatial-resolution panchromatic and multispectral imagery (esri.url.com/IGT43 for current version.)

Company	Urthecast	BlackSky Global
Constellation Name(s)	OptiSAR	Pathfinder
Status	Planned launch of 16 satellites (8 SAR and 8 optical) in 2019 and 2020	Planned launch of 6 satellites in 2017 with 54 more envisioned by 2019, 3 year lifespan per satellite
Orbit	two orbital planes	varied
Spatial Resolution (nadir)		
Panchromatic	0.5	NA
multispectral	2	1m
hyperspectral	NA	NA
SWIR	NA	NA
SAR	2 modes: (1) Stripmap 1m X-band and 5m L-band, (2) ScanSAR 10m x-band, 30m L-band	NA
Thermal	NA	NA
Spatial Accuracy	not stated	not stated
Bands	7 plus 0.5m RGB video	not stated
Spectral Resolution	B,G,Y,R,R-edge,NIR and SAR	not stated
Image Radiometric Resolution	12bits	less than 1 day when all 66 satellites are launched
Temporal Resolution (nadir to 25 degrees off nadir)	daily	more than once a day
Swath Width at nadir	12.29km optical; 2 modes SAR (1) StripMap 10km (2) ScanSAR 100km	not stated
Number of Satellites	16	60

Table 4.3 cont. Planned near-term launches of high- and very-high-spatial-resolution panchromatic and multispectral imagery

Hera Systems	OmniEarth (now part of EagleView)	Northstar	Astro Digital
not stated	OmniEarth LLC	Northstar	Landmapper-HD
Planned launch in 2017 of 9 satellites with an eventual total of 48 satellites	Planned launch of 18 satellites by 2018	Planned launch of 40 satellites sometime in the future	First planned launch in Q1 2018, remaining 19 by end of 2019. Lifespan of 5 years
varied	sun synchronous	not stated	sun synchronous at 10:AM
NA	2m	NA	NA
1m	5m	not stated	2.5m
NA	NA	not stated	NA
NA	NA	not stated	NA
NA	NA	NA	NA
NA	NA	NA	NA
not stated	not stated	not stated	not stated
not stated	5	not stated	5
not stated	standard visible bands	not stated	B,G,R, R-edge, NIR
not stated	not stated	not stated	not stated
not stated	daily	not stated	every 3 days
not stated	200km	not stated	25km
9	18	40	20

Chapter 4 : Choosing and Accessing the Right Imagery

- *Hexagon.* Esri and Leica Geosystems have partnered to serve 15- and 30-cm aerial imagery over much of the United States, Canada, and Europe. In the United States, the Hexagon imagery is a higher-spatial-resolution version of the NAIP imagery, with the same collection interval. Two products are available, the Basemap Service, which is cached true color imagery, and the Multispectral Imagery Service, which provides access to all four bands served across the web for use in ArcGIS software and applications. The imagery is available as a subscription service and is available from the ArcGIS Marketplace, Hexagon Geospatial's Power Portfolio, and Valtus.

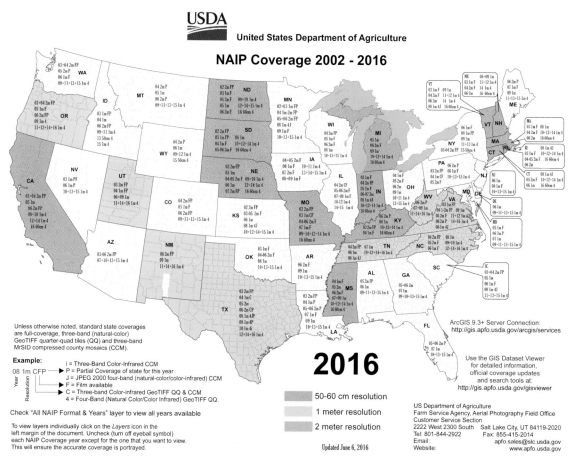

Figure 4.3. Current cycle of NAIP imagery by state (esriurl.com/IG43)

- *Commercial Satellite Imagery Providers.* DigitalGlobe and Airbus are the two largest and longest-established HVH-resolution commercial satellite imagery vendors. They sell licenses to worldwide 30-cm to 1-m panchromatic and 1-m to 4-m multispectral imagery from a constellation of satellites (see table 4.2) under licensing agreements. Both vendors also have extensive archives, especially DigitalGlobe, whose archive

dates back to IKONOS—the first high-resolution civilian satellite system, launched in 1999. Private companies currently operating HVH earth observing remote sensing systems include
- DigitalGlobe: https://www.digitalglobe.com/,
- Airbus Defense and Space: http://www.intelligence-airbusds.com/,
- Planet: https://www.planet.com/,
- UrtheCast: https://www.urthecast.com/,
- Earth-i: http://earthi.space/.

- *USGS EROS Center.* A full listing of EROS archive products can be found at https://lta.cr.usgs.gov/products_overview. HVH-resolution panchromatic and multispectral imagery offered for download from EROS are older datasets and include:
  - DigitalGlobe imagery purchased by federal agencies. IKONOS, GeoEye-1, QuickBird, and WorldView-1 to -3 imagery purchased by federal agencies is available to other US federal agency users only, either at no cost or for the cost of a license upgrade. https://lta.cr.usgs.gov/UCDP
  - Digital orthophoto quadrangles in panchromatic, true color, or color infrared with a 1-m spatial resolution. They are available to all users for download at no cost as quarter quadrangles (3.75 minutes) for 48 states and were completed in 2004. USGS also has black-and-white, full 7.5-minute quadrangles for much of Oregon, Washington, and Alaska. https://lta.cr.usgs.gov/DOQs
  - OrbView-3 imagery collected between 2003 and 2007. This 1-meter panchromatic and 4-meter multispectral data is available for download to all users at no cost and without license restrictions.
  - National High Altitude Program (NHAP) black-and-white (1:58,000) and color infrared photographs (1:80,000) of the conterminous United States collected from 1980 to 1989, which can be downloaded as scanned files. They are available to all users at no cost. The photos have not been terrain corrected. https://lta.cr.usgs.gov/NHAP
  - National Aerial Photography Program (NAPP) photography 1:40,000 black-and-white and color infrared images collected over the 48 conterminous states from 1987 to 2007 and can be downloaded as scanned files. They are available to all users at no cost. The photos have not been terrain corrected. https://lta.cr.usgs.gov/NAPP
- *USDA APFO.* The APFO has a vast amount of aerial photography of the lower 48 states, captured by USDA from 1955 to the present, including the NAIP imagery mentioned above. Until recently, the only way to gain access to its film archive was through requests made to APFO for an aerial photograph to be scanned. However, it is currently scanning much of the film archives. One of the most efficient ways to investigate APFO archive content is through the ArcGIS Online APFO Historical

Availability Tile Layer, which lists the type of imagery collected and the year of collection for every county in the 48 conterminous states.
- Bureau of Land Management. The location and date of BLM film collections can be researched using the BLM's interactive map (https://navigator.blm.gov/home). Copies of photos in the archive must be requested from the BLM.
- National Archives. This site allows registered researchers to gain access to historical maps and aerial photography. It is a rich resource for historical imagery and maps, going back to the 1930s (https://www.archives.gov/research/order/maps.html).
- *Declassified intelligence archives.* Both the USGS EROS Center and the National Archives and Records Administration maintain archives of more than 990,000 declassified black-and-white CORONA, ARGON, LANYARD, KH-7, and KH-9 satellite reconnaissance film images captured by the United States government worldwide from 1959 to 1980. USGS creates scans of the images on demand for $30.00 per frame. Most of this imagery was collected outside the United States. See https://lta.cr.usgs.gov/declass_1 and https://lta.cr.usgs.gov/declass_2.

Moderate- and Low-Spatial-Resolution Panchromatic and Multispectral Imagery

Table 4.4 summarizes the sources of currently operating panchromatic and multispectral moderate-spatial-resolution systems. Much of this imagery is collected from government satellite systems, and the imagery is usually freely accessible for download, or in the case of Landsat 8 and the USGS Landsat Global Land Survey imagery, is also dynamically served.

USGS EROS

The USGS EROS Center has one of the largest archives of moderate- and low-spatial-resolution imagery because they operate the US National Satellite Land Remote Sensing Archive, which includes the global Landsat archives as well as those for Advanced Very High Resolution Radiometer (AVHRR), Advanced Spaceborne Thermal Emission and Reflection Radiometer (ASTER), and other satellite systems. They also distribute the ESA Sentinel-2 imagery. Moderate- and low-spatial-resolution imagery available from EROS includes:
- All Landsat imagery from 1972 to the present. The archive distributes more than 7,000 terabytes of Landsat imagery annually. Starting with Landsat 1 (launched in 1972) with four bands (G, R, 2 IR) at 80-m spatial resolution, the Landsat constellation of satellites has evolved to the Landsat 8 system (launched in 2013) with 10

bands at 15-m (panchromatic); 30-m (visible, near-infrared, and short-wave infrared); and 100-m (thermal) spatial resolution. No other remote sensing system provides a 40+ year continuous record of global observations. Landsat 9 is currently under construction and will be identical to Landsat 8 with a planned launch date in the early 2020s.

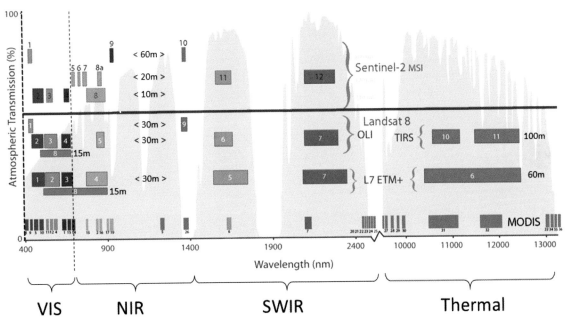

Figure 4.4. Comparison of the spectral resolutions of Landsats 7 and 8, Moderate Resolution Imaging Spectroradiometer (MODIS), and Sentinel-2. Source: USGS

- ESA Sentinel-2 imagery. Sentinel-2A was launched in June 2015 with 13 bands in the visible (10-m resolution), near-infrared (10 to 20-m resolution), and short-wave infrared bands (20 to 60-m resolution). The spectral resolution is very similar to Landsat's (figure 4.4) An identical satellite, Sentinel-2B launched in March of 2017. In a partnership with ESA, EROS distributes Sentinel-2 level 1C imagery.
- AVHRR imagery from 1979 to the present at a 1-km spatial resolution. AVHRR 1.1-km imagery is captured worldwide twice every day and has been collected by NOAA's polar-orbiting satellites from 1979 to the present.
- Multispectral data over the United States acquired by SPOT-4 and -5 satellites. This dataset of licensed commercial imagery is available only to US civil government agencies (2010 to 2013).
- The SPOT Controlled Image Base is an orthorectified product derived from panchromatic images with a 10-m ground sample distance. Coverage is limited to portions of

the United States, Europe, Middle East, Southeast Asia, North and South Korea, Central America, western Russia, and other smaller areas around the world (1986 to 1993).

- ASTER is one of a number of instruments on board the Terra platform, which was launched in December 1999. ASTER provides 14 spectral bands with 15- to 90-meter resolutions depending on the bands. ASTER does not acquire data continuously; its sensors are activated only to collect specific scenes upon request. The instrument consists of three separate telescopes; each provides a different spectral range and resolution. The visible and near-infrared sensor provides four bands at a 15-meter

Table 4.4. Sources of currently operating panchromatic and multispectral moderate-spatial-resolution systems. (esri.url.com/IGT43 for current version.)

Company/Country	UrtheCast	United States	
System Name	Deimos 1	Landsat 7	Landsat 8
Status	Launched in 2009 with 10 year projected life span	Landsat 7 launched in 1999. Scan line corrector problem occurs in 2003.	Launched in 2013 with design lifespan of 5yrs
Orbit	Sun synchronous	Sun synchronous @ 10:00am	Sun synchronous @ 10:00AM
Spatial Resolution (nadir)			
Panchromatic	NA	15m	15m
multispectral	22m	30m	30m
SWIR	NA	30m	30m
SAR	NA	NA	NA
Thermal	NA	120m	100m
Spatial Accuracy	10m	depends on DEM and ground control used	depends on DEM and ground control used
Bands	3	8	11
Spectral Resolution	R, G, NIR	pan, R,G,B,NIR, 2 SWIR, 1 Thermal	pan, R,G,B,NIR, 2 SWIR, 2 Thermal, 1 Cirrus
Image Radiometric Resolution	8 or 10 bits	8 bit	12 bit
Temporal Resolution (nadir to 25 degrees off nadir)	3 days	16 days	16 days
Swath Width at nadir	650km	185km	185km
Stereo	no	no	no
Number of Satellites	1	1	1

resolution. The short-wave infrared sensor provides six bands at a 30-meter resolution. The thermal infrared sensor provides five bands at a 90-meter resolution. The swath width for all sensors is 60 kilometers.
- The USGS and NASA have collaborated on the creation of the Global Land Surveys datasets. Each of these collections was created using the primary Landsat sensor in use at the time, which were processed for either 5-year or 10-year periods with best available images.

Table 4.4 cont. Sources of currently operating panchromatic and multispectral moderate-spatial-resolution systems.

India			European Union	
Resourcesat			Sentinel 1	Sentinel 2
Resourcesat 1 launched in 2003 and is operating with reduced capacity. Resourcesat 2 was launched in 2011 with a planned lifespan of 5 yrs. Both have 3 separate cameras with different resolutions and swath widths.			Sentinel 1A launched in 2014 with 7.25 design lifespan. 1B was launched in June of 2016 with a 7.25 year lifespan	Sentinel 2A launched in 2015 with a 7.25 yr design lifespan, 2B launched in 2016
Sun synchronous @ 10:30			Sun synchronous	Sun synchronous @ 10:30AM
LISS-4	LISS-3	AWiFS		
5.8m	23.5m	56m	NA	NA
5.8m	23.5m	56m	NA	10 and 20m
5.8m	23.5m	56m	NA	20 and 60m
NA	NA	NA	5-25m depending on mode	NA
NA	NA	NA	NA	NA
3	4	4	1	13
R,G, NIR	R, G, NIR, SWIR	R, G, NIR, SWIR	C band	R,G,B, NIR at 10m. 4 NIR bands at 20m. B, NIR & SWIR at 60m.
Resourcesat 1 LISS sensors are 7 bit, AWiFS is 10. All Resourcesat 2 cameras are 10bit.				12 bit
24 days	5 days		12 days with 1 satellite, 6 days with 2	10 days with 1 satellite, 5 days with 2
70	141	740	up to 770km depending on mode	290km
no			no	no
2			2	2

- WELD data. Global monthly and annual 30-meter global composites generated from contemporaneous Landsat 7 Enhanced Thematic Mapper Plus (ETM+) and Landsat 5 Thematic Mapper (TM) data for all non-Antarctic land surfaces are available for 3-year periods. Continental US and Alaska WELD data are weekly, monthly, seasonal, and annual 30-meter continental US and Alaska composites generated from Landsat 7 ETM+ data available for 10 years (2003 to 2012). Products are available through an interactive web interface or as hierarchical data format tiles from a direct download site.
- MODIS is one of a number of instruments carried on the Terra platform, which was launched in December 1999. MODIS provides continuous global coverage every one to two days and collects data from 36 spectral bands (band designations). Bands 1 and 2 have a resolution of 250 meters. Bands 3 through 7 have a resolution of 500 meters. The remaining bands, 8 through 36, have a resolution of 1000 meters. The swath width for MODIS is 2,330 kilometers.

ArcGIS Online

ArcGIS Online dynamically serves the pixel values for eight bands (no thermal or panchromatic) of all scenes of moderate-resolution Landsat 8 imagery worldwide at no charge. It also serves the Landsat Global Land Survey composites of Landsats 1 through 7, created by NASA and USGS, of decadal Landsat imagery. In addition, ArcGIS Online serves multiple cached moderate- and low-resolution imagery from a variety of international space agencies.

ESA

The ESA's Copernicus program is a combination of 30 multiple spectral resolution missions (Sentinels 1 through 6). Sentinel-1A and -1B (radar) and -2A and -2B (optical) have been launched and are operating successfully. https://sentinel.esa.int/web/sentinel/sentinel-data-access/access-to-sentinel-data.

Thermal Imagery—All Spatial Resolutions

Thermal sensors are a special type of multispectral scanner that sense only in the thermal wavelengths. Like radar imagery, thermal imagery is usually rendered in grayscale, with warmer areas represented by lighter shades and cooler areas in darker shades. The image can also be pseudocolored, with blue shades representing cool and red shades warm. Thermal sensors measure in wavelengths of electromagnetic energy that are naturally emitted from objects as a direct result of the objects' temperatures. Thermal sensors do not measure reflected energy, but rather emitted energy, and therefore can be acquired any time of day or night. Because of issues caused by the atmosphere, thermal imagery is collected with wavelengths of either 3 to 5 microns, 8 to 14 microns, or both.

Thermal imaging sensors can be flown as part of a multispectral scanner or in conjunction with one. For example, the Landsat TM sensor (on Landsats 4, 5, and 7) has a thermal band of 10.40 to 12.50 microns in addition to sensing in the visible, near-infrared, and middle-infrared wavelengths. The thermal band has a spatial resolution of 120 meters on Landsats 4 and 5 and 60 meters on Landsat 7. Landsat 8 also has a thermal sensor, called the Thermal Infrared Sensor (TIRS), which flies on the same platform as the Operational Land Imager (OLI). The OLI senses in the visible, near-infrared, and middle-infrared wavelengths. TIRS has two thermal bands (10.6 to 11.2 and 11.5 to 12.5 microns) and a spatial resolution of 100 meters. All Landsat imagery including the thermal data is freely available for download from a number of USGS websites including Earth Explorer (http://earthexplorer.usgs.gov), Global Visualization Viewer (GloVis) (http://glovis.usgs.gov/next/), and Landsat Look Viewer (http://landsatlook.usgs.gov).

Similarly, the Moderate Resolution Imaging Spectroradiometer (MODIS) senses in 36 bands at varying spatial resolutions depending on the wavelength. There are six thermal bands; four in the 3- to 5-micron range and two in the 10- to 12-micron range. The spatial resolution of the MODIS thermal imagery is 1 km. Additional information about the acquisition of MODIS imagery can be found at http://modis.gsfc.nasa.gov. Finally, the AVHRR also senses in the thermal wavelengths including one in the 3- to 5-micron range and two in the 10- to 12-micron range. AVHRR imagery is provided free by NOAA through the Comprehensive Large Array-Data Stewardship System at www.class.ncdc.noaa.gov/.

Thermal sensors on aircraft can also be a component of a multispectral scanner. However, it is more common for the sensor to be a thermal sensor with multiple wavelengths being collected just in the thermal range. These sensors have been flown on fixed-wing aircraft and helicopters and most recently on UAS's, otherwise known as drones. There are many applications for thermal imagery flown on an aircraft. Most notably, these applications include forest fire detection and measurement of heat loss from buildings. The US Forest Service and NASA have cooperated extensively on detecting forest fires using thermal imagery and have put thermal sensors on airplanes, helicopters, and most recently on UASs. Many utility companies employ thermal sensors to detect heat loss from houses to demonstrate where increased insulation or replacement of windows could significantly reduce heating costs. Other applications for thermal data include disaster mapping such as volcanic activity and sensing of large mammals (bears, moose, elk, etc.).

Hyperspectral Imagery—All Spatial Resolutions

Hyperspectral imagery is not generally available from remote sensing systems and is most often flown on aircraft for specific projects. While flown by both private and public agencies, the most well-known hyperspectral remote sensing system is NASA's AVIRIS, which operates an optical sensor capturing the spectral radiance in 224 contiguous spectral bands with wavelengths from 400 to 2500 nanometers. AVIRIS has been flown from 1992 to the present on four aircraft platforms mostly over the United States and Canada. The AVIRIS archive can be searched and data can be downloaded from http://aviris.jpl.nasa.gov/alt_locator/.

The only civilian source of hyperspectral satellite imagery is collected by the Hyperion sensor operating on the Earth Observing 1 (EO-1) Extended Mission. It is able to resolve 220 spectral bands (from 0.4 to 2.5 μm) at a 30-m spatial resolution. The EO-1 satellite was launched November 21, 2000, as a one-year technology demonstration/validation mission. After the initial technology mission was completed, NASA and the USGS agreed to the continuation of the EO-1 program as an extended mission. The EO-1 Extended Mission is chartered to collect and distribute Hyperion hyperspectral and Advanced Land Imager multispectral products according to customer tasking requests. Archived Hyperion imagery is available for search and download from the USGS EROS (http://eo1.usgs.gov/products/search).

Sources for Imagery Collected from Active Sensors

Lidar

Lidar imagery provides a comprehensive, detailed, and precise picture of the elevation of the earth's surface as well the height of vegetation, buildings, and other features. The data provides a detailed view of the vertical structure of our environment. Because lidar penetrates the canopy, it provides the ability to "see under trees," revealing roads, trails, manmade alteration of the landscape, and other landscape features that on an aerial photograph may be obscured by trees or vegetation. The lidar point cloud and its derivatives have myriad applications for land management, planning, archaeology, engineering design, hydrology, and other applications. High-resolution elevation data and forest structure metrics, such as tree height and canopy density, significantly enhance our ability to assess and monitor carbon stocks, document sea level rise, and map impervious surfaces and vegetation.

As reviewed in chapter 3, lidar data is collected as a "point cloud"; each point typically contains data representing the point's geographic location, elevation, and return intensity. Since most modern lidar missions produce high-density point clouds (multiple points per square meter), point cloud data is among the most space and resource consuming of remotely sensed datasets. The standard data format for the point cloud is the LASer (LAS) format, and lidar data is typically distributed as LAS files. The LAS format is a transfer standard but is not optimized for direct use because of the large size of its files. However, a number of lossless compression formats exist as alternatives to the LAS format, including zLAS and LAZ, which typically result in compression factors of 5×.

Lidar data is publicly available for a growing number of states, counties, and municipalities across the United States. Existing lidar data and derivatives can often be downloaded from many sources including Open Topography (http://www.opentopography.org/), NOAA's Digital Coast (https://coast.noaa.gov/digitalcoast/) and the USGS's Earth Explorer (http://earthexplorer.usgs.gov/).

Lidar data is typically provided in one of the formats below:
- Classified point cloud—this format is typically the least processed form made available. Access to the point cloud is needed to derive custom elevation models, to edit or change the point classification, or to visualize the point cloud directly. Working with the point cloud at a regional scale can be cumbersome and slow because the number of points in the point cloud is massive.
- Lidar-derived elevation models—derived elevation models are often the most useful products for the end user. Elevation models derived from the point cloud can depict the elevation of the ground (digital terrain models) or the elevation of the highest surface (digital surface models). Digital elevation models are provided in raster or three-dimensional model format and can be used for myriad types visualization, analysis, and modeling. Hillshades derived from elevation models are an excellent way to visualize lidar-derived topography. (For more information on elevation models, see chapter 8.)
- Elevation contours—contours are often derived from the point cloud and provided for download at the portals listed above.

If lidar data doesn't exist for your area of interest, or you require lidar data with different specifications or characteristics from what does exist, here are some very useful resources to help you with your planning:
- Lidar 101: An Introduction to Lidar Technology, Data, and Applications—https://coast.noaa.gov/digitalcoast/training/lidar-101.html
- USGS Lidar Base Specification—http://pubs.usgs.gov/tm/11b4/
- Manual of Airborne Topographic Lidar—http://www.asprs.org/Press-Releases/ASPRS-Launches-First-eBook-Manual-of-Airborne-Topographic-Lidar.html

- ASPRS Positional Accuracy Standards for Digital Geospatial Data—https://www.asprs.org/pad-division/asprs-positional-accuracy-standards-for-digital-geospatial-data.html.

Radar

Radar imagery has many uses. Because of its long wavelengths and use of active sensors, radar imagery can be acquired under almost any conditions. It is not impacted by clouds or fog and can even be acquired at night. Areas that experience constant cloud cover such as the Amazon jungle and coastal Alaska, which have very little optical imagery, can be regularly imaged with radar. In addition, depending on the wavelength, radar imagery has the ability to penetrate through the leaves/canopy of the forest and image more of the forest structure or to penetrate into the soil and reveal more about the soil characteristics and wetness. Two radar images of the same area can produce an image with parallax much like a stereo pair of aerial photographs. This parallax can be used in radargrammetry (photogrammetry for radar) to produce topographic/elevation data. Another topographic mapping method called interferometry uses two radar antennas that are spaced apart and can receive a single radar signal that is out of phase. Topographic information is extracted by analyzing the phase shift.

While radar systems have been and will continue to fly on aircraft, companies that provide these services change rapidly. Space-based systems have provided greater stability and sources of radar imagery over time even though many of these missions have ended as well. The imagery and products generated from these space-based missions are more readily accessible and available for use. The very first space-based radar mission was called Seasat, launched in 1978 by the United States with the goal of monitoring ocean conditions including the polar sea ice. The radar sensor was L band using HH polarization. Unfortunately, there were technical issues with the Seasat power system and the mission lasted less than 100 days. The next radar missions were part of the US shuttle program and called the Shuttle Imaging Radar (SIR) experiments. SIR-A (1981) and SIR-B (1984) used L-band radar with HH polarization, the same as Seasat. SIR-C (1994) was designed to experiment with multiband radar and used L, C, and X bands with multiple polarizations. After this, the United States has mostly disregarded radar imagery except for the Shuttle Radar Topography Mission in 2000. The goal of that mission was to use interferometric radar imagery to create topographic data for the majority of the inhabited earth's surface. The project was a joint one between NASA and the agency then called the National Imagery and Mapping Agency and now called the National Geospatial-Intelligence Agency (NGA). The mission was highly successful in collecting 1-arcsecond elevation data for most of the earth.

Since the early days of Seasat and the Shuttle Radar missions, many other countries have launched platforms with radar sensors into space. Of particular note are Canada,

Japan, and the European Space Agency (ESA). Table 4.5 presents a summary of some of the more important radar sensors, both historically and operating today (Lillesand et al., 2015). Because of the varying look angles used to collect radar imagery, the spatial resolution of these images also varies greatly and is not recorded in this table. Almaz-1 is listed in the table because it is the first commercial radar system, however it did not last long and is not as significant a data source as the other sensors in the list.

Table 4.5. Summary of important space-based radar sensors

Sensor	Source	Band	Polarization	Dates
Almaz-1	USSR	S	HH	1991–92
sERS-1	Eur. Space Ag.	C	VV	1991–2000
ERS-2	Eur. Space Ag.	C	VV	1995–2011
JERS-1	Japan	L	HH	1992–98
Radarsat-1	Canada	C	HH	1995–2013
Radarsat-2	Canada	C	Quad	2007–present
Envisat	Eur. Space Ag.	C	Dual	2002–12
ALOS	Japan	L	Quad	2006–11
ALOS-2	Japan	L	Quad	2014–present
TerraSAR-X	Germany	X	Dual	2007–present
COSMO-SkyMed 1 & 2	Italy	X	Quad	2007–present
COSMO-SkyMed 3	Italy	X	Quad	2008–present
COSMO-SkyMed 4	Italy	X	Quad	2010–present
TanDEM-X	Germany	X	Dual	2010–present
RISAT-1	India	C	Quad	2012–present
Sentinel 1A & 1B	Eur. Space Ag.	C	Dual	2014/16–present
SMAP	United States	L	Tri	2015–present

Soon, two new countries look to enter the radar data collection group. They are Spain, with Paz, a dual-polarization X band radar, and the United Kingdom with NovaSAR-S, a tripolarization S-band radar. In addition, Sentinel-1B was launched in April 2016. Launches are regularly postponed, and therefore it is important to check mission websites for the current status of any of these radar sensors.

Summary—Practical Considerations

This chapter has provided guidance on how to decide which imagery will best meet the requirements of your projects, as well as where to find it. The basic decision framework captured in the questions offered in this chapter do not change as new imagery products are created, because the questions are driven by your project requirements, not by changes in technology available. By asking and re-asking the questions with respect to each new project, you will be able to fairly easily determine the imagery that will best meet your needs.

However, sources of imagery change constantly as new sensors are invented and new platforms developed. As a result, the lists of imagery sources presented in this book will quickly become obsolete. Two decades ago, only two organizations provided global moderate-spatial-resolution satellite imagery of the earth to civilians (USGS and SPOT), and the first commercial airborne lidar system had just been introduced. Now, multiple countries and private companies offer a wide array of very-high-to-moderate-spatial-resolution imagery globally, and lidar technology has been adopted worldwide. Because there is no centralized resource for learning about new imagery sources, the remote sensing analyst must endeavor to continually keep abreast of new developments in remote sensing technology and to understand when that technology has matured enough to be adopted in an operational environment.

Section 2
Using Imagery

Chapter 5
Working with Imagery

Introduction

Chapter 3 presented a primer on imagery fundamentals, and chapter 4 provided instruction on choosing imagery appropriate for a given project. This chapter provides an overview of the fundamentals of working with the chosen imagery. The chapter begins by discussing the issue of scale in imagery and then covers a number of imagery processing and visualization topics including commonly used image storage and formats (pixel storage, image statistics, image compression, NoData, and image pyramids), image display fundamentals (histograms, image stretches), image enhancement and filtering, and image mosaics. The chapter concludes with a section on accessing imagery.

Image Scale

Using digital imagery and computers, it is easy to zoom in and out of an image, changing the scale of the display. Scale is the ratio of the length between two points on a paper or digital map to the actual distance between the same two points on the ground. It is expressed using a colon to differentiate between the distance of one unit on an image or map to the corresponding distance on the ground. For example, 1:15,840 is a scale commonly used in forestry applications where one inch on the image is equivalent to 15,840 inches (one-quarter mile) on the ground. This scale could also be represented with the units as 1 inch:0.25 miles.

The concepts of scale and spatial resolution are intertwined. Large-scale imagery has higher spatial resolution than small-scale imagery and, as a result, the terms "large" and "small" are a little counterintuitive. The term "large scale" refers to ratios with small denominators (e.g., 1:600) and images with high spatial resolution. The term "small scale" refers to ratios with a large denominator (e.g.1:60,000) and a corresponding lower spatial resolution.

Scale is an important consideration in both image analysis and image visualization. To use imagery effectively for mapping and analysis, the GIS analyst must choose imagery with a spatial resolution appropriate for the scale of the mapping and analysis that will be conducted during the course of the project (see chapter 3 for a more detailed discussion of spatial resolution).

An image's spatial resolution determines its maximum appropriate scale for display. As image spatial resolution increases, the maximum appropriate display scale also increases. When an image is displayed at a scale that exceeds its maximum appropriate display scale, the image becomes pixelated or fuzzy.

In addition to determining an image's maximum appropriate display scale, an image's spatial resolution determines the size of the smallest decipherable feature in the image. For an object or feature to be discernible in a digital image, the object's smallest x/y dimension must be at least two times the width of a single image pixel (Jensen, 2000). For example, in a 30-meter spatial resolution (30 × 30-meter raster cells) Landsat thematic mapper (TM) image, features with horizontal dimensions below 60 meters such as cars, small buildings, and houses are not discernible, as shown in figure 5.1. However, such features are readily discernible in 6-inch (.15-m) spatial resolution orthophotography, even when it is displayed (as it is in figure 5.1) at a smaller scale than the maximum recommended for 6-inch imagery.

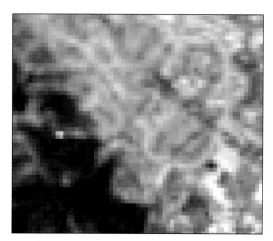

Landsat Imagery (true color) 30 meter spatial resolution

High Resolution Orthophotography (true color) 6 inch spatial resolution

Figure 5.1. Effect of spatial resolution on image appearance at 1:24,000 scale (esriurl.com/IG51)

It is important to note that increased spatial resolution leads to larger image file size for a given geographic area. Larger images consume more hard drive space and require longer times to analyze. As spatial resolution increases linearly, the number of pixels (and image size) grows arithmetically. Doubling spatial resolution (such as going from 2-foot to 1-foot spatial resolution) quadruples the number of pixels in the image and the image's file size. On a paper map, the scale defines the relation between paper units and ground units. On a computer screen, such a relation is more difficult to define because the sizes and resolutions of monitors can vary significantly. A definitive true screen scale is also not required, because you should not use a ruler to measure distances from a monitor. Some relation between screen pixels and map units is still required. The established convention is to assume each pixel on a screen is has a size of 1/96 inch (equivalent to 96 dots per inch)[1]. Native display scale (NDS) is the scale where one pixel on the ground represents one pixel on the screen. At display scales larger than NDS, the original pixels will be spread over more than one screen pixel and start to look blurry. At display scales smaller than this, some of the information in the pixel will be lost during display because multiple ground pixels will need to be interpolated to display a single display pixel.

As a result:

NDS = Raster resolution in meters × 96 dpi / 0.0254 inches per meter

For example, a Landsat scene with 30-meter pixels has an NDS of 30m × 96 dots per inch / 0.0254 meters per inch, or 1:113,386. If displayed at this scale, then each pixel on the screen would correspond to one pixel in the Landsat scene.

Due to the high resolution of screens, an image will remain looking sharp until about two times the NDS. If the scale is increased further the image will start to look blurry. Hence the following relationship:

Maximum Usable Display Scale = Raster resolution in meters × (96 dpi / 0.0254 inches per meter) / 2 times the NDS

= Raster resolution in meters × 1890

For example, for 30-meter Landsat data the maximum usable display scale for manual interpretation is 30 m times 1890, or about 1:56,700.

---

[1] On some screens, such as high-resolution retina screens, the pixel size is assumed to be 192 dpi.

Image Storage and Formats

Geospatial imagery is stored digitally as raster datasets. Raster datasets include several essential components: the image pixels themselves, image pyramids, and metadata (metadata includes image statistics and georeferencing information). There are many types of raster datasets, typically named for the format the pixels are stored in. This section discusses the most common types of raster dataset formats (see overview table 5.1), pixel storage, image properties such as georeferencing and image statistics, raster products, compression, image NoData, and image pyramids. This section also discusses image compression.

Pixels

Pixels are stored in a format that is often used as the name for the type of raster dataset (TIFF, JPEG, etc.). There are many different formats for storing pixels, each with different options for the types of pixel data they can store, the types of compression that they can use (see the discussion on compression below), how they support pyramids and NoData, and the ways that they store metadata.

Tiling Pixels

The simplest pixel formats store the pixels one after the other, row by row, while other formats break the pixels into tiles or blocks. The latter typically store the blocks of pixels in chunks of 128 × 128, 256 × 256, or 512 × 512 pixels. Pixels within a block that are geographically close to one another are stored physically close together on the computer's hard drive, improving image access speed.

Pixel and Band Ordering

In multispectral imagery—for example, a natural color orthophoto that contains a red (R) band, a green (G) band, and a blue (B) band—different file formats can store the pixels from the different bands in different orders. There are three ways to store multispectral pixel values in a raster dataset: band interleaved by pixel (BIP), band interleaved by line (BIL), and band sequential (BSQ).

Table 5.1. Common raster dataset formats

Format	Extension	Compression	Georeferencing	Pyramids	Statistics	NoData
TIFF	.tif	None, LZW, PackBits, JPEG	External (.TFW + .PRJ)	Internal or External (.ovr)	External (.AUX.XML)	Special Tag
TIFF – Very popular format for raster data with many options for compression, etc.							
geoTIF	.tif	None, LZW, PackBits, JPEG	Internal, Control points or External (.AUX.XML)	Internal or External (.ovr)	Internal or External (.AUX.XML)	Special Tag
geoTIFF – Enhancement to tagged image file format (TIFF) to define specific internal tags required for geospatial data. Should be the default format for most applications that use orthorectified imagery. The geoTIFF format is appropriate for the gamut of GIS and remote sensing imagery uses, ranging from storage of multispectral images for scientific analysis to the storage, management, and analysis of single-banded, thematic raster maps. Most GIS and remote sensing packages support geoTIFF imagery. The ability to use lossy JPEG compression also enables significant reductions in the file size when lossy compression is suitable.					
IMG	.img	None, PackBits	Internal or External (.AUX)	.rrd	External (.AUX)	Special Tag
IMG – Made popular by Erdas. Primarily used only by Erdas Imagine.					
MrSID	.sid	MrSID	Internal	Internal	External (.AUX.XML)	No Support
MrSID – Popular but proprietary format for delivery of pre-enhanced orthos for visualization. Lossy compression with very high compression factors. Security identifier (SID) is a highly compressed format developed by LizardTech. A license is required to compress imagery into this format.					
JPEG 2000	.jp2	Lossless or Lossy JPEG 2000	Internal	Internal or External (.ovr)	Internal	Special Tag
JPEG 2000 – Format with lossy and lossless wavelet compression. Its complexity can result in slow data access.					
ECW	.ecw	Lossy ECW	Internal	Internal	Internal	Special Tag (in V3)
ECW – Proprietary format used by Erdas for lossy compression of natural color imagery. A license is required to serve to the web.					
JPEG	.jpg	Lossy JPEG	.JPW	External (.ovr)	External (.AUX.XML)	No Support
JPEG – Popular web format with lossy compression. Should be used only for small natural color images with lossy compression.					
TileCache		JPEG, PNG	Internal	Internal	None	As PNG
TileCache – Format optimized for web distribution of rendered base images and natural color imagery					
MRF	.mrf	None, LERC, Deflate, PNG, JPEG	Internal	Internal	Internal	Internal Mask
MRF – Format optimized for cloud storage and access of satellite and aerial imagery					
CRF	.crf directory	None, LERC, LZW, PNG, JPEG	Internal	Internal	Internal	Internal Mask
CRF – Format optimized for cloud-based processing and analysis of large rasters					

BIP stores each band for each pixel in order. The red pixel value of the first pixel is stored, followed by the green pixel value of the first pixel, followed by the blue value of the first pixel. Then the values of the second pixel are stored, and so on. The example below illustrates BIP for a three-banded natural color (red, green, blue) image that is three pixels by three pixels:

R,G,B,R,G,B,R,G,B

R,G,B,R,G,B,R,G,B

R,G,B,R,G,B,R,G,B

BIL orders the pixels by lines (rows). All the red pixel values for the first row are stored first, followed by the green values of the first row, followed by the blue values of the first row. The same 3 × 3 three-banded image shown above would be stored like this using BIL:

R,R,R,G,G,G,B,B,B

R,R,R,G,G,G,B,B,B

R,R,R,G,G,G,B,B,B

BSQ stores the pixels by band. All the red pixel values for the entire image are stored first, followed by all the green pixel values and then the blue pixel values. Here is our 3 × 3 RGB image as BSQ:

R,R,R,R,R,R,R,R,R

G,G,G,G,G,G,G,G,G

B,B,B,B,B,B,B,B,B

BIP is typically the preferred pixel storage for natural color imagery with three bands, while BSQ is typically used for imagery with a large number of bands.

Image Properties

In addition to the pixel values, many other properties need to be stored within a raster dataset. These include general information about the pixel depth (8-bit unsigned, 16-bit unsigned, 32-bit float, etc.), the number of bands, georeferencing information, and image statistics.

Georeferencing

Most georectified imagery uses a simple affine transformation to define the relationship between pixels (columns and rows) and a ground coordinate system. The definition of the projection (and data) is also required. There are various ways to store this georeferencing information. Some formats, such as geoTIFF, JP2, IMG, and MrSID, embed the georeferencing information directly in the image files. For example, a geoTIFF file is a type of TIF that includes internal georeferencing information. Other formats, such as TIFF and JPEG, do not allow for the internal storage of georeferencing information. For those formats, georeferencing information is stored externally as two files that have the same prefix as the main image files. The .prj file stores information about the projection of the rasters, and the world file (.TFW for TIFF or .JPW for JPEG) stores the affine transformation parameters as a simple list of six values. Table 5.2 provides an overview of georeferencing information storage options for the most common raster dataset formats.

The world files can represent only simple affine transforms and are not suitable for handling more elaborate georeferencing transforms such as higher-order polynomial, warp, or RPC (Rational Polynomial Coefficient), which are often used to georeference satellite scenes. Some formats such as geoTIFF and JP2 can store those more advanced georeferencing parameters internally; otherwise, those parameters are stored in external .AUX or .AUX.XML files with the same prefix as the main image file. RPCs can be stored in a geoTIFF using a proposed extension to GeoTIFF 1.0 that is supported by the Geospatial Data Abstraction Library (GDAL) see geotiff.maptools.org/rpc_prop.html). While not officially part of the GeoTIFF standard, this enhancement is widely supported.

Image Statistics

For some types of imagery, image statistics may be required for the correct rendering of the imagery. This is primarily true for higher-bit-depth imagery such as satellite imagery and elevation data. The statistics typically computed are the minimum, maximum, mean, and standard deviation (see figure 5.2 for an illustration). The statistics are usually computed from the entire image, but image-wide statistics can be misleading if they contain anomalous areas such as clouds. The statistics are therefore generally used only as a rough guide for determining image display and image enhancement parameters. Image statistics are computed by image processing and GIS software and are either stored internally or added as external .AUX.XML files with the same prefix as the main image file. Table 5.2 shows how statistics are stored for the most common raster data formats.

Statistics	
imagery_c4	
Build Parameters	skipped columns:1, rows:1, ignored
Min	1
Max	255
Mean	113,7550341491982
Std dev.	54.020089979155
Classes	0

Figure 5.2. Image statistics

Raster Products

Raster products are designed to simplify adding imagery from sensors to an ArcGIS Desktop map. Raster products are defined in the imagery's metadata and provide a selection of ways to dynamically display the imagery as it was designed to be displayed by the vendor that produced it. Often—for example—multiple individual bands must be combined in a specific (and perhaps not so obvious) way to create a natural looking image. Sometimes, the data needs to be enhanced before one can understand certain features in the image. Sometimes, functions such as pan sharpening are required to view an image optimally in a certain way. Raster products allow the ArcGIS Desktop user to choose to display the imagery with a selected vendor-specified optimization applied; the system then applies the needed processing on the fly to the portion of the image that the user is displaying.

For example, selecting an image metadata (IMD) file for a Landsat scene will provide options for opening the scene as a multispectral, panchromatic, or pan-sharpened product, and the associated image attributes and band combinations will be correctly defined. Similarly, the ingesting of the parameters into a mosaic dataset (used to manage large collections of imagery as discussed below) is handled by raster types that extract the required parameters and set up suitable processing.

Compression

Large, high-resolution image files can have billions of pixels. Even with ever-increasing hard drive sizes, these vast amounts of imagery require massive amounts of hard drive space. Image compression reduces image size by more efficiently encoding the raw image data. Image compression is highly desirable, because it significantly reduces the size of imagery and the costs of image storage. Compression also improves imagery access speed, because less data needs to be transferred from storage to the CPU. However, if the decompression is CPU intensive, performance can be degraded. Table 5.2 shows compression options for the most common types of raster data formats.

One important consideration is whether to use lossy compression or lossless compression. Lossy compression provides much smaller file sizes than lossless compression, but the tradeoff is some reduction in image information. Lossy compression is best suited for image display (the degradation resulting from compression often isn't perceptible to the human eye). But because lossy formats change the original imagery's raw pixel values and remove information, this compression should be avoided for image interpretation, feature extraction, and image classification. For these applications, lossless compression should be used.

Lossless compression ensures that the data read is an exact match to the original data. Lossless methods include LZW, PackBits, Deflate, PNG, and some types of limited error raster compression (LERC) and JPEG 2000 compression. Lossless compression has significant value if the data has large areas with the same values, such as 1-bit scanned maps or rasters with large areas of NoData (see below for a discussion of NoData). For continuous tone imagery, lossless compression factors are typically around 1.5 to 3×.

Lossy compression results in some loss of image data but results in higher compression factors and, as a result, smaller file sizes than for lossless compressed imagery. There are two primary types of lossy compression: those that use the discrete cosine transform and those that employ wavelet compression. Compression algorithms such as the very popular JPEG compression use the discrete cosine transform. Note that JPEG is both a file format and a method of compression that can be included in other formats such as a TIFF with JPEG compression.

JPEG typically achieves compression factors of about 5 to 10× without significant artifacts. When artifacts do appear, they are often in the form of small blocks (8 × 8) of similar pixels with a sharp transition to the next block. JPEG compression is typically used only with 8-bit imagery, although a 12-bit version of JPEG is supported by ArcGIS.

The other primary lossy compression method—wavelet compression—uses wavelet transforms. Wavelet compression algorithms include JPEG 2000, ECW, and MrSID. Wavelet compressions typically achieve higher compression factors with good image quality; typically 30 percent to 50 percent better compression than the equivalent JPEG compression. However, wavelet compression often introduces artifacts such as smearing or loss of texture. Decompression of wavelet-compressed imagery can also be CPU intensive, resulting in longer access times. One of the advantages of wavelet compression is that it inherently includes lower-resolution pyramids of the imagery.

One additional disadvantage of lossy compression is that you cannot define the maximum amount that a pixel may vary from its original pixel value; lossy compression algorithms provide control only for a "quality factor" or "compression factor." Controlled lossy compression is a form of lossy compression that enables the maximum difference between the original and compressed pixel values to be defined. Its primary use is for storing elevation data or high-dynamic-range satellite imagery, which is often stored as floating point or 16-bit integer data. LERC is a controlled lossy compression in which the maximum error can be defined. Providing larger tolerances can result in significant compression factors (although not as high as for other types of lossy compression). Providing the LERC algorithm with a tolerance of zero results in fully lossless compression. One of the advantages of LERC is its very high compression and decompression speeds, making it a valuable compression algorithm for raster data used in science and analytics.

Using compression on imagery nearly always results in the imagery being internally tiled into blocks. The exception is a JPEG file. JPEG files use lossy JPEG compression but do not include tiling or pyramids. For this reason, it is typically not recommended to use JPEG files for larger images. JPEG file storage is inefficient in terms of access speed. To gain access to the last pixel in a JPEG file requires the complete file to be read. Note that using JPEG compressions in geoTIFF files circumvents this inefficiency and is recommended; in this case, the imagery is broken internally into tiles small enough to be read efficiently.

NoData

Image pixel values are numeric, but sometimes a pixel has no data associated with it because that data is missing or outside the extent of the frame of the imagery. These data-absent areas are stored with a NoData pixel value. NoData values can be given a specific display color or made transparent. Figure 5.3 shows an image of Sonoma County where the blue areas (where image data is absent) are NoData pixels. As is evident in figure 5.3, the structure of a raster dataset defines that it must be a rectangular grid; one reason that NoData needs to be specially defined. NoData is an important element of the raster data model because it provides a mechanism for separating invalid pixels, border areas, and background areas from valid pixel data.

Figure 5.3. NoData areas

It is important to recognize the difference between NoData pixels and pixels with a value of 0. To illustrate this difference, consider a digital elevation model (DEM) representing elevations ranging from 150 feet above sea level to 30 feet below sea level. In this example, the pixel value 0 is a valid value representing areas of mean sea level. In this case, NoData should be set to a pixel value outside the −30 to 150 range; −9999 would be a good choice. Table 5.2 shows the NoData options for the most commonly used raster data formats.

Areas of NoData are most commonly defined by a specific pixel value. For NoData defined by a pixel value, 0 or 255 is often used for 8-bit imagery. For 32-bit floating point imagery, −9999 (for elevation data) or the smallest possible value is often used.

NoData can also be stored as part of a raster dataset as a mask. This NoData mask can be either an internal band or a stand-alone .msk file. In ArcGIS, when working with mosaic datasets (discussed later in this chapter) it is also possible to define NoData areas using vector geometries (such as footprints). This approach is useful when working with large collections of overlapping images.

Pyramids

Image pyramids or pyramid layers speed up image display by displaying lower-resolution versions of an image when zoomed out. Pyramids are highly recommended to enable faster image display and are typically created for larger images (>3000 rows or columns). Figure

5.4 illustrates how pyramids reduce the number of pixels as display scale decreases. When zoomed out at 1:50,000, the reduced resolution image has only 1/16th of the pixels of the image that displays at 1:12,500. The 1:50,000 image renders much faster because of its small number of pixels but, because they are zoomed out to 1:50,000, the viewer will not be aware that they are viewing a reduced resolution version of the imagery.

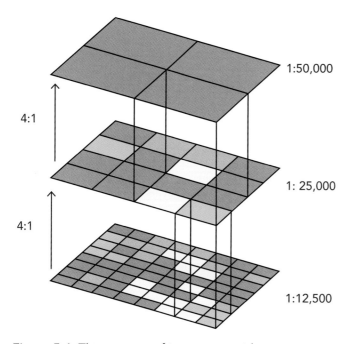

Figure 5.4. The concept of image pyramids

Pyramids are typically created with a factor of 2× or 3× the original data. For example, if the original data has a resolution of 1 meter and a factor of 2× is used, then the multiple reduced resolution pyramid layers would have pixel sizes of 2 m, 4 m, 8 m, 16 m, 32 m, 64 m, etc.

Pyramids are often created automatically when an image is created, but they can also be created manually by running a tool. Depending on the format of the image file, pyramids can be stored internally (this is the case for most wavelet-compressed files as well as some flavors of TIFF) or as external .ovr or .rrd files with the same prefix as the full-resolution image. Table 5.2 includes information about how pyramids are implemented in the most common raster dataset formats.

The existence of pyramids increases the size of an image file because both the image and the associated reduced resolution rasters must be stored on disk. For uncompressed imagery, pyramids with a factor of 2× would increase image file size by 33 percent, while a factor of 3× would increase the size by 15 percent. For compressed imagery, the factors are typically a bit larger as the pyramids do not compress as well.

When creating pyramids, the method of sampling the data must also be considered. The most common pyramid resampling methods are nearest-neighbor and bilinear/average interpolation. Nearest neighbor takes one value from the four input pixels (assuming a 2× factor is used to create pyramids). Bilinear interpolation takes the weighted average of the nearest four values. Nearest neighbor is appropriate for thematic rasters (an average value has no meaning for those types of rasters); bilinear interpolation should be used for continuous data such as remotely sensed imagery. Note that using nearest-neighbor resampling to create pyramids for continuous imagery will result in small-scale views of the images looking grainy.

It should be noted that creating or rebuilding pyramids can be time intensive, especially for large sets of high-resolution imagery—such as a countywide mosaic dataset of orthophotography—because the whole dataset needs to be processed.

Image Display for Continuous Raster Data

Continuous raster data, such as multispectral orthophotography and satellite imagery, is extremely useful for land management, planning, navigation, and general reference. High-resolution, multispectral imagery is now ubiquitous and often the preferred choice for a base image in GIS projects. By understanding the concepts of image display, you will be able to maximize the usability of imagery in your projects. This section discusses two fundamental image display concepts: histograms and image contrast stretches.

Histograms

Image histograms provide a graphic representation of the distribution of pixel digital number (DN) values in an image. This graphic summary is often a useful tool for image quality control and for detecting image problems or anomalies that might not necessarily be noticeable otherwise.

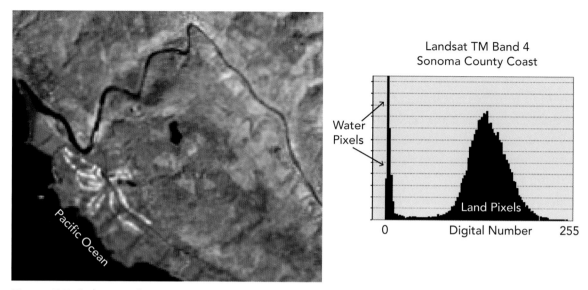

Figure 5.5. Relating a histogram to an image

Figure 5.5 shows the near infrared band histogram for an area of Landsat TM imagery over the Northern California coast. The values on the x-axis represent the DNs of the pixels in the image. In this case, pixel values range from 0 to 255. The y-axis values represent the number of pixels with a given DN.

This area contains a section of ocean and a section of land. Since water absorbs infrared energy, the reflectance of the water in the ocean, lake and rivers is minimal and the ocean pixels are very dark (low DNs). The water pixels are represented in the histogram in the spike on the left, at the low end of the DN scale (the x-axis). Since terrestrial cover types generally reflect infrared energy, the land pixels are much brighter; on the histogram, these land pixels are represented by the large hump in the middle of the histogram.

Image Stretch

Proper image display is critical to the utility of imagery for human interpretation. Displaying images at optimal contrast allows the viewer to resolve features of interest that might not otherwise be visible or perceptible. Often, the range of values that an image contains is not correctly stretched across the range that the computer can display, and as a result the image may appear dark or have little contrast. Modern satellite and aerial sensors often record with 14 bits of dynamic range resulting in 16,384 possible values for each band, while a computer monitor can display only 256 possible values for each band, and our eyes can perceive fewer than 256 shades of gray. Applying a contrast stretch takes full advantage of the range of values the computer can display by "stretching" the range of values in the image to the full range of values available to the computer display. Contrast stretching is performed for continuous image data and often temporarily for display purposes. A contrast stretch is most useful when applied on the fly to the pixels visible in the display to optimize the stretch for the particular area that an image analyst is viewing at a given time. Figure 5.6 demonstrates the usefulness of the contrast stretch in optimizing imagery for display and interpretation. Note how the histogram on the unstretched image on the left is compacted and how most of the left image's pixels are displayed in a small part of the radiometric resolution. On the other hand, the histogram for the stretched image on the right takes advantage of a much broader range of pixel values and, as a result has much higher contrast.

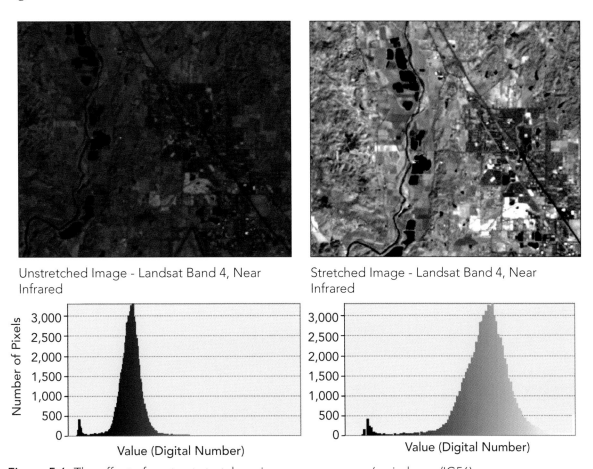

Figure 5.6. The effect of contrast stretch on image appearance (esriurl.com/IG56)

Many types of contrast stretches are available, and each stretches the histogram in a different way. A minimum–maximum stretch uses the entire range of image values (i.e., the dynamic range) and stretches them across the full radiometric resolution range of grayscale values that the software can display (0 to 255 for 8-bit imagery). If an image has a value range of from 21 to 210, and a minimum–maximum stretch is applied, an image pixel with a value of 21 will become 0 in the stretched image and an image pixel with a value of 210 will become 255 in the stretched image; everything between 21 and 210 in the image will stretch accordingly in a linear fashion. The minimum–maximum stretch is one of the most commonly used contrast stretches and is a good utilitarian stretch that will improve image contrast.

Standard deviation and percent clip stretches automatically set the pixels with very low and very high pixel values in an 8-bit image to 0 and 255 respectively, stretching the pixel values in between. The assumption for these types of stretches is that the DN values are normally distributed and that trimming off the relatively small number of pixels with extreme values will increase image contrast without detracting from the visual information displayed. Figure 5.7 illustrates the concept of a 2-σ standard deviation stretch. For this stretch, extreme pixel values are determined to be those further than two standard deviations from the mean pixel value of the input image (about 5 percent of input image pixels). These roughly 5 percent of input image pixels occupy the red areas to the left and right of the vertical 2σ lines in figure 5.7. Pixels in the stretched image will be set to 0 in the left red area and 255 in the right red area. The pixels between the 2σ lines—the pixels within two standard deviations of the mean—will be stretched linearly between 0 and 255.

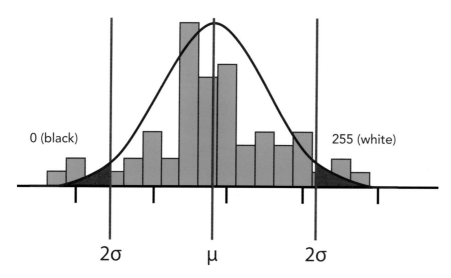

Figure 5.7. Standard deviation stretch (esriurl.com/IG57)

Image stretches are applied by the use of a lookup table. The lookup table contains the "from" and "to" values for the stretch. The "from" values include every possible image pixel value, while the "to" values represent the corresponding grayscale value (from 0 to 255) to be displayed for each pixel value in the software display.

Gamma

In conjunction with certain types of contrast stretches, you can also adjust an image's gamma value to further enhance the image for display. The eye responds to stimuli logarithmically. However, remote sensors respond to light linearly.

The gamma function is a logarithmic function that is used to stretch the image so that the eye sees gray level distinctions more naturally. If gamma is not applied to imagery from digital sensors, the images often appear too dark and bright features appear bleached. A gamma function is often performed when transforming imagery from a higher-bit source to 8 bits for visual display.

Adjusting the gamma changes a raster dataset's overall brightness by shifting the mid-level gray values. The gamma coefficient controls the degree of brightening or darkening. Higher gamma coefficients will result in the middle tones of an image appearing lighter, while lower gamma coefficients will result in a darker-toned image. Gamma does not affect the extreme values of the histogram (white or black areas). As with a linear stretch, gamma can also be applied at different levels to different bands, allowing you to adjust the ratios of red, green, and blue in the image display.

It should be noted that automated algorithms often expect a linear response from a sensor or simply work better when the image is still linear. Therefore, gamma correction should not be arbitrarily applied to imagery that will be used for automated classification or other types of analysis. As a general rule gamma should be applied if the output is to be viewed, but used with caution on imagery to be used for analysis.

In ArcGIS, images are enhanced for optimum display by setting appropriate stretch and gamma values. These enhancements can be saved as part of a new image (as is the case for cached basemaps) or defined as properties of a layer for display purposes. Stretch and gamma values can also be set on the fly (with each change of map extent) in dynamic range adjustment mode. In this mode, the stretch and gamma values are computed based on the pixel values of the current display extent. This ensures that the display is optimized for both dark and light areas in the image and will vary depending on the current display location and scale.

Image Enhancement and Filtering for Continuous Raster Data

Typically, imagery is provided to the GIS analyst after the image provider has already performed a long series of processing steps that optimize the imagery for general use. Often, not much needs to be done by the GIS analyst to use the data beyond setting up the appropriate display stretch (see above for a discussion of image stretch). However, there are times when additional image enhancement can highlight or "pull out" features of interest that aren't human-perceptible in the unenhanced version of the image. The most common method of image enhancement is to apply raster filters, which can be broken down broadly into two groups: high-pass filters and low-pass filters.

How Image Filters Work

Image filters work by assigning a new value to each pixel in an image based on that pixel's value and the values of neighboring pixels. During filtering, each pixel in the image is considered one at a time; the "neighborhood" (the blue area in figure 5.8) acts as a "moving window," stepping through the image pixel by pixel giving each pixel in the image a "turn" as the center processing cell.

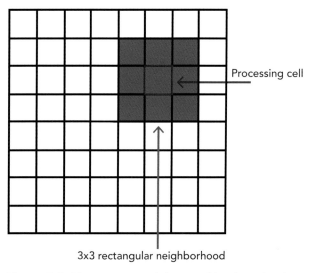

Figure 5.8. The concept of the neighborhood or kernel in image enhancement (esriurl.com/IG58)

Exactly which neighbors to consider is defined by the shape and the dimensions of the neighborhood. Figure 5.8 shows a three cell by three cell rectangular neighborhood. Other commonly used neighborhood shapes are a circle, an annulus, and a wedge. Many functions can be applied to the neighborhood. In figure 5.9, the neighborhood function is a simple summation: the output value represents the sum of all neighbor cell values. Other commonly used neighborhood functions include mean, majority, maximum, minority, median, variety, standard deviation, and range.

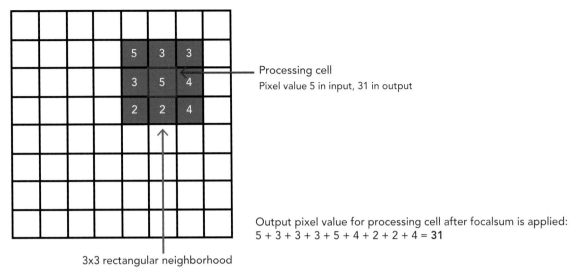

Figure 5.9. Neighborhood sum with equally weighted cells

In addition to setting the shape and dimensions of the neighborhood and the function that is assigned to the processing cell, most image filtering algorithms provide the ability to configure the weights that each neighbor has in the function being applied to the neighborhood as the "moving window" shifts across the image. The default is that all neighborhood pixels receive a weight of 1 and are equally weighted in the function applied (as in figure 5.9). Figure 5.10 illustrates a sum filter using the same example as above, but this time applying neighborhood cell weights.

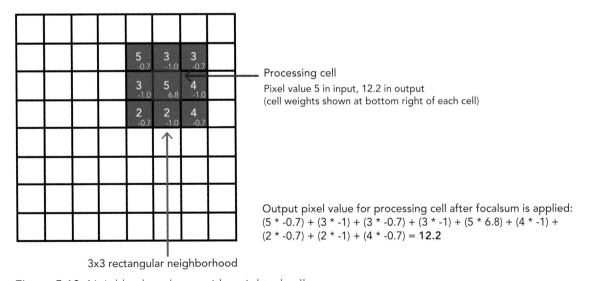

Figure 5.10. Neighborhood sum with weighted cells.

Chapter 5 : Working with Imagery 125

Smoothing Filters (Low-Pass Filters)

Low-pass filters are useful for removing anomalous ("noisy") pixels from an image and creating smoother, often more visually appealing images. If a pixel is slightly different from its neighbors in an image, a low-pass filter will decrease this difference and make the pixel in the output image appear more similar to its neighbors. In doing so, a low-pass filter reduces local variation and extreme values in an image. Low-pass filters typically apply a mean function to the neighborhood, with all cells in the neighborhood receiving equal weight. Figure 5.11 includes an example of a Landsat 8 panchromatic band with a low-pass filter applied. Notice the smoothed appearance of the filtered image.

Edge Filters (High-Pass Filters)

High-pass filters are useful for identifying edges in an image or sharpening an image. These types of filters enhance fine-scale, local details in an image and accentuate edges, such as the boundary between a forest and a meadow. If a pixel is slightly different from its neighbors, a high-pass filter will accentuate this difference in the output by adding contrast.

High-pass filters use the concept of the neighborhood and the moving window discussed above. Typically, high-pass filters use a neighborhood sum as their function with weights applied to the neighborhood. Figure 5.10 illustrates a high-pass filter. Figure 5.11 includes an example of a Landsat 8 panchromatic band with a high-pass filter applied. Note the sharpened appearance of the filtered image.

High pass filters can be implemented in ArcGIS using convolution filters. A typical example is the sharpening filter that can be used to improve the crispness of an image. This uses the following weights:

0	−0.25	0
−0.25	2	−0.25
0	−0.25	0

The sum of the weights is 1 so that, on average, the image does not become lighter or darker, but if a pixel is different from its closest neighbors the difference will be enhanced.

10	10	10
10	15	10
10	10	10

In this example, the center pixel will become
(−0.25 * 10) + (−0.25 * 10) + (2 * 15) + (−0.25 * 10) + (−0.25 * 10) = 30.

Landsat 8 panchromatic band with a low-pass filter applied

Landsat 8 panchromatic band with a high-pass filter applied

Landsat 8 panchromatic band with no filter

Figure 5.11. Low-pass and high-pass filters applied to Landsat 8 data (esriurl.com/IG511)

Image Mosaics and the Mosaic Dataset

Overview of Image Mosaics

Aerial photography and satellite imagery are collected in scenes. The scenes are then often combined into large, seamless image mosaics for use in GIS software as photo base images and for analysis. In some workflows, these mosaics are then cut up into smaller rectangular tiles for simpler distribution. Such tiles may have coincident edges or overlapping pixels. When working across large areas it's advantageous to work with a single large image versus many individual scenes or tiles. It is therefore necessary to merge the scenes or tiles into a single physical image or a single virtual image. The creation of a single physical image is referred to as creating a mosaic and results in duplication of the scene data in the new mosaicked image file. Within ArcGIS, there is also the ability to create a mosaic dataset that combines the multiple inputs into a virtual image, but does not require additional storage and provides greater flexibility.

Figure 5.12 illustrates the concept of an image mosaic. Chapter 12 provides more in-depth information about creating mosaic datasets.

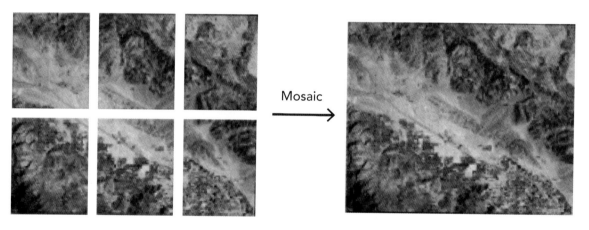

Figure 5.12. The concept of an image mosaic

Mosaic Datasets—an Introduction

Until recently, image mosaics were typically created by merging the pixels of all the input images together and creating a new image containing pixels from all the inputs. However, this process is inefficient in terms of storage, because each pixel may be stored twice on the hard drive; once as a pixel in the input and once as a pixel in the mosaic. Moreover, in the past multiple mosaics were often created with different stretches or derivatives (e.g., NDVI), further multiplying storage requirements for a collection of mosaicked images.

Esri developed the mosaic dataset and associated functionality to simplify the creation, management, and maintenance of mosaics and to enable distribution of large collections of imagery as web services. The mosaic dataset does not store pixels of imagery. Instead, it is an information model that catalogs imagery and performs on-the-fly processing. At its core, the mosaic dataset is simply a catalog or index of metadata that points to the original imagery, along with a framework for processing and displaying the imagery on the fly. The mosaic dataset does not result in the creation of new image files; instead, it dynamically mosaics the pixels in the original image scenes or tiles without altering or converting the original imagery. As a result, mosaic datasets require a small fraction of the space that traditional mosaics require and can scale to massive collections of imagery.

Components of a Mosaic Dataset

Mosaic datasets are defined in a geodatabase and reference the original imagery. Mosaic datasets have the following components:
- A catalog of metadata about each image (each of the individual scenes or tiles of imagery) as well as the footprint of each image.
- A feature class that defines the boundary of the entire mosaic dataset.
- Mosaic dataset properties that include mosaic methods that define rules for the required order of overlapping images as well as a range of other properties.

A mosaic dataset is a well-defined geodatabase structure optimal for working with large collections of imagery and rasters. Mosaic datasets are stored as file geodatabases, or use an enterprise geodatabase such as Microsoft SQL, Oracle, or PostgreSQL. The imagery and raster data that composes a mosaic dataset is not stored in the database but instead is referenced by the database and stored across a file system or in the cloud. A single mosaic dataset can reference millions of images and make them appear as single virtual dataset or enable quick access to any individual or group of referenced images.

Once a mosaic dataset has been created, image references (again, not the images themselves, but references to where the images reside on the file system) can be added to it or removed from it. Mosaic datasets support spatially noncontiguous inputs to easily model images along linear features such as roads. Through the catalog, mosaic datasets track the spatial, temporal, spectral, and radiometric details of their component images. As a result, a single mosaic dataset can contain various types of overlapping imagery (even images with different spatial resolutions and projections). What the end user of the image mosaic sees is controlled by the mosaic methods and mosaic operators, which are discussed in the following section.

Dynamic Mosaicking and Mosaic Methods

There are often areas where mosaic dataset input images overlap with their neighbors. "Dynamic mosaicking" is the ability to define or refine the order in which images should be displayed or blended in a mosaic dataset. The user or application using a mosaic dataset can use a number of mosaic methods to handle these overlaps. Mosaic methods essentially control the drawing order of the input images and are applied dynamically as the user pans and zooms around. For example, the "closest to center" method resolves overlaps by displaying topmost the image that is closest to the user's view center at any given time. Another mosaic method, "by attribute," sorts the input images by an attribute and resolves overlaps this way. For example, the by attribute method could use an acquisition date attribute in the mosaic dataset to resolve overlaps by displaying the most recently collected scene on top. The "seamline" mosaic method can be used to create seamless mosaics, where the images are blended along predefined seamlines. Since the mosaic dataset's catalog contains the metadata and extents of its component images, it is simple to define query filters in the form of "where clauses" to display or work with a subset of the imagery referenced by the mosaic dataset. Query filters are easily applied in ArcGIS Desktop by setting definition queries on the footprints layer (catalog) of a mosaic dataset. The combination of mosaic methods and query filters allows the user to control the pixels that are actually displayed in areas of overlapping images.

Mosaic Operators for Mosaic Datasets

Working side by side with mosaic methods and query filters, which control the drawing order of input images (controlling which image is on top of another), mosaic operators control how overlapping pixels are displayed. Typically, the "first" mosaic operator is used, which results in an output pixel value equal to the topmost overlapping image's pixel value. Other operators include "mean," which results in the average of the overlapping pixel values, or "blend," which gives a weighted average of the overlapping pixel values with the weights dependent on the distance to the image edge. Figure 5.13 illustrates the mean mosaic operator.

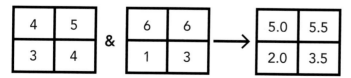

Figure 5.13. Mean mosaic operator

Mosaic Dataset Raster Functions

Esri raster functions allow for the on-the-fly rendering of derivative image products from a mosaic dataset, transforming the pixels from data values stored on disk to values required by the end user or application. For example, by applying a slope raster function to a mosaic dataset of DEM images, the mosaic dataset will appear as a slope image instead of a digital elevation model (see figure 5.14). Changing the slope raster function to an aspect function will change the appearance to an aspect image. Removing the raster function altogether will revert the mosaic dataset's appearance to a DEM. Multiple functions can be associated with a single mosaic dataset by defining processing templates. In this way, a single mosaic dataset can have multiple representations, and the user can quickly select the most appropriate with the required processing applied on the fly as the data is accessed, while the only pixels stored on disk are those of the original images.

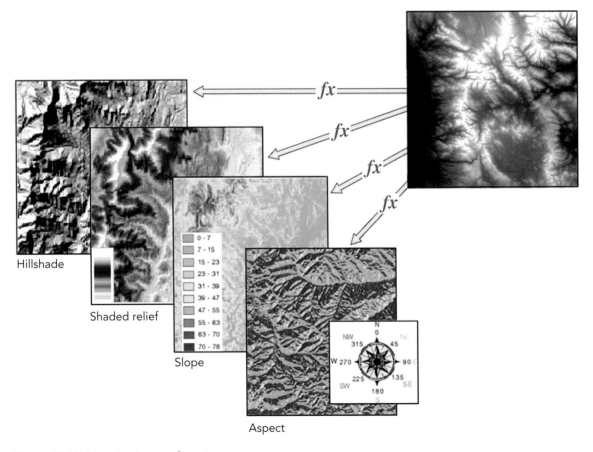

Figure 5.14. Mosaic dataset functions

Mosaic datasets are changing the way imagery is accessed and processed by enabling the source image data to be stored once and having multiple derivative products generated on demand. The processing speed is typically much higher and the storage requirements much lower than the traditional approach of passing imagery through multiple processing steps with multiple inputs and outputs.

Mosaic datasets are also used for publishing collections of images and rasters as web services, which provide access to imagery across the Internet. The following section introduces web services; a more detailed discussion of web services and mosaic datasets is presented in chapter 13.

Accessing Imagery as Web Services

Introduction

This section provides an overview of using and accessing imagery that exists as Esri web services (for more detail, see chapter 13). Web services enable data to be stored on servers and be quickly accessible to anyone with access to those servers over the web. There are many types of web services for imagery outside the scope of our discussion—this chapter and this book focus on Esri's technology for serving and consuming imagery over the Internet.

Organizations collect and maintain increasingly large volumes of imagery and raster data. These data come from a wide range of sources, from sensors on platforms ranging from satellites to unmanned aerial systems, to scanned maps and scanned film-based aerial photography. In many cases, an organization's raster data may be the output from an image processing or analytical workflow. These imagery and raster datasets are important to organizations and are used to make important decisions. Traditionally, there have been challenges in managing such data and making it accessible to a wide range of users for various applications. Those challenges have been exacerbated by the ever-increasing number and size of those datasets.

The massive volumes of data mean that it is not practical to move the data to the end user's application for processing. Instead, it is advantageous to serve the imagery as web services that simultaneously provide access to the imagery and derivative image products that are processed on the fly by the server.

Web services for imagery can be used for visual display and for analysis by desktop applications such as ArcMap and ArcGIS Pro, by web maps displayed in a web browser, and by applications running on computers, tablets, or smartphones (see figure 5.15).

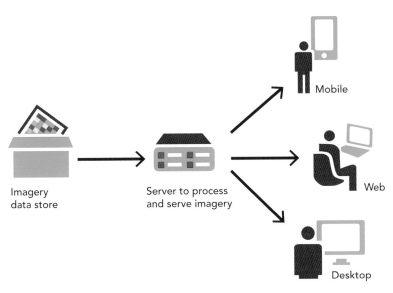

Figure 5.15. ArcGIS web services for imagery

Types of Web Services for Raster Data

Two different types of web services provide access to raster data and imagery, each with advantages and disadvantages. The two types, dynamic image services and tile cache services, are discussed below.

Dynamic Image Services

Dynamic image services provide full access to imagery. Requests are made by applications to servers that quickly gain access to and process the imagery (if requested by the user), and deliver the imagery and metadata to the end user's application. Image services enable applications and their users to gain access to individual images or massive collections of imagery, with the server returning unprocessed raster data or applying processes such as image enhancement, vegetation index computation, or image classification. Imagery requests to the server are synchronous, with the imagery being returned by the server nearly instantaneously. ArcGIS provides a robust set of desktop tools and a rich web service protocol that enables the end user or application to define the processing to be applied on the server. Dynamic image services provide full access to image data, making them useful for analysis and scientific applications. Client applications can even export or download pixels from the service, storing tiles or portions of the imagery locally.

Chapter 5 : Working with Imagery 133

Tile Cache Services

Tile cache services provide background imagery for many applications. Tile cache services deliver highly compressed, preprocessed imagery from servers to applications and end users. Before serving as a tile cache, the imagery must first be processed and transformed into large collections of small tiles. Tile cache services are the most efficient way to serve background imagery (such as image basemaps), because they put minimal load on servers and maximize the use of caching on both the server and the client. Tile cache services are designed for providing background imagery, and should not be used for analysis and scientific applications.

Geoprocessing Services

Though they don't typically provide access to raster data, geoprocessing services are a third type of service worth noting here. Geoprocessing services provide a client application or end user access to a task running on a GIS server that can process data submitted by the client. Geoprocessing services take location-based input from a client application, apply a geoprocessing workflow to the user-provided information, and return a result or some data to the end user. An example of a geoprocessing service: a web map user clicks on a point on a web map (submitting the point to a geoprocessing service) and the geoprocessing service calculates the watershed that the point falls within and returns to the user a vector polygon of the watershed's boundary.

Examples of Publicly Available Web Services for Raster Data

ArcGIS Online (www.arcgis.com) provides access to large collections of web services through its Living Atlas. The Living Atlas includes world basemap imagery, providing global 1-meter or better resolution imagery. Other global datasets include multispectral and multitemporal Landsat 8 image services, as well as global DEMs. Subscriptions to imagery directly provided by satellite and aerial imagery providers are also available. Access to these online services enables organizations to immediately add imagery to their GIS maps and applications.

Many public agencies have made their imagery available as services. For example, the US Department of Agriculture's National Agricultural Imagery Program provides image services of 1-meter resolution imagery for the entire country for many dates of collection.

Using Image Services in ArcGIS Desktop

A user of ArcMap can gain access to web services in one of two ways: by adding services from ArcGIS Online or by setting up a connection to an ArcGIS server directly. For the first method, click the Add Data icon and select "Add Data from ArcGIS Online" (see figure 5.16). Signing in provides access to both public services and those available to you through your organization; not signing in will provide access only to public services. Once logged in, you can browse ArcGIS Online or use search terms to find imagery, and then load the image service or tile cache service directly into ArcMap. To connect directly to an ArcGIS server from ArcMap to gain access to web services, click Add Data and change "Look In" to "GIS Servers." Select "Add ArcGIS Server" from the list and go through the wizard to connect to the server. You will need the URL of the ArcGIS server. In ArcGIS Pro, adding web services is done by clicking the Add Data icon and navigating to the Portal icon on the left. Searching with "All Portal" selected returns a list of public services from ArcGIS Online that fit your search term. If you have configured private portals in ArcGIS Pro, services on those portals that fit your search will be listed as well.

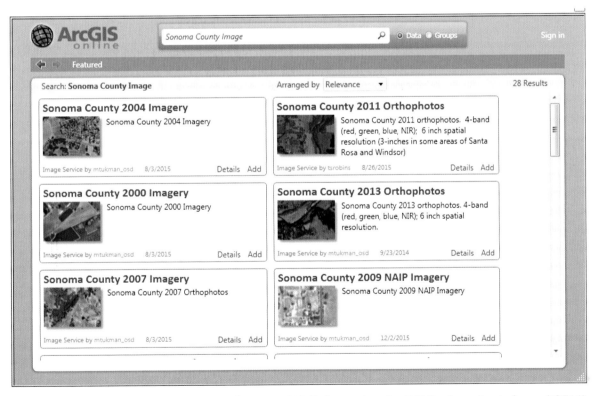

Figure 5.16. Accessing image services from ArcGIS Online using ArcGIS Desktop (esriurl.com/IG516)

Once you've loaded a raster web service into ArcMap or ArcGIS Pro, you can pan around, zoom in and out, and visualize the service in the context of other raster and vector data. If the service is a tile cache service (see "Types of Web Services for Raster Data" above), then access to the imagery is limited to viewing the imagery in its preconfigured appearance. You cannot, for example, apply custom stretches, gain access to pixel values, download portions of the imagery, or use the data for processing or analysis. However, if you're using a dynamic image service, then you have full access to the raster data as if it were stored locally. For example, you can use the "Identify" tool to query pixel values, apply a custom stretch to the imagery, use the service as an input to a geoprocessing or spatial analyst tool, or save a portion of the service's raster data to your local computer (if export capabilities are enabled on the service).

One of the most powerful aspects of image services is that they can contain thousands or millions or overlapping images, but since the information about each is contained in a searchable catalog (accessible by right-clicking the service in ArcGIS Desktop and selecting "Open Attribute Table"), you have the ability to control which images display on top by adding filters, also known as definition queries, to the mosaic dataset's attribute table. Let's use the example of an Esri-hosted image service of Landsat 8 pan-sharpened imagery. This mosaic dataset comprises hundreds of thousands of Landsat scenes—ArcGIS Desktop will display only the ones in its view extent. If you load the mosaic dataset and use it, the service's default rules will decide which scenes to display on top. However, you can reorder the imagery by changing the mosaic method or set a specific query by opening the image service's attribute table and applying a definition query. Figure 5.17 shows two such definition queries (applied in ArcGIS Pro) to the Landsat 8 service. The query on the left side of the figure filters the scenes displayed by the image service to winter scenes with less than 30 percent cloud cover, while the query on the right side limits scenes displayed by the image service to summer scenes with less than 30 percent cloud cover.

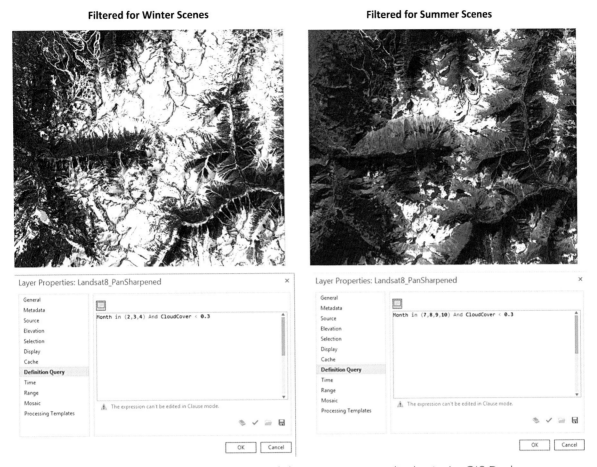

Figure 5.17. Using attribute queries to control the image service display in ArcGIS Desktop (esriurl.com/IG517)

Exporting and Downloading Imagery from Dynamic Image Services

In most cases, images from image services can be used directly without the need to make local copies of the data. If required however, users can export or download sections of raster data from image services using ArcGIS Desktop, creating a local copy of a subset of the service's original pixel data and metadata. In ArcGIS Desktop a user can export raster data from services in a number of ways, such as by using the Copy Raster tool, or by right-clicking on the service and selecting "Export Data" (see figure 5.18) or by using the "Clip Management" tool to clip the image service to a vector area of interest and save it locally.

Chapter 5 : Working with Imagery   137

In ArcGIS Desktop, the intact source images that compose an image service can be downloaded directly to the file system using the Download Mosaic Dataset Rasters tool. Downloading source files can also be done manually by selecting source rasters in the image service's attribute table, and then right-clicking the image service and selecting "Data" and then "Download Selected Rasters." During the download, the server extracts the original pixel values and metadata and returns these without any resampling. Optionally, the server can also clip the pixels to a specified extent.

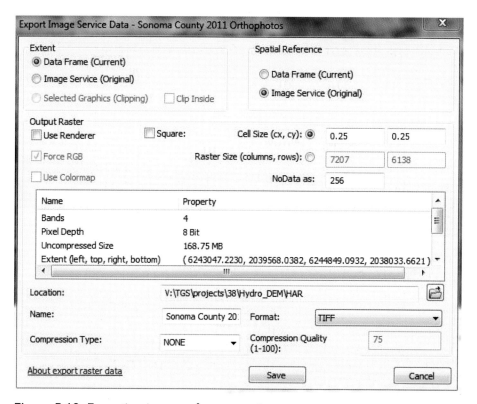

Figure 5.18. Exporting imagery from a service

Since raster data exported from an image service is bandwidth intensive, it is usually recommended to limit image service downloads to relatively small areas, especially for high-resolution imagery. The publisher of an image service may restrict the maximum size of the image that may be exported or the maximum number of images downloaded. They may also disable download functionality entirely.

Using Web Services in ArcGIS Web Maps

ArcGIS Online web maps provide a browser-based mapping environment that is easy to use, even for the non-GIS user, and useful for creating maps to share online, provide an interactive display of geospatial information, or tell a story. Web services are easily added to an ArcGIS Online web map. Figure 5.19 shows an ArcGIS Online web map with an image service of high-resolution orthophotography.

To add image services to an ArcGIS web map, go to http://arcgis.com, create a web map, and click the Add icon. From there, you can use keywords to search from among the thousands of image services available in ArcGIS Online. If you log in to your organization, you will also have access to your organization's web services for use in your web map.

Figure 5.19. Using image services in ArcGIS Online web maps (esriurl.com/IG519)

Using Raster Web Services in Story Maps

Using web services in Esri story maps is a powerful way to provide nontechnical end users with access to imagery and raster data. Story maps are built from a rich set of customizable templates, and can also be extended with custom code. Figure 5.20 shows an example of a story map (http://sonomavegmap.org/1942) that allows the viewer to swipe a historical image service (1942 air photos) on top of a modern image service (2011 air photos). The comparison of the two image services provides a visual illustration of land-use change between the 1940s and the present day. The story map also includes a series of vignettes, so that the user can zoom to different areas of Sonoma County, compare the images, and read about localized changes in the landscape.

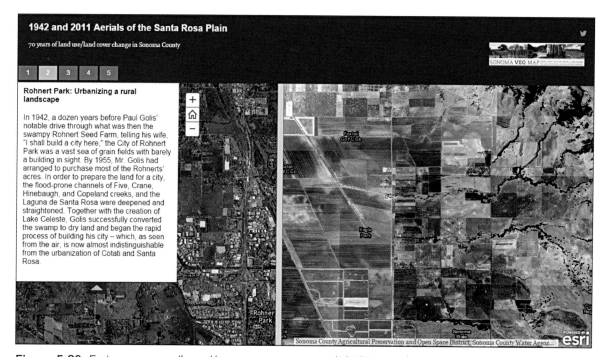

Figure 5.20. Esri story maps (http://sonomavegmap.org/1942) (esriurl.com/IG520). Source: Sonoma County Agriculture Preservation and Open Space District

Using Raster Web Services in Apps

Esri provides a rich development platform and set of APIs for developers to integrate web services into their applications and apps. Developers can create a wide range of applications that gain access to and use raster web services, ranging from focused apps running on mobile phones, to web applications, to desktop GIS applications optimized for geospatial analysis.

ArcGIS provides a robust platform for application development. Using ArcGIS Runtime, native applications running on mobile devices such as Android, iOS, or Windows phones can gain access to imagery locally on the device or by connecting to raster web services. An application for field data collection may use an image service only as a backdrop. A more advanced application may use the image service for analysis such as performing line-of-sight computations for military situational awareness.

For the development of web applications, ArcGIS offers an extensive JavaScript API that provides access to imagery for both visualization and analysis. Web applications can act as thin clients that make requests to servers that perform all processing and return the results quickly for display. The ArcGIS APIs enable extensive filters and queries for displaying the image service on the client application. The result is that the end user of an application can have at their disposal massive collections of imagery, such as collections of many years of multispectral satellite imagery. Applications can also interact with geoprocessing services running on servers. Geoprocessing services take data submitted by the user of an application, such as a point digitized by the user, and return information or data about the user's submission, such as a list of parks closest to the user-digitized point.

The image service APIs enable a wide range of functionality. By providing access to client-defined image processing (through the use of raster functions), the APIs promote imagery uses that go far beyond visualization, enabling applications to perform advanced image processing and image analysis.

A good example of a web application that deeply integrates web services is the US Geological Survey's Landsat Look Viewer (http://landsatlook.usgs.gov/). (See figure 5.21.) This web application provides quick search and access to the complete archive of over 4 million Landsat scenes. Users can zoom to any location and perform a search for Landsat scenes corresponding to specific criteria, and then immediately see full-resolution versions of the scenes and apply image enhancements. Users can add scenes to their cart and download them.

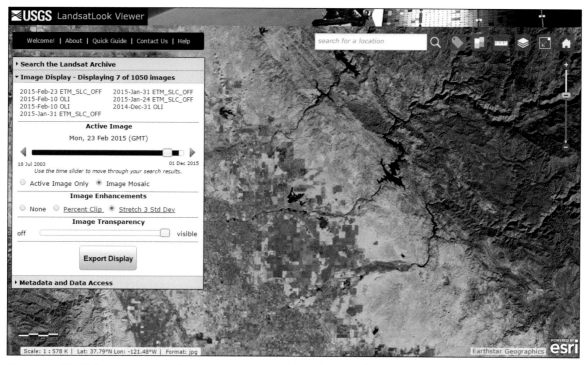

Figure 5.21. Using image services in apps—USGS Landsat Look Viewer (http://www.esri.com/landsatonaws) (esriurl.com/IG521)

Summary—Practical Considerations

This chapter reviews fundamental concepts that are required to work with imagery. The concept of scale is discussed—scale in the context of imagery is critical to understand both for selecting the right type of imagery for the task at hand and for understanding the appropriate uses for types of imagery with varying spatial resolutions.

Chapter 5 discusses common raster data format and raster storage. Image compression, introduced in this chapter, is an important consideration. Lossy compression algorithms can greatly reduce the storage requirements for image datasets, but their use in compressing imagery should generally be limited to imagery used for visual reference only, not for rasters that will be used for image interpretation, feature extraction, or image classification. For these applications, lossless compression should be used.

The concept of NoData is an important one for GIS analysts. Choosing an organizational standard for dealing with NoData areas in imagery will reduce confusion and inconsistency for data users and analysts.

Chapter 5 introduces the concepts of image enhancement and filtering of continuous raster data. These techniques are worth understanding—they are commonly used by the GIS analyst to optimize imagery for various uses. Often, with a little work, these techniques will increase the value of imagery to your organization.

Mosaic datasets offer a revolutionary approach to managing, storing, organizing, querying, analyzing, displaying, and serving imagery. Using mosaic datasets can result in massive increases in productivity and large savings in disk space. Understanding the myriad capabilities of the mosaic dataset and the ability to apply on-the-fly functions to the mosaic dataset's underlying raster data are among the most important practical pieces of knowledge contained in this chapter. Using mosaic datasets to their full capabilities will reap wide-ranging practical benefits.

Chapter 6
Imagery Processing: Controlling Unwanted Variation in the Imagery

Imagery is a valuable tool in geospatial analysis because what you see on the imagery is indicative of ground conditions. In other words, the imagery is highly correlated with the ground. Thus, imagery can aid us in learning about the ground without having to visit it. Imagery also provides a different perspective of the landscape than can be seen from the ground. Imagery's "bird's-eye view" provides a comprehensive panorama, allowing us to study how different landscape features interact.

The stronger the correlation between what can be seen on the imagery and what is actually on the ground, the more useful the imagery becomes. This is true whether the image is going to be used simply as an image layer as part of a geospatial analysis or if the image is going to be processed to create a layer of biophysical attributes or a thematic map. Therefore, it is important to control the variation in the imagery that is not correlated with variation on the ground to ensure that the imagery best represents those ground conditions. If the imagery is simply used for visualization and the variation is left uncontrolled, then the image will be less representative of the ground than it could be. However, if the image is to be processed to create a map and the variation is left uncontrolled, then the resulting map will be less accurate than it could be.

This chapter begins by reviewing some important concepts of electromagnetic energy. It then presents the issues and solutions for the three factors that most often introduce spurious variation into the imagery: atmosphere, clouds, and terrain. These three factors can be controlled with radiometric correction, cloud removal, and geometric correction. A final section is included on image mosaicking because variation between images must be controlled to combine them in a mosaic (i.e., a collection of images merged to create a single combined image).

A key assumption made in this chapter is that very often the processing done to control unwanted variation (i.e., radiometric and geometric correction and cloud removal) is performed as a service before the geospatial analyst obtains the imagery. For example, Landsat imagery downloaded from USGS EROS comes corrected for reflectance and is registered to the ground. Therefore, it is often unnecessary for analysts to be able to perform these corrections themselves. However, it is critical that an analyst have a strong appreciation and understanding of the usefulness, assumptions, and application of these techniques. A presentation regarding general knowledge and discussion of the importance of these techniques follows. The reader who wishes to study any of the processes in more detail can find each of these topics thoroughly described in most remote sensing textbooks (e.g., Jensen, 2016; Lillesand, Kiefer, and Chipman, 2015; and Campbell and Wynne, 2011).

Review of Electromagnetic Energy

As discussed in chapter 3, electromagnetic energy has many wavelengths. Not all are equally important for remote sensing. To human beings, the most important wavelengths are those in the visible portion of the spectrum because our eyes sense those to see. Wavelengths in the infrared (near, middle, and thermal) are also very important for vegetation analysis and land-cover mapping because variations in vegetation and land cover are often more highly correlated with variations in infrared wavelengths than in visible wavelengths. Electromagnetic energy that passes through our atmosphere is said to pass through atmospheric "windows." Not all electromagnetic energy passes through an atmospheric window; some of it is blocked. For example, gamma rays are blocked by the atmosphere and do not reach the earth's surface. In the late 20th century, humans unintentionally created the ozone hole; a new atmospheric window allowing additional ultraviolet light to reach the earth's surface over the Antarctic.

Interactions

To control unwanted variation in imagery, we need to understand how electromagnetic energy interacts with the objects we are interested in imaging. When electromagnetic energy reaches an object, four interactions are possible: 1) absorption, 2) reflection, 3) transmittance, and 4) emittance, as shown in figure 6.1. One possible interaction is that the energy is absorbed by the object. For example, the energy could be converted to heat that warms up the cool sand early in the morning on a beach. In plants, the blue and red wavelengths are absorbed by the chlorophyll in the plant and converted into chemical energy through a process called photosynthesis. A second interaction is reflection. Electromagnetic energy reflected by an object can be detected by a remote sensor (including our eyes) and is the most important interaction for the majority of remote sensing systems. For example, while the blue and red wavelengths are absorbed by a plant leaf, the green light is reflected, which is why humans see the leaf as green and why it appears green on natural color imagery. A third possible interaction is transmittance. In this case, the energy is allowed to pass through the object; it is transmitted. When electromagnetic energy passes through objects of different densities (e.g., from air to glass), the energy is refracted (i.e., bent). The amount of refraction depends on the difference in density. Lastly, electromagnetic energy can be emitted or given off by the object. Volcanoes and forest fires emit electromagnetic energy. In the beach example, the sand absorbs electromagnetic energy throughout the day and continues to heat up. Sand early in the morning is cool and feels great on our bare feet. However, by the middle of the day, the sand can be quite hot and may burn our feet. After the sun sets, the sand emits the electromagnetic energy and begins to cool down again, so by late evening the sand is safe to walk on with bare feet.

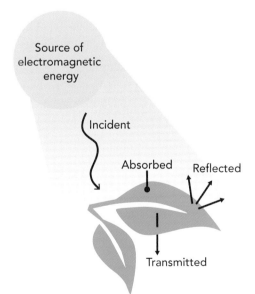

Figure 6.1. Three of the four possible interactions between electromagnetic energy and an object.

All objects interact in one or more of the four ways described here. Understanding these interactions is critical to the effective use of any imagery. The majority of remote sensors rely on sensing reflected energy.

Reflectance versus Radiance

While it is not uncommon to hear the terms reflectance and radiance used interchangeably, there are important differences between them that should be understood by anyone using remotely sensed imagery. Remote sensing devices sense reflected energy that is measured in radiance. The sensor records values called digital numbers (DNs) that are easily converted into radiance values using the calibration information provided in the image metadata about that scene. Because electromagnetic energy must pass through our atmosphere, the measured radiance is actually a combination of the energy reflected and/or radiated from the object, minus the energy absorbed by the atmosphere on the upwelling path (i.e., from the object back to the sensor), plus the energy scattered by the atmosphere into the path of the sensor (figure 6.2). Scattering is caused by particles and water vapor in the atmosphere that change the trajectory of the energy so that some energy reflected from an object does not reach the sensor, while other energy reflected from other objects might. The amount of scattering is wavelength dependent. For example, the reason that the sky appears blue to us is that blue light has the shortest wavelength of the visible light and is most scattered by the atmosphere, hence the sky is blue. The sensor is in a specific location, and therefore the radiance is a measure of the energy (a combination of interactions) that reaches the sensor and is recorded by it. Radiance is measured in watts.

Reflectance is a unitless ratio of the amount of energy reflected from an object divided by the amount of energy reaching that object. Reflectance is a rather stable characteristic of a material, hence for object identification it is better to measure reflectance than the upwelling radiance coming from an object because that varies due to many factors. As discussed later in this chapter, methods are available to compensate for atmospheric effects and, therefore, use the radiance to compute the reflectance properties of the object of interest. As a result, the reflectance values then represent the true spectral response of that particular object and can be used in the identification of that object.

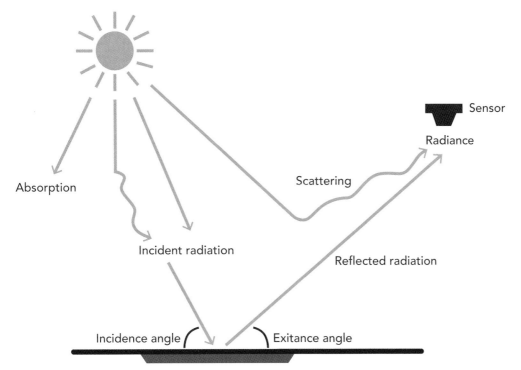

Figure 6.2. Graphic representation of the radiance reaching the sensor as a combination of electromagnetic energy reflected by the object, absorbed by the atmosphere, and scattered into the path of the sensor.

In summary, starting with an uncalibrated image (DN values), the image is then calibrated to a radiance image. This radiance image is then atmospherically corrected, resulting in a surface reflectance image. The final reflectance image is then ready to be used to extract quantitative information about features on the surface.

The example below helps to demonstrate the importance of not simply using the radiance (what the sensor sees), but actually correcting the values to reflectance. A few leaves of a plant could be collected, ground up, and examined in a lab using a spectrometer to find out the amount of energy they reflect in the visible and infrared portions of the electromagnetic spectrum. Those results could be shown in a spectral pattern analysis graph as shown in chapter 3 (see figure 3.1). A sensor on a high-spatial-resolution satellite could image a pixel of a tree having the exact same type of leaves as collected and analyzed in the lab, but the instruments' measures of reflectance would most probably differ because of the interference of the earth's atmosphere. For the spectral pattern of the remotely sensed pixel to be directly compared to the one from the lab, the radiance collected by the sensor would need to be corrected to reflectance by compensating for atmospheric effects occurring where and when the image is captured by the sensor. Failure to do so would make it impossible to compare this same object (i.e., leaves) collected using the two different sources or to accurately track change in the object over time.

For some remote sensing applications, correcting radiance to reflectance does not matter. For any comparison of image feature values that are within one image, and therefore relative to each other, radiance comparisons will be sufficient and reflectance is not required because the atmospheric effects will be the same between the features. As discussed in more detail later in the chapter, correction to reflectance is not necessary when a single image is being used to create a thematic map as long as atmospheric conditions do not vary across the image. For example, if a single Landsat image is used to make a land-cover map, only the information from that image will be used in the image processing and it will not be necessary to correct the image. However, if two dates of imagery or more than a single scene is used, correction should be applied, because the atmosphere may have caused issues between the dates or scenes. It has also been widely accepted that ratios/indices in which one band of imagery is divided by another (e.g., the Normalized Vegetation Index or the NIR/red ratio) adequately compensates for using radiance rather than reflectance. However, the most recent research results indicate that since reflectance represents the exact properties of the object, more reliable and repeatable results will be achieved when using reflectance.

Bidirectional Reflectance Distribution Function

The bidirectional reflectance distribution function (BRDF) is an easy concept to understand, but in practice it is quite complex. BRDF is simply what we observe every single day when looking at objects. That is, objects look different to us when viewed or illuminated from different angles or locations. The same is true when we remotely sense these same objects. The amount of energy received by the sensor from the object of interest is highly dependent on the viewing and illumination geometry. The actual BRDF varies based on wavelength and is influenced by the physical properties of the object, including factors such as the shadow cast by the object; the scattering of the object based on smoothness/roughness of the surface; and the ability of the object to reflect, absorb, or transmit energy.

An easy example to demonstrate this concept is water. If a lake were very calm and still, the lake's water would act as a specular reflector, and the light from the sun would bounce off it in a very coherent form related directly to the angle of the sun in the sky. If the sensor were at this exact angle, a high percentage of the light would be recorded for those wavelengths that are reflected by water. If the sensor were not at this angle, then much less reflectance would reach the sensor. The amount would depend on the angle. However, if this same lake were experiencing some very strong winds that caused the water to become choppy, then the surface of the lake would no longer be smooth, and the reflectance of the light off the water would be altered significantly.

Correcting for variable view and illumination angle effects is especially important in remote sensing when multiple images are to be mosaicked together to form a larger image, or when some standardization of the image is necessary. These corrections are part of the next section in this chapter: radiometric correction.

Radiometric Correction

All remotely sensed imagery has radiometric errors. The process of correcting radiometric errors is key to the effective use of the imagery. Failure to radiometrically correct the imagery means failure to control unwanted variation, thereby weakening the correlation between what is on the ground and what is being imaged. Radiometric errors can be grouped into three categories: 1) issues related to the sensor, 2) issues related to sun angle and topography, and 3) issues related to the atmosphere.

Sensor Correction

Sensor issues and errors tend to be either minor, predictable, and easily corrected, or catastrophic resulting in the sensor being decommissioned. Minor issues include such things as bad pixels, partial line or column dropouts, optical corrections for color shift, fall off, and others. Also, line start problems can occur where data at the beginning of a scan line is not collected. Some sensor issues are more serious, such as striping when instead of a detector failing completely, it just gets out of calibration with the other detectors. The result would be a stripe in the image when the sensor is looking at the same object with multiple detectors at the same time. For example, if a sensor has 16 detectors that swipe across the landscape to collect an image, and one of those detectors is not in calibration, then when an image is captured of a large area of the same land-cover type, there will be a stripe in the image where the uncalibrated detector collected data. This issue is commonly seen over water areas, where it is easy to see the striping. One method for correcting a striping problem involves taking some average of the pixels surrounding the stripe and substituting this average for the miscalibrated data. Other modeling and convolution methods can also be used.

A notable example of a sensor issue is the problem with the Scan Line Corrector (SLC) on Landsat 7. The SLC is used to correct for forward motion of the satellite, and without it significant gaps occur, as shown in figure 6.3. The gaps in the imagery make the image less useful for many applications. Luckily, the gaps occur in different portions of a scene over time and many organizations have developed processes for filling the gaps using multitemporal imagery.

Figure 6.3. A portion of a Landsat 7 scene showing Scan Line Corrector gaps

Sun Angle and Topographic Correction

As discussed in the section above introducing bidirectional reflectance, differences in illumination and viewing angles (created by varying sun location, sensor location, and topography) can have significant impacts on the radiance recorded by the sensor. Areas with steep terrain may be in complete shadow, affecting the response recorded by the sensor. For example, dense coniferous forest on steep terrain can exhibit a spectral response similar to that of water, if not properly corrected. Clearly, labeling a forested area on the side of a mountain as water would be a serious and unacceptable error. This issue could be fixed easily by including a slope layer in the mapping process to make sure that water only occurs on flat terrain. Additionally, the process could be as complicated as using a variety of mathematical algorithms to correct the imagery for topography before classification.

All the mathematical algorithms used to radiometrically correct imagery for sun angle and topography are based on computing the proportion of electromagnetic energy striking that particular place on the ground as defined by the cosine of the incidence angle. In other words, knowing where the sun is and the slope of the terrain, the proportion of light recorded at that area on the ground (pixel) can be determined. These values can then be used to correct the imagery to normalize it so that the corrected value represents the pixel without the effect of the slope. There are a number of additional adjustments made to this cosine correction empirically by including a factor that compensates for the land surface not being a perfect reflector (i.e., it does not reflect all light incident upon it). The two most common of these methods are called the Minnaert correction and the C-correction.

Atmospheric Correction

As the name suggests, atmospheric correction involves correcting the imagery to deal with the unwanted image variation created by the atmosphere. If imagery is collected close to the earth's surface (i.e., from aircraft) there is less chance for variation caused by the atmosphere to occur. As the imagery is collected from higher and higher altitudes (and especially from space), the atmosphere can have a greater and greater impact. Before beginning a discussion of atmosphere correction, it is important to determine when such a correction must be performed and when it is not needed.

As mentioned earlier in the chapter, a sensor records radiance, not reflectance. The radiance values measured by the sensor are a combination of the electromagnetic energy being reflected by the objects minus the energy absorbed by the atmosphere plus the energy scattered into the path of the sensor by the atmosphere. Reflectance, on the other hand, has been corrected to compensate for these atmospheric effects and better represents the physical properties of the objects being imaged. Therefore, it would seem that it is always better to correct for atmospheric impacts and convert radiance into reflectance. While it is true that this conversion always results in better imagery, it is not always necessary. For example, if a single date of a single image is being used to create a land-cover map (e.g., a Landsat 8 image), then all the information needed to classify that image (e.g., training data or other statistical information) will be generated from that image, and hence no corrections are necessary. No comparisons between images have been done, so no normalization or correction to reflectance is required. However, if derivative bands such as ratios or vegetation indices are created, then better results will occur with imagery corrected to reflectance, because the reflectance values better represent the physical properties of the objects of interest. It is important to note that in some situations, atmospheric correction cannot be performed, as the information needed to do the correction is not available. In this situation, ratios and vegetation indices are useful as they do tend to have some normalization effect on the imagery.

Atmospheric correction is needed in many remote sensing applications. Any time multiple images from different dates are being used, the imagery should be corrected. Any time biophysical parameters are being determined from the imagery, correction is essential. When creating derivative bands such as ratios and vegetation indices, correction provides for better results.

Historically, atmospheric correction of imagery was a difficult process, and despite the development of many sophisticated algorithms in recent years, the process has not gotten much easier. A key factor in atmospheric correction is the requirement of detailed knowledge of the atmospheric profile at the time the image was acquired. Obtaining this profile is the most difficult component of the entire process.

Atmospheric correction can be divided into two general types: relative correction and absolute correction. Relative correction is easier and requires less information, but only normalizes the image or images. Normalization makes the images directly comparable to each other rather than completely removing the atmospheric effects. The most common single image relative atmospheric correction is known as dark object subtraction (DOS). This correction is based on the concept that objects in an image that have very low reflectance should appear to be near zero when plotted on a histogram (i.e., frequency distribution) of the objects in an image by wavelength. Figure 6.4 provides an example. Identifying objects on the image that have very low reflectance for a given wavelength (i.e., dark objects) means that these objects should appear very near zero on a histogram of the imagery. However, as shown in figure 6.4, these dark objects have a value near 30. The reason this value is at 30 and not 0 is that atmospheric scattering added light into the area. Subtracting the value of 30 from the entire histogram (see figure 6.4) will then normalize the image and remove this atmospheric effect.

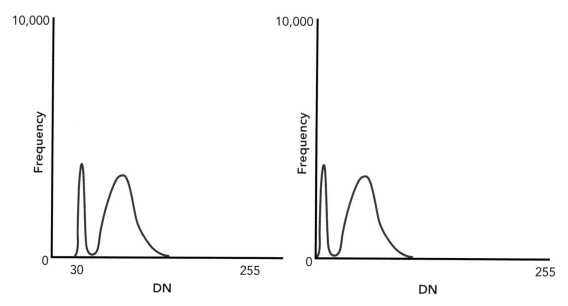

Figure 6.4. Histograms for a given wavelength (in this example, NIR) showing the original distribution of the image data without dark object subtraction (DOS) on the left and the corrected histogram after DOS on the right

A similar process can be applied to multiple images to normalize each one so that they can be directly compared without the impact of the atmosphere. In this case, a base image (usually the one with the least atmospheric issues or one that has been previously atmospherically corrected) is selected and the spectral characteristics of the other images are transformed to match those of the base image. This multidate image procedure relies on using regression analysis in which the selection of objects on the imagery that remain the same over time can be used as effective radiometric "control points" to normalize the images. Figure 6.5 shows an example of how this normalization looks graphically.

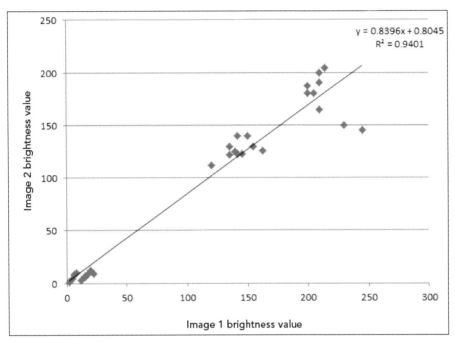

Figure 6.5. Graphic representation of multidate image normalization using regression analysis

As previously discussed, absolute atmospheric correction generally requires an atmospheric profile obtained on the same day and at the same location as the imagery, which is used in conjunction with an algorithm to compensate for the atmosphere. As such, absolute atmospheric correction can be difficult to perform because this information is often unavailable: relative correction must be performed instead. Recently, some of the information required in the atmospheric profile can be obtained from specific atmospheric absorption bands available in some image products, especially hyperspectral imagery, which has more and narrower spectral bands than multispectral imagery. The availability of this information within the imagery itself, along with the development of more hyperspectral sensors, has sparked a resurgence of algorithm development to produce better absolute atmospheric correction methods.

Most algorithms used for absolute atmospheric correction are based on one of two very robust and time-tested radiative transfer models: MODTRAN (MODerate resolution atmospheric TRANsmission) or 6S (Second Simulation of the Satellite Signal in the Solar Spectrum). Information gained from either the atmosphere profile or the image absorption bands along with the exact location, date, time, and other image information such as average ground elevation, altitude of the sensor, and band wavelength ranges are used to determine the amounts of atmospheric absorption and scattering present when the image was acquired. These characteristics are then used to convert the radiance of the image to atmospherically corrected reflectance. The most common algorithms used to perform this correction are Atmospheric CORrection Now (ACORN), Fast Line-of-sight Atmospheric Analysis of Spectral Hypercubes (FLAASH), and Atmospheric CORrection (ATCOR).

One final method available to perform absolute correction does not rely on information about the atmospheric profile. Instead, this method, called empirical line calibration, requires on-the-ground spectral reflectance measurements collected at the same time and place as the image acquisition. Areas are selected on the ground that can be clearly identified on the image. Some bright areas such as pavement or sand are selected as well as dark areas such as deep, clear water. Measurements are collected using a spectroradiometer and are used to calibrate the imagery separately for each band (wavelength).

Clouds and Cloud Shadows

Clouds and cloud shadows affect most imagery and introduce unwanted variation that can be challenging to control. Even if the platform carrying the sensor flies below the clouds so that they are not part of the image, clouds still have an overall effect on the imagery by reducing the amount of sunlight available during the image collection. If the platform is above the atmosphere (e.g., a satellite), then clouds and cloud shadows are a constant consideration.

Some clouds are easy to see and simply block the view of the sensor to the ground. Other clouds, such as cirrus clouds, tend to act more like the haze in the atmosphere because of their wispy, translucent nature and can be corrected as already described in this chapter. The latest Landsat 8 Operational Land Imagery sensor has a band dedicated to detecting cirrus clouds. Perhaps the biggest issue with these types of clouds that obscure but do not block the ground is that they are not uniformly distributed across the imagery, and therefore the same correction cannot be uniformly applied throughout the image.

In images with snow, detecting clouds can be a challenge. Clouds versus snow are indistinguishable in the visible wavelengths but can be separated using other wavelengths of electromagnetic energy. It is important when identifying clouds to also find their shadows (see figure 6.6). Cloud shadows can be difficult to detect. Depending on the illumination and view angles, the cloud shadow can be anywhere from directly below the cloud to some distance from it. Cloud shadows appear very dark on the imagery and are easily confused with other dark objects such as water and dense forests. When conducting a change analysis by comparing two or more images of the same area to one another, clouds and cloud shadows are even more problematic because both will appear as changes from one image to the other, while neither represents actual change on the ground.

Figure 6.6. Image showing clouds and cloud shadows

In most situations, little can be done about clouds and cloud shadows. The most common correction to fix a single date of imagery is to use pieces of cloud-free images from other image dates as close as possible to the base date for those areas impacted by the clouds and their shadows. While this is a growing practice, it comes with its own problems, especially if the sun or viewing angle of the image patches are significantly different from the base image, as shown in figure 6.7. If the number of areas or the total area covered by clouds is large, it may be best to classify multiple cloudy images of the area separately, and then select the cloud-free areas from each thematic map to create one complete map of the area.

Clearly, the best approach is to acquire imagery with few or no clouds, if possible. In some regions of the world such as the American West, this is a realistic goal, while in others such as the tropics and Southeast Asia it is extremely rare to find a cloud-free satellite image. When clouds are persistent, alternative image sources such as radar or airborne collects may be the only way to remotely sense the ground.

Figure 6.7. An example of different dates of QuickBird imagery mosaicked together to create a cloud-free image of a portion of Kalaupapa National Historical Park in Hawaii. Notice the distortions in the mosaic caused by the different sun angles and view angles of the imagery used to create the mosaic. The mountains of Hawaii are often shrouded in clouds, and it is nearly impossible to collect cloud-free imagery at the same time for both the coastal plains and the mountains.

Geometric Correction

Geometric correction corrects errors introduced into the position (geometry) of the imagery. The goal is to fix the image so that it can be used with other spatially accurate imagery and geospatial data layers. In other words, all the layers in our GIS must line up with each other. An image that is not properly registered introduces unwanted variance as the imagery is not truly indicative of what is on the ground. This variation must be controlled to obtain the most effective imagery for use in any geospatial analysis.

Before any errors can be corrected, they first must be recognized as errors. There are two general types of errors: systematic and random. Systematic errors occur throughout the image and present the same issue everywhere. They tend to be much easier to correct, because once the issue has been identified it can be corrected throughout the image. Random errors can occur anywhere in the image and are more difficult to correct. For geometry, random errors are generally the result of poor digital elevation model data or the lack of an accurate sensor model, which causes small errors in the attitude and position information.

As in radiometric correction, the altitude of the platform plays a role in what errors are introduced. For example, image scanning systems that are long distances from their targets tend to have small view angles that are nearly vertical, and therefore the ground area covered by each pixel in the image remains fairly constant. A similar scanning system flown in an airplane close to the ground will have a significantly greater scan angle, resulting in greater geometric distortion because the ground area covered in a single pixel toward the edges of the scan will be much larger than the areas covered toward the center. This issue is analogous to a vertical photograph versus an oblique photo. The vertical photo looks straight down on objects and the scale of the photo remains relatively constant except for changes in topography. The oblique photo changes scale and tends to look more at the sides of objects as the image moves toward the horizon. If the scanning system is pointable, as in the case of most high-spatial-resolution systems such as the DigitalGlobe and Pleiades constellations, then there are additional issues similar to what we have discussed in an oblique photo (i.e., the images become oblique) as the view angle increases.

The type of sensor is also important. An example of sensor type is the difference between a scanner and a camera. A camera acquires an image that contains a principal point: the geometric center of the image. If that image were vertical, then by definition, the nadir (the point directly below the center of the camera) and the principal point (the geometric center of the image) are in the exact same spot. An important concept known as topographic displacement is the displacement, caused by a sensor that is not perfectly horizontal, which radiates from the nadir. Objects with positive elevation are displaced outward from the center of the image (nadir) while objects with negative elevation are displaced inward. A quick look at any aerial photograph or high-spatial-resolution image will demonstrate the impact of topographic displacement. For example, one of the first IKONOS images was of Washington, DC. Topographic displacement is obvious in this image, as the Washington Monument is displaced outward from the center of the image (see figure 6.8). For this reason, an image is not a map (not planimetrically correct) and geometric correction is required to correct for this displacement. In a scanning system, the principal point (or geometric center of the image) moves along with the scanner. There is no one center point, but many as the scanner moves along. This sensor system results in a one-dimensional topographic displacement that also must be corrected if the image is to be properly registered to the ground. Other factors such as curvature of the earth, atmospheric refraction, and the earth's rotation must also be accounted for depending on the altitude and type of sensor used to collect the imagery.

Figure 6.8. An image of the Washington Monument showing topographic displacement. Source: DigitalGlobe

The following sections provide an overview and discussion of various components of geometric correction. It is important to remember that the goal of geometric correction is to correct the issues in the imagery so that the imagery best fits the ground. Remotely sensed imagery is the source of much geospatial data. Without effective geometric correction, this imagery would be far less useful. It is actually the ability to make these corrections that has dramatically increased the use of imagery in GIS.

Again, most geometric correction of imagery is conducted as a service and is not commonly performed by the geospatial analyst. However, some coordinate transformation and reprojection are done by the analyst depending on the geospatial data being used for a particular project. The following section begins with these topics and then provides an overview of the other issues and considerations most helpful to the geospatial analyst.

Coordinate Systems and Map Projections

For geospatial data to be useful it must have a coordinate system and a map projection. Coordinate systems are used to define the spatial location of an object on the earth (i.e., to allow you to know where you are). However, three factors complicate the use of coordinates. When most of us think of coordinates, we think of Cartesian coordinates with straight lines in the x and y directions. However, the earth is not flat, but rather curved, so flattening an image of it into a 2D representation causes some distortion. Second, not only is the earth not flat, but it is also not a regular sphere or even an ellipsoid. Instead, the shape is quite irregular (a geoid). This fact makes locating the coordinates even more difficult. Finally, methods are constantly being improved to measure the shape of the earth and also the position of objects on it (e.g., think about the advances in GPS over the last 20 years). However, none of these measurements are perfect, and this introduces errors in our coordinates.

Map projection renders the locations from a 3D earth to a 2D map (flat surface). The process most often uses a mathematical algorithm to accomplish this rendering. It is impossible to avoid introducing some distortions during the projection process. An infinite number of map projections distort one or more of the following characteristics during the projection process: 1) shape, 2) distance, 3) direction, 4) scale, and 5) area. Selection of the appropriate projection takes into account the size of the area being studied and the objectives of the analysis to decide which characteristics should have minimum distortion and which do not matter. For example, a pilot using a map to aid in navigation would wish to use a map with a projection that maintains distance and direction so to fly to the right place. The pilot would not care if the shapes or areas of the objects were distorted. In another example, everyone has probably seen a Mercator projection map in which Greenland seems to be the size of North America. The Mercator projection distorts area.

Latitude and longitude compose the geographic coordinate system that has an origin at a line we call the equator (0 degrees latitude) and a line we call the Prime Meridian going through Greenwich, England (0 degrees longitude). Latitude and longitude are components of an unprojected coordinate system that is not associated with any specific map projection. Instead, everything is measured in degrees north or south of the equator and east or west of the Prime Meridian. It does not use a Cartesian coordinate system because it measures position on the curved earth where the lines of longitude converge at the north and south poles. Therefore, the distance covered by a degree of longitude varies from the largest value at the equator to zero at the poles where the lines converge.

The most commonly used projected coordinate systems in the United States include the state plane coordinate system and the universal transverse Mercator (UTM) coordinate system. The state plane system is a standard set of projections using Cartesian coordinates for each state. Depending on the size of the state, each one might have more than a single zone. In addition, the projection used for each state depends on the shape of the state. States that are wide use the Lambert conformal projection, while states that are tall use the transverse Mercator projection (Bolstad, 2012).

A state plane coordinate system works well if working in a single state, but requires using multiple state coordinate systems if the project incorporates multiple states. Therefore, another common projected coordinate system (the UTM coordinate system) is a global system that is not limited to a single state. The UTM system is based on the transverse Mercator projection and is used commonly throughout all of North America and the world.

One final consideration that may cause a geometric/positional error in geospatial layers is the datum. A datum is the reference surface used as the starting point for all other measurements. Our ability and technology to accurately define this reference surface have changed over time, and, therefore datums have also changed over time. In addition to knowing the coordinate system and map projection, the datum and if necessary the transformation between datums must also be known for all the geospatial data in a project. The use of different datums with unknown transforms will result in positional errors.

Converting between geographic coordinates and projected coordinate systems and datums is quite simple, in that the mathematical conversions are readily available to allow this process to occur. The conversions are part of most image analysis and GIS software packages such as ArcGIS. The goal is to have all the geospatial layers, including those that are imagery-based, in the same coordinate system and map projection and datum so that all the layers directly coincide with each other. Image analysts should generally be aware of the metadata of each spatial layer to confirm these.

Image Registration

To enable two images to be compared, they need to be registered to each other so that for each pixel in one image there is a corresponding pixel in the second image. To do this, the geometric transform to warp one image to the other needs to be determined and applied. The geometric correction of imagery can be performed in two ways. Either an image can be registered to another image (image-to-image registration) or an image can be registered to a map (i.e., the ground). The processes to conduct this registration are similar, with the exception that two images that are registered to each other will not necessarily be registered to a map. To overlay an image on a map, the transform between the image and the map needs to be known. If the transforms for multiple images to the same map are known, then it becomes possible to overlay the images on the same map and perform the required comparison. ArcGIS handles both with the image coordinate system (ICS). It is best to use image–ground transforms, as this results in the image being registered to the ground.

Image-to-map registration, or geometric correction, is the process by which the image is made planimetric. That is, geometric distortion in the image is removed so that the image aligns directly with a map taking on the same coordinate system and map projection. However, this process may not remove all the topographic displacement in the image (see the discussion on orthorectification later in this chapter). The process of geometric correction or image-to-map registration includes two important components. The first is determining the geometric relationship between the coordinates of the pixels in the image and the coordinates of the exact same place on the map (or other image in image-to-image registration). The second component is determining the pixel values at the new pixel locations as a result of applying the geometric correction and resampling.

If the transformation from an image to a map is known, then that transform may also be inverted, resulting in the map being transformed to the image (i.e., the image remains unchanged and the vectors of the map are overlaid on the image). This process is referred to as working in image space rather than working in map (ground) space. In a GIS, most image analysis is done in map space with the image transformed to the map. This difference between working in map space and image space was one of the primary differences between a GIS and traditional imagery applications. ArcGIS, by default, displays all image and vector data in map space and includes tools to determine the appropriate transformation if the image is not georeferenced. ArcGIS also has the ability to work in image space (or the ICS) by using the "Focus on image" option.

Most images that analysts receive will be orthorectified to map space, but this is not always the case, and therefore it is good to understand how these transformations take place. In a GIS, each raster dataset has an associated geometric transform that defines the transform between the rasters and the ground in a specified projection. Depending on the level of processing applied to the image, this may be a simple shift and scaling (as is the case if the input is an orthoimage in the same projection) or other transforms such as affine, projective, warp, or more complex orthorectification transforms. Orthorectification is the process of transforming an image so as to take into account the elevation changes and is explained in more detail later.

The transformation of an image to fit another image or a map requires that the image be resampled so that the pixels align. This resampling is also necessary if two images are in map space, but not with the exact same projection, pixel size, and/or pixel alignment. In all cases, the source imagery must be resampled for display or analysis such that pixels are displayed in the correct location on a screen or align with other images.

Resampling

Resampling of an image occurs whenever any transform is applied for display or analysis. Resampling an image will always have some effect on either the radiometry or geometry of the image. There are many different resampling techniques and each has advantages and disadvantages. For optimum results, it is important to understand what resampling has taken place on the imagery and also to mitigate the number of resampling steps that take place.

Image resampling is performed by defining a raster in the required output coordinate system (or display) and for each output pixel and transforming the x,y coordinates at the center of each pixel to determine the corresponding column and row values of the input image. The output pixel value is then determined from the nearest values of the pixel or pixels from the input image. There are various resampling methods; the three most common are: 1) nearest neighbor interpolation, 2) bilinear interpolation, and 3) cubic convolution interpolation.

The nearest neighbor interpolation approach is the simplest of the three methods. The pixel value from the original image that is closest (nearest) to the transformed pixel in the registered image is assigned to that pixel (figure 6.9). Figure 6.10 shows the same approach using a mathematical example. The new pixel value is simply the closest pixel value from the image before registration. The biggest advantage of this approach is that all the original pixel values are preserved. In the other approaches, some averaging is performed and the original values are lost. Therefore, if there are particular pixel values indicative of a certain characteristic or property in the image, those values remain to be discovered and explored by the analyst. The biggest disadvantage of nearest neighbor interpolation is that there can be significant degradation in the visual quality of the imagery as well as artifacts such as step effects, especially on the edges of features. These effects are also known as spatial aliasing. Looking at figure 6.9, one could see that if a road or river was represented by the raw pixels (black squares) at the top of the figure, this linear feature would be broken and no longer be linear in the corrected image (open circles). In the first part of the image, the road or river pixel values would be pushed upward and be in the first row, while in the second half of the image, they would go downward and be in the second row. This result would cause an unfavorable visual rendition of the image, and if the image were being used for visualization purposes would cause some issues.

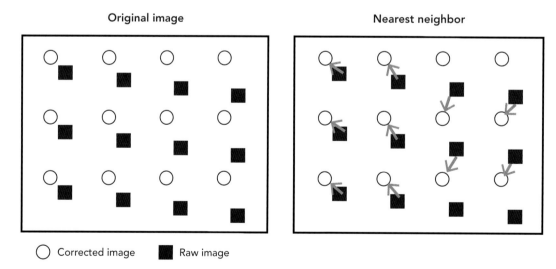

Figure 6.9. A diagram showing the nearest neighbor interpolation process

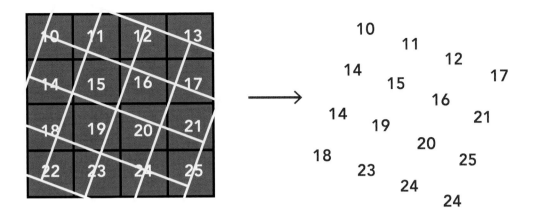

Figure 6.10. A mathematical example of the nearest neighbor interpolation approach

Bilinear interpolation is a compromise between the nearest neighbor and the cubic convolution approaches. In this method, instead of a single value being selected for the new pixel location, the distance-weighted average value for the surrounding four pixels are selected (figure 6.11). If the four pixels are equal distances from the center, then the result is simply the average (mean). Figure 6.12 shows a mathematical example for a single pixel. The closest four pixels from the original image are averaged, and that value is placed in the new pixel location in the transformed image. The advantage of this approach is that linear features tend to be maintained, unlike in nearest neighbor interpolation. Also, the image is more visually pleasing without artifacts such as jagged edges or linear features that end abruptly. The biggest disadvantage is that the original image values are lost in the averaging process (i.e., they can be smoothed). The impact of this averaging/smoothing is that the high and low values in the image are removed. If these values were indicative of a specific feature or property in the image, those values are now reduced (if not eliminated) by the interpolation. Bilinear interpolation resampling is the most commonly used and generally provides good results.

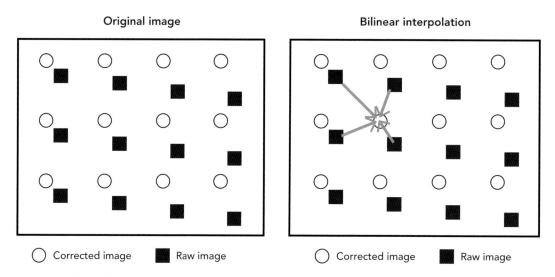

Figure 6.11. A diagram showing the bilinear interpolation process

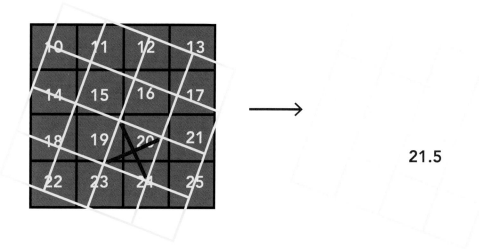

Select pixel with X in it. Result is average of the 4 nearest pixels.
(19 + 20 + 23 + 24) / 4 = 21.5

Figure 6.12. A mathematical example of the bilinear interpolation approach

    The third approach is called cubic convolution interpolation and uses a weighted averaging of the closest 16 pixels. The weighting is performed in such a way as to maintain the sharpness of the imagery. Figure 6.13 shows an example of the pixels selected to compute the new value for a single pixel. The result of this interpolation is some smoothing of the imagery despite the weighting chosen to minimize the smoothing. In other words, the high and low pixel values (the variation) in the original image are reduced, resulting in a more uniform image. While this image can be aesthetically and visually pleasing, it has lost some of the information in those pixel values that may be useful in image analysis. Therefore, it is important to understand the purpose of the imagery before any resampling is applied. Cubic convolution is often the most suitable resampling method selected for visual interpretations. Those who wish to use the imagery as a backdrop in their geospatial analysis should be quite satisfied with this resampling approach. However, it is not necessarily the best for those analysts wishing to digitally process the imagery for land-cover mapping and other related purposes.

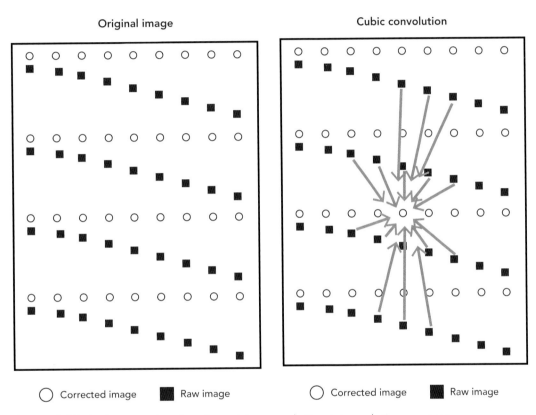

Figure 6.13. A diagram showing the cubic convolution interpolation process

In many cases, the analyst has a choice and can order imagery geometrically corrected with the resampling approach appropriate for their purposes. While the geospatial analyst will typically not perform resampling, knowledge of which approach is applied to the imagery is important for deciding the future use of the imagery. In other cases (as in Landsat imagery, which is processed using cubic convolution), the imagery is always resampled using a single resampling technique. However, the geospatial analyst must still be aware of which approach was used as a reason for explaining the results of their analysis.

Georeferencing

Images are typically georeferenced to the ground, which means that a transform from ground coordinates to image coordinates has been determined. The new image is created by applying the transforms with the appropriate resampling either for displaying or analysis of the image. If the image has already been rectified, such as is the case with an orthoimage, then to produce an output that has a different pixel alignment, the original image will be resampled at least twice: once during the original orthorectification process and a second time to resample it for display or analysis. Remember that resampling alters the radiometric and/or the geometric properties of an image and should be performed a little as possible. Therefore, it is optimum to orthorectify the image directly to the pixel alignment to be used for analysis. Alternatively, it is possible to create the full orthorectified image by storing the original image and the required transform, and then applying the orthorectification and resampling on the fly such that only a single defined resampling takes place. The disadvantage to such on-the-fly processing is that the different transforms need to be maintained and applied, which increases the complexity of the process. Hence for most GIS applications, the images are preorthorectified to ground coordinates, and it is acknowledged that some loss of information will have taken place because of the additional resampling. When comparing two images, no additional resampling needs to take place so long as the two images are in the same projection, pixel size, and alignment.

2D Transforms

Multiple transforms can be computed between image space and ground space. The appropriate transform to use depends on the distortion that needs to be modeled to correct the imagery, and the accuracy required. The transformation parameters also need to be determined. These may be implicit or may need to be determined based on measuring control points.

Control points (or ground control points [GCPs]) are locations in an image for which the ground coordinates are known. They may be measured by a manual process in which the user selects the locations in the image and the corresponding locations on a map. In some cases, the measured ground locations are determined by field surveys. It is imperative that the control points be accurately located on both the image and the map (ground). GCPs need to be distinct and easily identifiable on both the map and the image. Usually, it is far easier to identify the exact location of a road intersection or a boundary between different land-cover types on a map than it is on the image. Depending on the spatial resolution of the imagery (pixel size), many features that are clear on a map or the ground appear to be fuzzy or indistinct on the imagery. However, effective modeling of the geometric relationship between the image and the map is highly dependent on very good GCP selection.

The number and distribution of the GCPs are also very important. A sufficient number of good points must be selected across the entire image. If the image has lots of road intersections and other easily distinguishable points, it is relatively easy to collect points throughout the image, including near each of the corners as well as in the middle. Often, points are more distinguishable in one or two sections of the image, and more work must be devoted to obtaining good GCPs in the other parts of the image. The key is to get sufficient good GCPs to accurately determine the relationship between the image and the map to perform the appropriate coordinate transformation.

With the rapid increase of available accurate basemap imagery, a process of automated image-to-image registration is also often used by which the system searches for locations in a reference image that has already been accurately georeferenced. This process works well if the images are relatively similar in appearance and resolution, but there are often issues when imagery from different seasons are used or if significant changes have taken place between the two dates.

The simplest transform is just a shift and scale change. The determination of such a transform requires a minimum of two known points in ground and image spaces to determine the parameters. Such a transform is defined using four parameters and is applicable only to images that have already been orthorectified. Figure 6.14 shows an example of how a shift and scale can be applied to an image based on two control points.

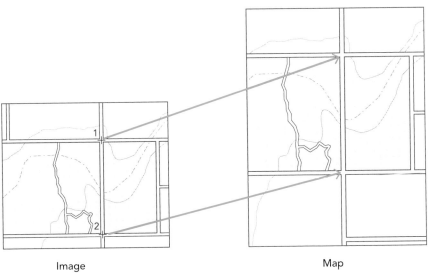

Figure 6.14. An illustration of shift and scale in x and y

The affine transform is common and applies scale and rotations for each axis. It has the property that parallel lines remain parallel. A minimum of three known points in ground and image spaces must be collected to determine the parameters. It is defined using six parameters and referred to as a first-order polynomial transform. Figure 6.15 shows an example of how an affine transform can be defined by three points and that the lines remain parallel.

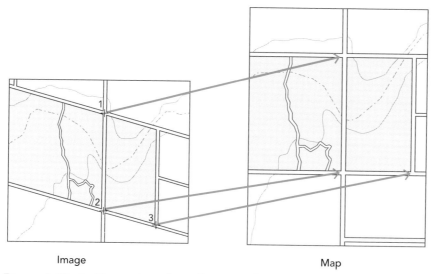

Figure 6.15. An illustration of an affine transform

The projective transform is useful because it is a good, simple representation of a how an image that is not horizontal gets projected to the ground. Lines that were straight remain straight, but need not remain parallel. A minimum of four known points in ground and image spaces is necessary to determine the parameters. It is defined using eight parameters. In figure 6.16 a projective transform with 4 points is shown. Such transforms are common for oblique imagery.

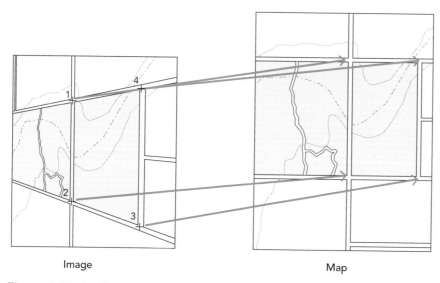

Figure 6.16. An illustration of a projective transform

The second-order polynomial provides additional parameters to enable the modeling of some nonlinearity in the transforms. It is used relatively rarely. A minimum of six ground and image points are required to determine the parameters. It is defined using 12 parameters. Figure 6.17 shows how a second-order polynomial can account for curved features in the input image.

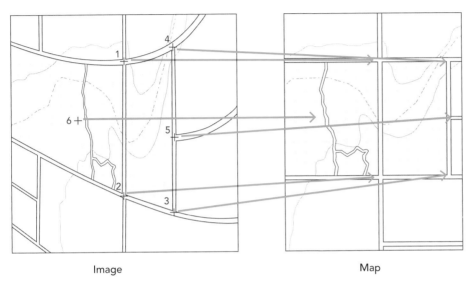

Figure 6.17. An illustration of a second-order polynomial transform

Higher-order transforms are also possible, but rarely used. The advantage of using these parameter-based transforms is that if more than the minimum number of well-distributed ground and image points are defined, then residual errors can be computed to give an estimate of the transformation accuracy.

Other methods of defining such transforms include corrections grids, modeling using splines, or the tinning methods where any number of irregular ground and image points can be used and the transforms are determined using the nearest three control points for any pixel. In all these cases, the aim is to determine a transform such that for any x,y output coordinate the appropriate image column and row can be determined.

Finally, the quality of any transformation can be tested and reported by collecting an independent (not used in the building of the model) set of GCPs and comparing these transformed image coordinates to the ground coordinates to see whether they agree.

Orthorectification

Orthorectification is the process of rectifying an image by taking into consideration the impacts of terrain. When images are taken from aerial sensors, the transform from image to ground space cannot be completely and accurately represented without taking into account the terrain model (i.e., elevation), and so a 2D transform is not sufficient. With aerial imagery, two locations at the same ground x,y coordinates but at different heights would have different image coordinates. If the terrain model is not taken into consideration during rectification, significant errors will result. The size of these errors is dependent on the angle of a ray of light to the ground and the height difference. If the height is not taken into consideration, ground points that vary in height will have associated ground locations that are in error.

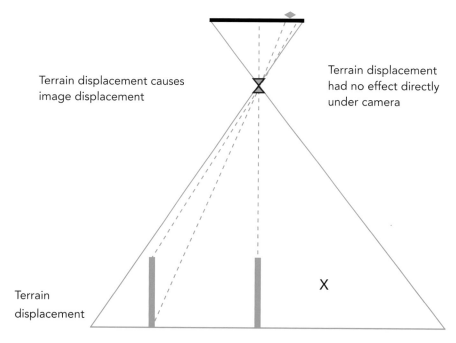

Figure 6.18. Image displacement due to terrain difference

Whereas the 2D transforms described above required only values of x and y, a 3D transform for orthorectification requires x, y, and z values. The process of orthorectifying an image is the same as rectifying an image with the addition of a z coordinate that needs to be determined and used in the transform. The corresponding z value for each x,y value is typically obtained using a digital terrain model that defines a z value for any x,y value. The accuracy of the orthorectification is dependent on not only the accuracy of the transform but also the accuracy of the digital terrain model.

A number of different transforms can be defined for orthorectification. The most appropriate transform to use is dependent on the physics of the sensor that was used to collect the imagery and the method used to determine the parameters. The most common models used by sensors are the standard frame model and rational polynomial coefficients (RPCs).

The standard frame camera model defines the physical characteristics of a camera in terms of the camera location (x0, y0, z0), the orientation angles (omega, phi, kappa), the focal length of the camera, and parameters that define the interior orientation of the CCD sensor in relation to the camera location and camera distortion parameters.

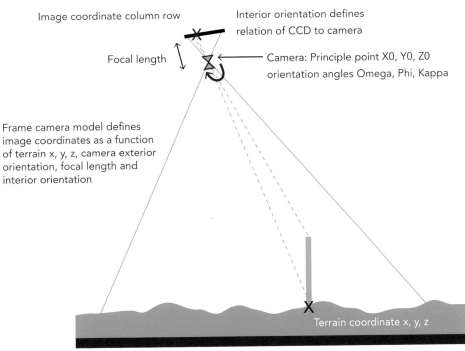

Figure 6.19. A frame camera model defines image coordinates as a function of terrain x,y,z, camera exterior orientation, focal length, and interior orientation.

The advantage of the frame camera model is that for a collection of images taken with the same camera, only the six exterior orientation parameters (x0, y0, z0, omega, phi, kappa) change from image to image. With the appropriate modern sensors, global positioning satellites, and inertial measurement units attached to the camera, these exterior orientation parameters can be determined without the need for ground control. The focal length and interior orientation parameters remain constant and need to be determined only once. Such a process is referred to as "direct georeferencing." Technologies to enable direct georeferencing have improved significantly in the last 15 years, but in many cases some GCPs are still required to check the achieved accuracies.

If the focal length and interior orientation parameters are known, then the exterior orientation parameters can be determined using three GCPs for each image. For a large collection of images, collecting a large number of GCPs would be very expensive, so a process of aerial block triangulation is used to reduce the number of points needed. Aerial block triangulation works by first determining tie points between images. Tie points are image coordinates that define the same location in two or more images and can be used to determine the spatial relationship between overlapping images. Image processing can be used to quickly determine a large number of such tie points between images. Once the tie points are computed, a process called bundle block adjustment is used to compute the exterior orientation parameters for the images while requiring significantly fewer control points. The process determines the unknown parameters by an iterative process that reduces errors between all measurements to a minimum. If there is sufficient redundancy in the parameter determination, then other parameters such as the camera focal length or camera distortion parameters can also be determined.

The standard frame camera model is not normally used for satellite imagery, because the satellites are typically not framing sensors, the focal lengths are too long, and the physics of the sensor and the atmosphere are not sufficiently defined. Although other physical camera models exist, the equations become complex and require the accurate knowledge of camera parameters that satellite vendors often consider proprietary. The RPC model is therefore often used instead to define the sensor orientation.

The RPC is defined as a rational polynomial of 20 terms for x,y,z in the numerator and denominator (i.e., it is defined by a total of 80 parameters). It provides a good approximation to most airborne and satellite sensor models. One of the advantages of the RPC model is that the parameters can be determined by the satellite vendors based on the physical characteristics of the sensors, but provided in a generic form that can be used by most image processing applications.

Depending on the accuracy of the satellite orientation parameters, these RPC parameters can be determined without the need for ground control to an accuracy of a few pixels. However, in many cases the provided RPCs are not sufficiently accurate, and they are refined by using GCPs or, for collections of imagery, using tie points, control points, and a block adjustment process.

As with the 2D georeferencing of imagery, once the 3D transforms for an image are known, and assuming that a digital terrain model exists, then orthorectification can be performed to determine for each pixel of an output image what the corresponding input pixel is and then perform the resample. Errors are related to the accuracy of the transform and the accuracy of the available digital terrain model. In areas where the terrain is smoothly changing, terrain modeling is relatively easy. However, in cases where the terrain is undulating or has sharp changes caused by cliffs, bridges, or buildings, terrain modeling can cause shifts or artifacts in the images. It is also important that the digital terrain model (DTM) uses the same datum as is used for the exterior orientation. For example, many RPCs for satellite imagery are given using ellipsoid heights, while most DTMs are stored with orthometric heights, so these corrections need to be made. If there are inaccuracies in the DTMs or there are any shifts caused by the use of incorrect vertical datums, then this will have an effect on the resulting accuracy.

In cases where a suitable DTM does not exist, it can often be created photogrammetrically from the overlapping images. There are different photogrammetric methods for extracting terrain models. As above, the sensor model defines the transform from x,y,z to a specified row and column. If two images overlap, then for any x,y point there should be a specific z value that results in the transform returning the same image point in the overlapping images. The different algorithms work to find the best solution.

Mosaicking

In most large projects, a single image from the satellite or aerial sensor may not be sufficient to cover the entire project area. Often tens, hundreds, or even thousands of individual images are required depending on the extent covered by each image. The combining of many overlapping orthoimages into a single large image is referred to as mosaicking or creating a mosaic. As part of this process, one needs to determine what part of the overlapping image to use. Mosaics can be very large and are handled in different ways. An introduction to mosaicking was provided in chapter 5, and more detail on creating mosaics is covered in chapter 13.

Summary—Practical Considerations

This chapter presents a discussion on controlling the factors that cause unwanted variation in imagery. Since remote sensing works because the variation in the imagery is highly correlated with what is happening on the ground, it is very important to minimize any unwanted variation that reduces this correlation. To understand this process, it is important to know about electromagnetic energy and how it interacts with the objects being imaged. While the sensor recording the imagery measures the amount of electromagnetic energy reaching that sensor, a number of factors actually must be considered before using the imagery. Digital imagery is recorded in DNs based on the radiometric resolution of the sensor. These values can be used to process the imagery, but are more often converted to radiance using the calibration information about that particular image and then finally to reflectance using ATCOR.

ATCOR is one type of radiometric correction used to control unwanted variation in the imagery. The other types of radiometric correction control either issues with the sensor or issues caused by the angle of the sun or the topography of the ground. Other factors that cause unwanted variation in the imagery are clouds and cloud shadows. Clouds are usually easy to detect, while their shadows can be more difficult. Both cause the information recorded in the imagery to be unrepresentative of what is actually on the ground. Failure to deal with clouds can be a large source of unwanted variation on the imagery. It is always better, but not always possible, to use cloud-free imagery.

In addition to corrections needed to the radiometry to control unwanted variation in the imagery, geometric corrections are also very important. Given that the power of a GIS is to be able to very accurately overlay various data layers, the use of imagery in a GIS relies heavily on registering the imagery to the ground. Because the earth is a sphere spinning on its axis, and its surface is not flat, the geometric correction of imagery is not trivial. The ability to register an image depends on a technique called resampling that accounts for these factors. Different resampling methods have varying advantages and disadvantages depending on how the imagery will be used in the GIS. Nearest neighbor resampling preserves data values that may be important in creating a thematic map but is not the best choice when resampling imagery for a base image because of the artifacts it can introduce. Bilinear interpolation and cubic convolution both solve the linear feature problem, but both compute some type of averaging of the data, which may remove some information. Finally, correcting the imagery for changes in the earth's surface (elevation) is especially complicated and requires a process called orthorectification.

Section 3
Extracting Information from Imagery

Chapter 7
Understanding Variation on the Ground—the Importance of the Classification Scheme

Introduction

Creating an accurate map requires understanding the variation on the ground and the variation on the imagery and then relating them to one another. The greater the correlation between what varies on the ground and what varies on the image, the better the map. In addition, other spatial data layers (e.g., slope, aspect, elevation, soils) can aid in understanding this variation. The development of an effective classification scheme that distinguishes and defines the variation to be mapped is vital to successful mapping.

This chapter focuses on how to create a robust classification scheme. First, we examine the requirements of robust classification schemes. Next, we review common classification schemes in use today. The chapter ends with a real-world example of a classification scheme.

Definition and Why Classification Schemes Are Important

Classification schemes are systems for logically organizing and categorizing information or data (Cowardin et al., 1979). They enable the map producer to classify a complex landscape into classes and allow the map user to readily recognize and use those classes (Congalton and Green, 2009). Therefore, creating a map requires the development of an appropriate and effective classification scheme that clearly defines and distinguishes the classes of objects to be mapped. The scheme must be

- constrained by minimum mapping units (MMUs),
- defined by labels and rules,
- totally exhaustive,
- mutually exclusive, and
- hierarchical.

The lack of a well-defined scheme will doom a project to failure. The inability to develop and agree on the scheme very early on in the project will result in cost overruns and significant inefficiencies.

Constrained by Minimum Mapping Units

The MMU is the smallest area to be mapped on the ground. Anything smaller than the MMU is grouped into a neighboring map class because it has been determined to be too small to be of use or to discern on that map. The choice of the MMU is determined by the scale or spatial resolution of the imagery and the objectives of the mapping. If the imagery used to make a map has a spatial resolution (i.e., pixel size) of 30 × 30 meters, it would not be appropriate to try to map any objects smaller than that size. In fact, the MMU is typically larger than a single pixel, and a grouping of pixels is usually a more meaningful level of detail appropriate for a map. The smaller the MMU, the more effort may be required to create the map. Sometimes, a map will have multiple MMUs, where one minimum area is set for some map classes and another is set for the other classes. While multiple MMUs are not common, they are possible depending on the objectives of the mapping. Therefore, MMU must be carefully considered so that the map's level of detail is consistent with project needs and budget constraints.

Defined by Labels and Rules

Despite what is often seen in practice, a robust classification scheme must be more than just a list of the map class labels (e.g., water, forest, urban, agriculture). A valid classification scheme includes rules that define the map classes. Without defined rules, different map users and producers will make assumptions about the map classes based on their own experiences, resulting in significant confusion about the classes. For example, it is not sufficient to simply list water as a class, but rather water must be defined so that map producers and users know what is meant by the water map category. While water might sound like a self-explanatory class, failing to define the class will cause confusion. For example, is a swimming pool considered water? In some maps, such as those used by a fire department, identifying swimming pools as water may be very important because swimming pools can provide a source of water for firefighting. In other maps, a swimming pool may be considered part of an urban class and not mapped as water. Besides swimming pools, examples of classes where a "water" label not accompanied by a rule defining water include reservoir and ocean edges, swamps, wetlands, and vernal pools, all of which are covered with water part but not all of the year because of tides, seasonal rainfall, or the draining of reservoirs.

Figures 7.1 and 7.2 illustrate the importance of rules being included in classification schemes. Figure 7.1 is a hierarchy of labels for a very simple map. However, without definitions or rules for distinguishing these very simple map classes, it is easy to see how confusion and uncertainty may quickly arise. The first decision is to determine whether the area is water or land. We have already discussed the need for defining even something as simple as water. In this case, we will create a rule that defines water as an area that is 80 percent or more covered in water all year.

If we determine we are on land, then we need to decide whether the land is vegetated or nonvegetated. If the land is mostly just soil but has some tufts of grass every so often, is the area vegetation or nonvegetation? Again, it is clear that we need a rule of vegetation density to make this determination. If we state that the area must have at least 10 percent vegetative cover to be labeled vegetated, then we can determine whether the area is vegetated or nonvegetated. If we decide it is vegetated, then we next need to decide whether it is woody or herbaceous vegetation, which requires a definition of woody vegetation. If we decide it is woody vegetation we need another rule that determines whether the woody vegetation is forest or shrub. Figure 7.2 provides the map class hierarchy of figure 7.1 with the addition of rules and MMUs. It is now a classification scheme. Hopefully, this very simple example demonstrates the need for clear and concise definitions as a critical part of any classification scheme.

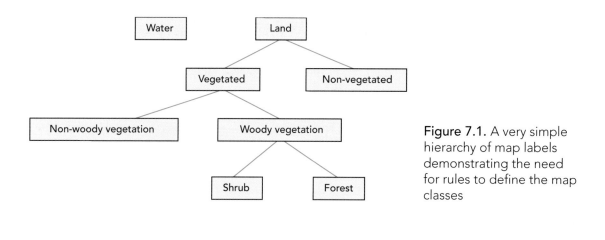

Figure 7.1. A very simple hierarchy of map labels demonstrating the need for rules to define the map classes

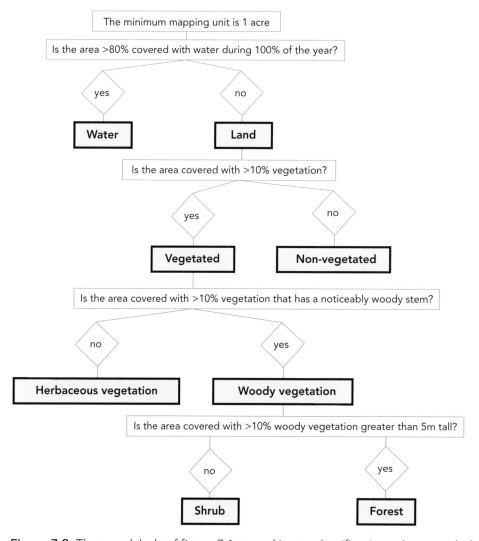

Figure 7.2. The map labels of figure 7.1 turned into a classification scheme with the addition of map label rules and minimum mapping units (MMUs)

Another example of the importance of a class definition can be found when mapping a forest map class. Most people believe they know what a forest looks like. However, many characteristics define a forest such as how tall trees must be, how large an area must contain trees, or how dense (i.e., how close together) the trees must be. Many schemes use tree height to define a forest, such as an area composed of trees at least 5 meters tall (Congalton et al., 2014). Using this definition alone, a group of recently planted 3-meter tall trees would not be called a forest even though the trees would be taller than we are when standing in the middle of them.

Totally Exhaustive

A totally exhaustive classification scheme labels all the meaningful variation found on the ground, resulting in every single area on the map falling into one of the map classes in the scheme. Careful consideration of what map classes are appropriate for your map will usually result in an exhaustive classification. However, it is not unusual to create a classification scheme in the office only to later discover unique classes in the field that need to be added to the scheme.

It is always useful to include the class "other" in a classification scheme to label any area on your map that does not fall into one of your defined map classes. In feature extraction and identification, most of the map will be in the "other" class because only specific features are being identified and mapped. In thematic mapping, a small proportion of a final thematic map should fall into the other class. A large portion of a thematic map labeled as other often indicates that the classification scheme is not capturing all of the variation on the ground and it may need improvement.

Mutually Exclusive

Another characteristic of a good classification scheme is that the map classes are mutually exclusive. That is, each and every object is clearly labeled by one and only one map class. Here again, having clear and effective definitions or rules of the map classes should result in a mutually exclusive scheme. In other words, if the map classes are well defined so that field personnel and image analysts can easily understand each class and the differences between the classes, then each object ideally should fall into only one of these classes.

While the goal is to have a mutually exclusive classification scheme, in reality this can be very difficult to achieve. For a simple scheme with only a few (e.g., 6 to 12) general map classes, one would expect to be able to use good definitions to separate the map classes and ensure that they are mutually exclusive. However, as the complexity of the scheme and the number of map classes increases, it becomes much more difficult to guarantee that the scheme is mutually exclusive.

Hierarchical

A hierarchical classification scheme is one that is divided into levels that increase in detail from level to level. Figure 7.2 is a hierarchical scheme. Figure 7.3 is another example of a hierarchical organization of map labels.

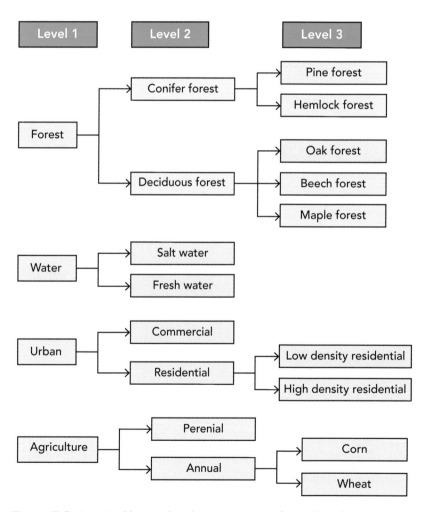

Figure 7.3. A typical hierarchical organization of map labels

As can be seen in this simple example, level 1 contains the broadest map classes, while level 3 contains the most specific. There is great power in organizing classes hierarchically, for a number of reasons. First, not every map class may need to be mapped to the same level of classification detail, and therefore the analyst can use a higher level of the hierarchy where necessary while using more-general classes where detail is not needed. In addition, when assessing the accuracy of the map, it may not be feasible or cost effective to assess the map at its most detailed level. In this case, the hierarchical structure of the classification scheme can be used to collapse the map classes down to fewer more-general classes that can then be assessed for accuracy. Figure 7.4 is another example of a hierarchical classification scheme, this time showing different levels of detail based on the objectives of the map.

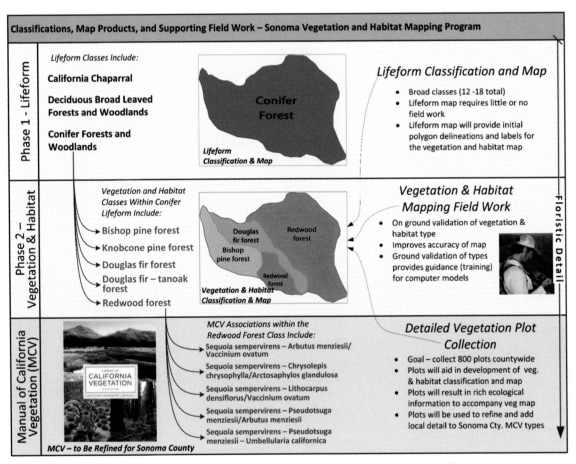

Figure 7.4. An example of a hierarchical classification scheme for forest classification in Sonoma County, California. Source: Sonoma County Agriculture Preservation and Open Space District

Which Classification Scheme to Use

Which classification scheme to use is usually a decision negotiated between the map producers and users. There are always trade-offs between the detail of the map desired by users versus the costs and schedule required to produce the map. The more detail, the higher the costs, and the longer the time required to make the map.

Once the characteristics of a good classification scheme are adopted, it is possible to select or build the appropriate scheme for a project. The first decision will be whether to adopt an existing scheme or to build a new one.

Existing Classification Schemes

There are many well-developed, widely used classification schemes for various types of thematic and feature maps. If an existing scheme is well suited to a project, it makes sense to use it (or adapt it to the particular project), rather than creating a completely new scheme. Using or adapting an existing classification scheme saves the considerable work of creating a new one and has the added benefit of ensuring that the map will integrate well with other existing maps that use the same or similar classification scheme.

Many times an existing scheme can be used. For example, for land-cover and land-use mapping, the Anderson classification scheme (Anderson et al., 1976) is often chosen as a starting point. The Anderson scheme was developed by scientists at the US Geological Survey (USGS) particularly for use with remotely sensed imagery. There are other schemes for land-cover and vegetation mappings such as the one used for USGS's 2011 National Land Cover Database and the National Vegetation Classification (NVC) scheme, which is the result of collaboration between federal agencies through the Federal Geographic Data Committee and the private sector (FGDC, 2008). The Food and Agricultural Organization (FAO) of the UN has also developed a robust land-cover classification scheme. There are also schemes designed specifically for other applications such as wetlands, wildland fuels, transportation, hydrology, coastal environments, and soils. Table 7.1 presents a list of existing classification schemes commonly used today. One should always search to see whether a robust and appropriate classification scheme exists before deciding to create a new scheme.

Table 7.1. Commonly used classification schemes

Feature Type	Classification Scheme Title	Author	Link
Land Use–Land Cover	National Land Cover Database 2011	MRLC Consortium	http://www.mrlc.gov/nlcd11_leg.php
Land Cover	Land Cover Classification System (LCCS)	FAO	http://www.fao.org/docrep/003/x0596e/x0596e00.HTM
Coastal and Marine	Coastal and Marine Ecological Classification Standard	FGDC	https://coast.noaa.gov/digitalcoast/training/cmecs-pub.html
Hydrology	The National Hydrography Dataset	USGS	http://nhd.usgs.gov/chapter1/chp1_data_users_guide.pdf
Wetlands	Classification of Wetlands and Deepwater Habitats of the United States	FGDC	http://www.fws.gov/wetlands/Documents/Classification-of-Wetlands-and-Deepwater-Habitats-of-the-United-States-2013.pdf
Wildland Fuels	Standard Fire Behavior Fuel Models: A Comprehensive Set for Use with Rothermel's Surface Fire Spread Model	USDA–FS	http://www.fs.fed.us/rm/pubs/rmrs_gtr153.pdf
Vegetation	NVC Standard	FGDC	https://www.fgdc.gov/standards/projects/FGDC-standards-projects/vegetation/NVCS_V2_FINAL_2008-02.pdf
Soils	Keys to Soil Taxonomy	USDA–NRCS	http://www.nrcs.usda.gov/wps/portal/nrcs/detail/soils/survey/class/taxonomy/?cid=nrcs142p2_053580
Transportation	Highway Functional Classification Concepts, Criteria and Procedures	DOT	http://www.fhwa.dot.gov/planning/processes/statewide/related/highway_functional_classifications/fcauab.pdf

Building New Classification Schemes

Sometimes it is necessary to build a new or more detailed classification scheme than is available in existing schemes, especially for vegetation management. For example, the National Park Service builds new classification schemes when each national park is mapped for the first time, and the California Department of Fish and Wildlife develops new schemes when performing detailed vegetation mapping of a section of the state. Both agencies rely on the conceptual framework of the NVC, which uses the hierarchy of vegetation grouping displayed in figure 7.5. To determine how vegetation in an area should be grouped, multiple vegetation samples are collected throughout the areas to be mapped. Determination of the area's vegetation classes is accomplished through multivariate analysis of the sample species cover data using techniques such as cluster analysis and ordination (Green et al., 2015). Once the samples are grouped into vegetation associations, a robust classification scheme is developed that includes a rigorous key and vegetation class descriptions.

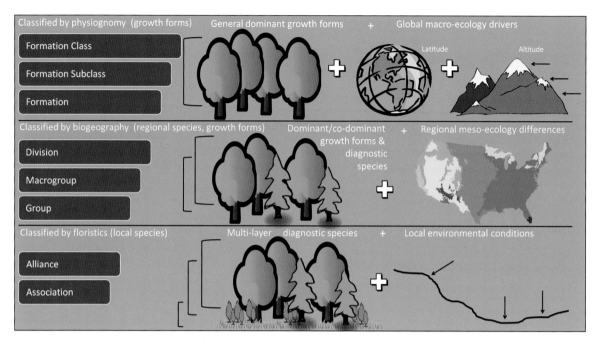

Figure 7.5. National vegetation classification (NVC) hierarchy

When using any classification scheme, and especially when creating a new one, it is important to remember that it is possible that not all the map classes to be mapped can be distinguished using the remotely sensed imagery. Thinking of a Venn diagram is helpful here (figure 7.6). One circle of the diagram represents the map classes of interest in the classification scheme, while the other circle represents what can be identified on the remotely sensed imagery. If these circles completely overlap, then the imagery should be sufficient to create the thematic map. However, if these circles do not completely overlap, then other information including other geospatial data layers would be necessary to create the map. At this point, the power of using GIS becomes important in the creation of the map. This synthesis of GIS and imagery is the central theme of this book.

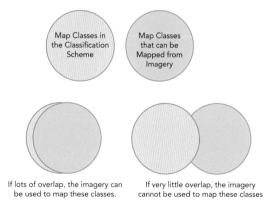

Figure 7.6. Relationship between the map classes of interest in the classification scheme and the ability of the remotely sensed imagery to distinguish these classes

Case Study—Sonoma County, California, Vegetation Map Classification Key

A good example of a robust map classification scheme is the one created for Sonoma County, California, by the California Department of Fish and Wildlife in 2015. The NVC provides the hierarchy for the scheme, and the scheme is expressed as a classification key. To create the key, more than 1,400 detailed vegetation samples were collected throughout the county. Next, personnel with the California Department of Fish and Wildlife used ordination analysis to classify the samples into NVC vegetation macrogroups, groups, and alliances. They then created a key to distinguish the vegetation alliances from one another. Finally, the vegetation alliance key was transformed into a land-use land-cover mapping key by including life-form rules and combining some alliances into groups or macrogroups. The following are important elements of the mapping key:

- The map key is more general than the vegetation alliance key. The map key was developed by a team composed of vegetation analysts, map users, and map producers who made explicit trade-offs between the level of map detail desired versus the mapping technologies, budget, and schedule available.
- The key employs the nationwide standard NVC hierarchy, which means that the resulting map can be compared with other areas mapped within the NVC hierarchy, and map users from throughout the nation in many different organizations will be familiar with the hierarchy.
- Terms used in the key, such as relative cover, dominance, and codominance are clearly defined. If the terms of a scheme are not well defined, different field personnel and image analysts will define the terms differently, resulting in map confusion.
- Unlike the NVC, the key starts with simple life-form classification rules that enable map producers and users to clearly decide whether they are in a forest, shrubland, perennial cropland, etc.

The field key follows, at the end of the chapter.

Summary—Practical Considerations

This chapter is wholly dedicated to understanding the importance of developing a robust classification scheme when using imagery to create a map. This concept is so important and yet so often overlooked that it deserves its own chapter. When thinking about a classification scheme it is most important to remember the four characteristics of any good scheme: defined by labels and rules, totally exhaustive, mutually exclusive, and hierarchical. This chapter really emphasizes the importance of having accurate definitions of each thematic map class in the classification scheme. While this seems obvious, it is the factor that seems to be most neglected when using a classification scheme. Each class must be well defined so that the mapmakers and the map users both know exactly what each class means.

The MMU is also strongly emphasized in this chapter. While MMU is a well-known and common concept for those conducting photo or image interpretation, it is not as common in the digital processing community. However, deciding the smallest unit of significance on the map is important no matter how the map is created.

Finally, this chapter provides a practical example of a classification scheme. All the issues, challenges, considerations, and problems presented in the chapter are demonstrated through this example. A poor classification scheme can cause serious problems with the entire mapping project, from training site collection, to data exploration, to image classification, to map accuracy assessment. A quickly developed and nonrigorous classification scheme can more than double the anticipated costs of a mapping project.

Hierarchical Field Key to the Vegetation Alliances of Sonoma County

This key is for the vegetation types found in Sonoma County, based on the classification developed by analyzing survey data collected for this and other relevant projects. It is intended as a guide to field-based and image interpretation-based identification of vegetation. This key is not dichotomous; instead it follows the hierarchy of the United States National Vegetation Classification (USNVC) as of the publication of the Manual of California Vegetation (Sawyer et al., 2009). The USNVC hierarchy is promoted by the Survey of California Vegetation (SCV), Federal Geographic Data Committee (FGDC) and the Ecological Society of America's Vegetation Panel (FGDC 2008, Faber-Langendoen et al. 2014).

This key lists vegetation types starting at the USNVC macrogroup level and proceeding down to the association level. The complete hierarchy for this classification is listed in Table 1, Final Vegetation Classification for Sonoma County, California.

Due to the high diversity of the vegetation types in the area, this is a complex key. Follow the instructions in a section carefully and sequentially to arrive at the correct vegetation type. You will need to collect or refer to plant composition data that includes not only those species that are dominant but also those "indicator" or characteristic/diagnostic species, whose presence may cause a stand to key to a particular vegetation type. If it seems that a stand of vegetation could key to more than one type, review the descriptions (e.g., stand tables, environmental information) for each type to determine which one fits best. Note that this vegetation key may include types that are not accurately detectable in remotely-sensed imagery.

Terms and Concepts Used throughout the Key

Stand: The basic physical unit of plant communities in a landscape. It has no set size. Some vegetation stands are very small, such as certain wetland types, and some may be several square kilometers in size, such as certain forest types. A stand is defined by two main unifying characteristics:

1. It has compositional integrity. Throughout the stand, the combination of species is similar. The stand is differentiated from adjacent stands by a discernible boundary that may be abrupt or occur indistinctly along an ecological gradient.

2. It has structural integrity. It has a similar history or environmental setting that affords relatively similar horizontal and vertical spacing of plant species. For example, a hillside forest originally dominated by the same species that burned on the upper part of the slopes but not the lower would be divided into two stands. Likewise, a sparse woodland occupying a slope with very shallow rocky soils would be considered a different stand from an adjacent slope with deeper, moister soil and a denser woodland or forest of the same species.

The compositional and structural features of a stand are often combined into a term called homogeneity. For an area to meet the definition of a stand, it must be homogeneous at the scale being considered.

United States National Vegetation Classification (USNVC): A central organizing framework for how all vegetation in the United States is inventoried and studied, from broad scale formations (biomes) to fine-scale plant communities. The purpose of the NVC is to produce uniform statistics about vegetation resources across the nation, based on vegetation data gathered at local, regional, or national levels. The latest classification standard was published in by the FGDC (2008).

The hierarchy units in the USNVC from highest to lowest (i.e., broadest to finest) are:
1. Formation Class
2. Formation Subclass
3. Formation
4. Division
5. Macrogroup
6. Group
7. Alliance
8. Association

Alliance: Plant communities based on dominant/diagnostic species of the uppermost or dominant stratum. Accepted alliances are part of the USNVC hierarchy. For the Sonoma County Vegetation Map (SVM), map classes are typically at the alliance level of the USNVC hierarchy.

Association: The most botanically detailed or finest-scale plant community designation based on dominant species and multiple co-dominant or subdominant indicator species from any stratum. Associations are also part of the USNVC hierarchy. The SVM map classes are not typically defined to the association level.

Plant community nomenclature: Species separated by "–" are within the same stratum; species separated by "/" are in different strata.

Cover: The primary metric used to quantify the importance/abundance of a particular species or a particular vegetation layer within a stand. It is measured by estimating the aerial extent of the living plants, or the bird's-eye view looking from above, for each category. Cover in this mapping project uses the concept of "porosity" or foliar cover rather than "opacity" or crown cover. Thus, field crews are trained to estimate the amount of light versus shade produced by the canopy of a plant or a stratum by taking into account the amount of shade it casts excluding the openings it may have in the interstitial spaces (e.g., between leaves or branches). This is assumed to provide a more realistic estimate of the actual amount of shade cast by the individual or stratum which, in turn, relates to the actual amount of light available to individual species or strata beneath it. However, as a result, cover estimates can vary substantially between leaf-on versus leaf-off conditions. Stands dominated by deciduous species (e.g., *Aesculus californica, Toxicodendron diversilobum*) should be sampled during leaf-on since they will have substantially less cover when leaves are absent and may key to another type.

Absolute cover: The actual percentage of the surface area of the survey that is covered by a species or physiognomic group (trees, shrubs, herbaceous), as in "tan oak covers 10% of the area being surveyed." Absolute cover of all species or physiognomic groups, when added together, may total greater than 100%, because this is not a proportional number and plants can overlap with each other. For example, a stand could have 25% tree cover in the upper layer, 40% shrub cover in the middle layer, and 50% herbaceous cover when surveyed on the ground. However, when aerial interpretation is being used, the maximum absolute value is 100%, since lower levels of vegetation cannot be seen through the overstory on aerial photographs.

Relative cover: The percentage of surface area within a survey area that is covered either by one species relative to other species within the same physiognomic stratum (tree, shrub, herbaceous) or one stratum relative to the total vegetation cover in a polygon. Thus, 50% relative cover of *Quercus douglasii* in the tree layer means that *Q. douglasii* composes half the cover of all tree species within a stand, while 50% relative shrub cover means that shrubs make up half the cover of all vegetation within a stand. Relative cover values are proportional numbers that, when added together, total 100% for each species within a stratum or each stratum within a stand of vegetation.

Dominance: Dominance refers to the preponderance of vegetation cover in a stand of uniform composition and site history. It may refer to cover of an individual species as in "dominated by tan oak," or it may refer to dominance by a physiognomic group, as in "dominated by shrubs." When we use the term in the key, a species is dominant if it is in relatively high cover in each stand. See "dominance by layer," below, for further explanation.

Strongly dominant: A species in the dominant life-form stratum has 60% or greater relative cover.

Co-dominant: Codominance refers to two or more species in a stand with similar cover. Specifically, each species has between 30% and 60% relative cover. For example in a coastal scrub stand with 5% *Baccharis pilularis*, 4% *Frangula californica*, and 3% *Rubus ursinus* (total 13% shrub cover), technically only the *Baccharis* (5/13 = 39% relative cover) and the *Frangula* (4/13 = 31% relative cover) would be co-dominant because *Rubus* would only have 23% relative cover (3/13 = 23%).

Characteristic/Diagnostic species: Should be present in at least 80% of the stands of the type, with no restriction on cover. Relatively even spacing throughout the stand is important, particularly in vegetation with low total cover, since an even distribution of the diagnostic species is a much better indicator than overall cover. Characteristic species that are evenly distributed are better indicators of a type than species with higher cover and patchy distribution.

Dominance by layer/stratum: Tree, shrub, and herbaceous layers are considered physiognomically distinct. Alliances are usually named by the dominant and/or characteristic species of the *tallest characteristic layer* (see tree-characterized, shrub-characterized, and herb-characterized vegetation definitions below). Average covers within the dominant layer reflect the "modal" concept of the health/age/environment of a particular vegetation type. For example, a higher average cover of woody plants within a stand not recently affected by disturbance reflects a mode of general availability of water, nutrition, and equitable climate, while lower average cover under similar conditions would reflect lower availability of these things.

Woody plant: A vascular plant species that has a noticeably woody stem (e.g., shrubs and trees). It does not include herbaceous species with woody underground portions such as tubers, roots, or rhizomes.

Tree: A one-stemmed woody plant that normally grows to be greater than 5 meters tall. In some cases, trees may be multistemmed (ramified due to fire or other disturbance) but the height of mature plants typically exceeds 5 meters. If less than 5 meters tall, undisturbed individuals of these species are usually single-stemmed. Certain species that sometimes resemble shrubs but may be trees in other areas (e.g., *Aesculus californica*) are, out of statewide tradition or by the USNVC, called trees. It behooves one to memorize which species are "traditionally" placed in one life-form or another. We use the accepted life-forms in the USNVC or the PLANTS Database (USDA NRCS 2015) to do this.

Tree-characterized vegetation: Trees are evenly distributed throughout the stand. In the Mediterranean climate of the North Coast, tree-dominated alliances typically have >10% absolute tree cover, providing a consistent structural component.

Forest: In the USNVC, a forest is defined as a tree-dominated stand of vegetation with 60% or greater absolute cover of trees. Most forest alliances tend to have an average cover of trees >60%, but individual stands under certain conditions may drop lower than 60%.

Woodland: In the USNVC, a woodland is defined as a tree-dominated stand of vegetation with between 25% and 60% absolute cover of trees. Most woodland alliances tend to have an average cover of trees with 25-60%, but individual stands under certain conditions may drop higher or lower than this range.

Emergent: A plant (or vegetation layer) is considered emergent if it has low cover and rises above a layer with more cover in the stand. For example, individual *Pseudotsuga menziesii* trees may compose an emergent tree layer of 2% cover over dense *Gaultheria shallon* and *Rubus parviflorus* in the shrub understory; the stand would be considered within the *Gaultheria shallon – Rubus* (*ursinus*) Shrubland Alliance because the total tree cover is <10% and the shrub cover is >10%. Medium to tall shrubs are not considered emergent over shorter shrubs, but short trees are considered emergent over tall shrubs.

Shrub: A multistemmed woody plant that is usually 0.2-5 meters tall. Definitions are blurred at the low and high ends of the height scales. At the tall end, shrubs may approach tree size based on disturbance frequencies (e.g., old-growth resprouting chaparral species such as *Cercocarpus montanus*, *Fremontodendron californicum*, *Prunus ilicifolia*, and so on, may frequently attain "tree size," but are still typically multistemmed and are considered shrubs in this key). At the short end, woody perennial herbs or subshrubs of various species are often difficult to categorize into a consistent life-form (e.g., *Eriogonum latifolium*, *Lupinus chamissonis*); in such instances, we refer to the PLANTS Database or "pick a lane" based on best available definitions.

Subshrub: A multistemmed plant with noticeably woody stems less than 0.5 meter tall. May be easily confused with a perennial herb or small shrub. We lump them into the "shrub" category in stand tables and descriptions of vegetation types.

Shrub-characterized vegetation: Shrubs, including subshrubs, are evenly distributed throughout the stand, providing a consistent (even if sparse) structural component; the stand cannot be characterized as a tree stand; and one or both of the following criteria are met: 1) shrubs influence the distribution or population dynamics of other plant species; 2) shrubs play an important role in ecological processes within the stand. Shrub alliances typically have at least 10% absolute shrub cover.

Herbaceous plant: Any species of plant that has no main woody stem development; includes grasses, forbs, and perennial species that die back each year.

Herb-characterized vegetation: Herbs are evenly distributed throughout the stand, providing a consistent (even if sparse) structural component and playing an important role in ecological processes within the stand. The stand cannot be characterized as a tree or shrub stand.

Nonvascular vegetation: Nonvascular organisms characterize a stand, providing a consistent (even if sparse) structural component and playing an important role in ecological processes within the stand.

Botanical nomenclature: We use the PLANTS Database (USDA NRCS 2015) as our standard for botanical names, except in two cases. When a more current name has been assigned in *The Jepson Manual, second edition* (Baldwin et al. 2012), that name is frequently used and a code beginning with "2JM" is assigned. General vegetation types, such as moss and lichen, have codes beginning with the number 2 (e.g., 2MOSS).

KEY TO NATURAL AND SEMINATURAL VEGETATION OF SONOMA COUNTY

Class A. Vegetation dominated, co-dominated, or characterized by an even distribution of overstory trees. The tree canopy is generally greater than 10%, but may occasionally be less than 10% over a denser understory of shrubs and/or herbs **= Tree-Overstory (Woodland / Forest) Vegetation.**

Class B. Vegetation dominated, co-dominated, or characterized by woody shrubs in the canopy. Shrubs usually have at least 5% cover. Tree species, if present, generally total less than 10% absolute cover. Herbaceous species may have higher cover than shrubs **= Shrubland Vegetation**

Class C. Vegetation dominated, co-dominated, or characterized by nonwoody, herbaceous species in the canopy, including grasses, graminoids, and broad-leaved herbaceous species. Shrubs, if present, usually compose less than 5% of the vegetation cover. Trees, if present, generally compose less than 5% cover **= Herbaceous Vegetation.**

Class A. Tree-Overstory (Woodland / Forest) Vegetation

Section I: Woodlands and forests dominated or characterized by needle or scale-leaved conifer trees. Includes *Abies*, *Hesperocyparis*, *Pinus*, *Pseudotsuga*, and *Sequoia*.

1. Temperate rainforest dominated or co-dominated by *Sequoia sempervirens* or *Abies grandis*. Found in maritime climates with summertime fog.

Vancouverian Rainforest Macrogroup

Vancouverian Hypermaritime Lowland Rainforest Group

1a. *Sequoia sempervirens* dominates, co-dominates, or characterizes (rarely with as little as 5% cover) stands near streams, along all slopes and aspects, or on ridges. Associated trees include *Acer macrophyllum*, *Notholithocarpus densiflorus*, *Pseudotsuga menziesii*, *Torreya californica*, and *Umbellularia californica*, which are typically sub- to co-dominant but may occasionally exceed *Sequoia* in cover. *Vaccinium ovatum*, *Oxalis oregana*, and *Woodwardia fimbriata* may intermix in the understory.

Sequoia sempervirens **Alliance**
Sequoia sempervirens – Acer macrophyllum – Umbellularia californica Association
Sequoia sempervirens – Notholithocarpus densiflorus / Vaccinium ovatum Association
Sequoia sempervirens – Pseudotsuga menziesii – Notholithocarpus densiflorus Provisional Association
Sequoia sempervirens -– Pseudotsuga menziesii – Umbellularia californica Association
Sequoia sempervirens -– Umbellularia californica Association
Sequoia sempervirens / Oxalis oregana Association
Sequoia sempervirens / Woodwardia fimbriata Riparian Provisional Association

1b. *Abies grandis* has strong dominance in the tree overstory, with *Pinus muricata* and *Sequoia sempervirens* intermixing locally as subdominants. Stands are rare in the county. One stand, found on a convexity running along a middle slope up to the ridgetop, was sampled for this project.

Abies grandis **Alliance**

2. Cool-temperate coniferous forests and woodlands influenced by warm, relatively dry summers and cool rainy winters. Stands are dominated or co-dominated by *Pinus ponderosa*, *Pseudotsuga menziesii,* or *P. menziesii* in combination with *Notholithocarpus densiflorus* in the tree overstory.

Californian–Vancouverian Montane and Foothill Forest Macrogroup

2a. Vegetation characterized by a mixture of *Pseudotsuga menziesii* and *Notholithocarpus densiflorus* in the canopy. *Pseudotsuga* is typically dominant to co-dominant with *Notholithocarpus*, but may occasionally be slightly subdominant.

Vancouverian Evergreen Broadleaf and Mixed Forest Group

Pseudotsuga menziesii – Notholithocarpus densiflorus **Alliance**
Pseudotsuga menziesii – Notholithocarpus densiflorus Association

2b. Vegetation characterized by *Pinus ponderosa* and/or *Pseudotsuga menziesii*. If *Notholithocarpus densiflorus* is present, it is subdominant with relatively low cover.

Upland Vancouverian Mixed Woodland and Forest Group

2b1. *Pinus ponderosa* is dominant to co-dominant with *Pseudotsuga menziesii*. Stands with significant *Pinus ponderosa* were only encountered twice for this project – in the higher elevation, eastern portion of the county in The Geysers. In both instances, *Arbutus menziesii*, *Arctostaphylos manzanita*, and *Quercus chrysolepis* were present.

Pinus ponderosa – *Pseudotsuga menziesii* **Alliance**
Pinus ponderosa – Pseudotsuga menziesii Association

2b2. *Pseudotsuga menziesii* not as above, but instead dominant or co-dominant with *Arbutus menziesii*, *Quercus agrifolia*, *Q. chrysolepis*, or *Umbellularia californica*. When *P. menziesii* co-dominates with hardwoods, key to *P. menziesii*, except when with *Quercus garryana*, *Q. kelloggii*, or *Notholithocarpus densiflorus* (see *Q. garryana* (step 4a3) or *Q. kelloggii* Alliance (step 5c4) below, or *P. menziesii* – *N. densiflorus* Alliance above, step 2a).

Pseudotsuga menziesii **Alliance**
Pseudotsuga menziesii – Arbutus menziesii Association
Pseudotsuga menziesii – Quercus agrifolia Association
Pseudotsuga menziesii – Quercus chrysolepis Association
Pseudotsuga menziesii – Umbellularia californica Association
Pseudotsuga menziesii – Umbellularia californica / Polystichum munitum Association

3. Closed-cone or xerophyllic conifers, including *Hesperocyparis* spp., *Pinus attenuata*, *Pinus muricata*, *Pinus radiata*, or *Pinus sabiniana* is dominant, co-dominant, or characteristic in the overstory.

California Forest and Woodland Macrogroup

Californian Evergreen Coniferous Forest and Woodland Group

3a. Stands dominated by a native or planted species of *Hesperocyparis*.

3a1. Planted *Hesperocyparis macrocarpa* dominates in patches or along roads. In this region of California, stands are considered seminatural since they are not naturally occurring.

Hesperocyparis macrocarpa **Special Stands and Seminatural Alliance**
Hesperocyparis macrocarpa Provisional Seminatural Association

3a2. A native cypress species, *Hesperocyparis macnabiana* or *H. sargentii*, dominates or characterizes stands on serpentine, volcanic, or other ultramafic substrates. *Adenostoma fasciculatum*, *Arctostaphylos* spp., *Ceanothus jepsonii*, and *Quercus durata* are commonly found in stands.

3a2a. *Hesperocyparis macnabiana* characterizes the tree canopy (sometimes with <10% cover) and may be similar in height to surrounding shrubs. Found on open slopes and ridges and only known locally in the eastern part of the county.

Hesperocyparis macnabiana **Alliance**
Hesperocyparis macnabiana / Arctostaphylos viscida Association

3a2b. *Hesperocyparis sargentii* dominates on slopes, ridges, or along stream benches and terraces. Sites are known near Harrison Grade or The Cedars.

Hesperocyparis sargentii **Alliance**
Hesperocyparis sargentii / Ceanothus jepsonii – Arctostaphylos spp. Provisional Association
Hesperocyparis sargentii / Quercus durata (mesic) Provisional Association
Hesperocyparis sargentii Riparian Association

3b. Stands dominated by *Pinus attenuata*, *P. muricata*, *P. radiata*, or *P. sabiniana*.

3b1. *Pinus attenuata* dominates in the tree overstory, sometimes with a moderately dense cover of shrubs such as *Adenostoma fasciculatum*, *Arctostaphylos* spp., and *Ceanothus cuneatus* in the understory.

***Pinus attenuata* Alliance**
Pinus attenuata / Arctostaphylos (*manzanita, canescens*) Provisional Association
Pinus attenuata / Arctostaphylos viscida Association

3b2. *Pinus muricata* is the sole dominant or may co-dominate with *Hesperocyparis pigmaea* in the tree overstory. The understory may include moderate to dense cover of shrubs such as *Arctostaphylos nummularia*, *Gaultheria shallon*, and *Vaccinium ovatum*.

***Pinus muricata* Alliance**
Pinus muricata Provisional Association
Pinus muricata – Hesperocyparis pigmaea Provisional Provisional Association
Pinus muricata / Vaccinium ovatum Provisional Association

3b3. *Pinus sabiniana* dominates or co-dominates with *Umbellularia californica* in the tree overstory. *Adenostoma fasciculatum*, *Arctostaphylos viscida*, *Quercus durata*, and other shrubs may exceed *P. sabiniana* in cover.

***Pinus sabiniana* Alliance**
Pinus sabiniana / Quercus durata Provisional Association
Pinus sabiniana /Arctostaphylos viscida Association

3b4. Planted stands of *Pinus radiata* are found along roadsides or on slopes where they were introduced after fires in the 1960's.

***Pinus radiata* Alliance**
Pinus radiata Provisional Seminatural Association

Section II. Woodlands, forests, and riparian vegetation characterized and/or dominated mainly by native and nonnative broad-leaved evergreen and deciduous trees. Includes species of *Aesculus*, *Acer*, *Alnus*, *Arbutus*, *Fraxinus*, *Juglans*, *Notholithocarpus*, *Populus*, *Quercus*, *Salix*, and *Umbellularia*.

4. Vegetation dominated, co-dominated, or characterized by one or more of the following broadleaf trees: *Acer macrophyllum*, *Arbutus menziesii*, *Notholithocarpus densiflorus*, or *Quercus garryana*.

Californian–Vancouverian Montane and Foothill Forest Macrogroup

4a. Broadleaf trees such as *Arbutus menziesii*, *Notholithocarpus densiflorus*, or *Quercus garryana* dominate, co-dominate, or characterize moist, coastal, mixed evergreen forests and woodlands. Stands of *Quercus garryana* may also occur in more interior settings, where the winters are cooler and the summers are warmer.

Vancouverian Evergreen Broadleaf and Mixed Forest Group

4a1. *Arbutus menziesii* is either dominant with subdominant *Quercus agrifolia* or is dominant to co-dominant with *Quercus kelloggii* and/or *Umbellularia californica*. *Pseudotsuga menziesii*, *Heteromeles arbutifolia*, and *Toxicodendron diversilobum* are often present. If *Arbutus* is sub- to co-dominant with *Quercus agrifolia*, *Q. chrysolepis*, or *Notholithocarpus densiflorus*, key to one of these alliances instead of *A. menziesii*.

***Arbutus menziesii* Alliance**
Arbutus menziesii – Quercus agrifolia Association
Arbutus menziesii – Umbellularia californica Provisional Association *Arbutus menziesii – Umbellularia californica – Quercus kelloggii* Association

4a2. *Notholithocarpus densiflorus* is strongly dominant in the tree canopy or co-occurs with subdominant to co-dominant *Arbutus menziesii*.

Notholithocarpus densiflorus Alliance
Notholithocarpus densiflorus Provisional Association
Notholithocarpus densiflorus – *Arbutus menziesii* Association

4a3. *Quercus garryana* dominates or co-dominates with other broadleaf trees or *Pseudotsuga menziesii*. Stands are of two types: 1) relatively dense woodlands without a significant understory herb component or 2) open woodlands over moderate to dense native and nonnative herbs (e.g., *Cynosurus echinatus* and *Festuca californica*). *Pseudotsuga menziesii*, *Umbellularia californica*, *Quercus agrifolia*, and/or *Q. kelloggii* commonly intermix, typically as subdominants. If two or more species of *Quercus* are present and, collectively, they are dominant or co-dominant with *Q. garryana*, key to the *Quercus* (*agrifolia, douglasii, garryana, kelloggii, lobata, wislizeni*) Alliance (step 5c1).

Quercus garryana (tree) Alliance
Quercus garryana – *Umbellularia californica* – *Quercus* (*agrifolia, kelloggii*) Provisional Association
Quercus garryana / (*Cynosurus echinatus* – *Festuca californica*) Provisional Association

4b. *Acer macrophyllum* dominates or co-dominates with *Umbellularia californica* or, occasionally, *Fraxinus latifolia* in riparian or, occasionally, upland stands. *Pseudotsuga menziesii*, *Quercus agrifolia*, and *Q. chrysolepis* may intermix. *Acer* stands were found farther than 15 miles from the coast or closer to the eastern boundary of the county, usually in low-lying, rocky, steep canyons.

Upland Vancouverian Mixed Woodland and Forest Group

Acer macrophyllum Alliance
Acer macrophyllum Association

5. Vegetation dominated or co-dominated by the following broadleaf, primarily upland tree species: *Aesculus californica, Quercus agrifolia, Q. chrysolepis, Q. douglasii, Q. kelloggii, Q. lobata, Q. parvula* var. *shrevei, Q. wislizeni*, and/or *Umbellularia californica*.

California Forest and Woodland Macrogroup

Californian Broadleaf Forest and Woodland Group

5a. *Aesculus californica* dominates in open to moderately dense woodlands. If *Umbellularia californica* is present, it is subdominant. A variety of herbs may be found in the understory.

Aesculus californica Alliance
Aesculus californica / *Toxicodendron diversilobum* / Moss Association

5b. *Umbellularia californica* is either dominant or co-dominant with *Quercus agrifolia* in open to dense woodlands. Found in a variety of settings, such as streamsides, valley bottoms, coastal bluffs, inland ridges, steep north-facing slopes, rocky outcrops, and postfire landscapes. If *U. californica* is co-dominant with *Arbutus, Acer,* or *Pinus sabiniana* on serpentine, or *Pseudotsuga menziesii, Quercus garryana, Q. kelloggii,* or *Sequoia*, key to one of these other hardwood or conifer alliances instead.

Umbellularia californica Alliance
Umbellularia californica – *Acer macrophyllum* Association
Umbellularia californica – *Notholithocarpus densiflorus* Association
Umbellularia californica – *Pseudotsuga menziesii* / *Rhododendron occidentale* Association
Umbellularia californica – *Quercus agrifolia* Provisional Association
Umbellularia californica (Pure – Coastal) Provisional Association
Umbellularia californica / *Polystichum munitum* Association

5c. One or more species of *Quercus* listed above (step 5), other than *Quercus garryana* (step 4a3), dominates or co-dominates in the tree overstory or *Quercus garryana* co-dominates with two other oak species.

5c1. *Quercus agrifolia*, *Quercus garryana*, and/or *Quercus kelloggii* are present and at least two of the oak species co-dominate. Other oaks such as *Q. chrysolepis*, *Q. douglasii*, and *Q. lobata* may also be present. This mixed type is for stands where multiple *Quercus* tree species intermix and it is difficult to assign to an alliance defined by one oak species – read steps to key to individual oak alliances below.

Quercus (agrifolia, douglasii, garryana, kelloggii, lobata, wislizeni) **Alliance**
Quercus agrifolia – *Quercus garryana* – *Quercus kelloggii* Provisional Association

5c2. *Quercus chrysolepis* is dominant or co-dominant with *Arbutus menziesii* in the tree overstory. *Quercus wislizeni* is occasionally found as a subdominant tree.

Quercus chrysolepis (tree) **Alliance**
Quercus chrysolepis – *Arbutus menziesii* Provisional Association *Quercus chrysolepis* – *Quercus wislizeni* Association

5c3. *Quercus douglasii* or *Quercus* ×*eplingii* (the hybrid between *Q. douglasii* and *Q. garryana*) dominates or co-dominates with *Quercus agrifolia* or *Arbutus menziesii* in the tree overstory. The understory herbaceous layer is often moderately dense to dense, with a mixture of native and nonnative forbs and grasses.

Quercus douglasii **Alliance**
Quercus × *eplingii* / Grass Provisional Association
Quercus douglasii – *Quercus agrifolia* Association
Quercus douglasii / *Arctostaphylos manzanita* / Herbaceous Association
Quercus douglasii / Grass Association

5c4. *Quercus kelloggii* dominates or co-dominates with *Pseudotsuga menziesii*, *Q. agrifolia*, and/or *Umbellularia californica* in the tree overstory. *Arbutus menziesii* is often present as a subdominant species. Stands in Sonoma County are found inland, above maritime influence, on northern exposures.

Quercus kelloggii **Alliance**
Quercus kelloggii – *Arbutus menziesii* – *Quercus agrifolia* Association
Quercus kelloggii – *Pseudotsuga menziesii* – *Umbellularia californica* Association

5c5. *Quercus lobata* dominates or co-dominates with *Fraxinus latifolia* and/or *Quercus agrifolia* in the tree overstory. Stands are typically found along valley bottoms, lower slopes, and summit valleys on seasonally saturated soils that may flood intermittently. Common understory shrubs include *Rosa californica*, *Rubus* spp., and *Toxicodendron diversilobum*.

Quercus lobata **Alliance**
Quercus lobata – *Fraxinus latifolia* / (*Vitis californica*) Association
Quercus lobata – *Quercus agrifolia* / Grass Association
Quercus lobata / Grass Association
Quercus lobata / *Rubus ursinus* – *Rosa californica* Provisional Association

5c6. *Quercus parvula* var. *shrevei* dominates as a tree or shrubby regenerating tree, co-occurring with *Umbellularia*, *Adenostoma*, and a variety of other shrubs that prefer more mesic, northerly exposures. One stand was sampled and classified in Sonoma County, and likely further variation will be seen.

Quercus parvula var. *shrevei* **Provisional Alliance**

5c7. The tree form of *Quercus wislizeni* dominates or co-dominates in the tree canopy, often with *Arbutus menziesii*, *Pseudotsuga menziesii*, and/or *Umbellularia californica*. If *Q. wislizeni* has a shrubby habit or is a regenerating tree intermixing with a variety of other shrub species, key to the *Quercus wislizeni* (shrub) Alliance, step 9b.

Quercus wislizeni (tree) **Alliance**
Quercus wislizeni – *Arbutus menziesii* / *Toxicodendron diversilobum* Association

5c8. *Quercus agrifolia* dominates or co-dominates with *Arbutus menziesii* in the canopy. If *Q. douglasii* (or hybrid *Q. ×eplingii*), *Q. lobata*, or *Umbellularia californica* is co-dominant, key to one of these other alliances instead of *Q. agrifolia*. The understory herbaceous layer often contains a mixture of native and nonnative herbs and/or shrubs.

Quercus agrifolia **Alliance**
Quercus agrifolia – Arbutus menziesii – Umbellularia californica Association
Quercus agrifolia / Grass Association
Quercus agrifolia / *Toxicodendron diversilobum* Association

6. *Acer negundo, Juglans hindsii, Populus fremontii*, or *Salix laevigata* is dominant, co-dominant or characteristic in permanently moist or riparian settings, where subsurface water is available all year. Nearby upland vegetation is often dominated by broadleaf evergreen or deciduous trees, as opposed to conifers.

Southwestern North American Riparian, Flooded, and Swamp Forest Macrogroup

Southwestern North American Riparian Evergreen and Deciduous Woodland Group

6a. *Acer negundo* dominates in the tree overstory, often along major streams and rivers, with other riparian plants such as *Fraxinus, Populus, Rubus*, and *Salix*. Stands are considered rare in the state and may be small and monospecific.

Acer negundo **Alliance**

6b. *Juglans hindsii* or hybrids dominate in naturalized stands along riparian corridors, floodplains, stream banks, and terraces. Other riparian species may be present, including *Acer, Fraxinus*, and *Rubus*.

Juglans hindsii **and Hybrids Special Stands and Seminatural Alliance**

6c. *Populus fremontii* dominates or co-dominates with *Acer negundo, Juglans*, and/or *Salix*, sometimes with *Populus* having as little as 5% absolute cover. If *Juglans hindsii* is dominant, but *Populus* has at least 20% relative cover in the tree layer, key to this alliance.

Populus fremontii **Alliance**
Populus fremontii – Acer negundo Association
Populus fremontii / *Salix exigua* Association

6d. *Salix laevigata* dominates along streams, rivers, ditches, floodplains, and lake edges. Associated trees and shrubs include *Alnus rhombifolia, Populus fremontii, Quercus agrifolia, Rubus, Salix*, and others.

Salix laevigata **Alliance**
Salix laevigata / *Salix lasiolepis* Association

7. *Alnus rhombifolia, Fraxinus latifolia*, and/or *Salix lucida* are dominant, co-dominant, or characteristic of broadleaf riparian tree vegetation. Stands are more likely to occur near cool-temperate coniferous forests, unlike vegetation of the Southwestern North American Riparian, Flooded, and Swamp Forest Macrogroup described above. Found along riparian corridors, incised canyons, seeps, stream banks, midchannel bars, floodplains, and terraces.

Western Cordilleran Montane–Boreal Riparian Scrub Macrogroup

Vancouverian Riparian Deciduous Forest Group

7a. *Alnus rhombifolia* dominates or co-dominates with *Acer macrophyllum* or *Umbellularia californica* in the tree overstory. If *Fraxinus latifolia* is co-dominant, key to the *Fraxinus latifolia* Alliance below. A variety of shrubs and herbs may be found in the understory, including *Carex, Rubus, Toxicodendron, Xerophyllum*, and *Woodwardia*. Careful identification of alder stands closer to the coast is necessary to differentiate from *A. rubra* stands.

Alnus rhombifolia Alliance
Alnus rhombifolia Association
Alnus rhombifolia – Acer macrophyllum Association
Alnus rhombifolia / Carex (*nudata*) Association

7b. *Alnus rubra* dominates in the tree canopy in riparian settings, typically within a few miles of the coast. The understory often comprises one to many species of *Rubus*, which may exceed *Alnus* in cover. *Alnus rubra* stands were encountered usually less than 10 miles from the coast in riparian or swampy bottomlands, but can occur along rocky streambeds in similar settings to *A. rhombifolia* stands. Careful identification of the species of Alnus is important closer to the coast.

Alnus rubra Alliance[1]
Alnus rubra / Rubus spp. Provisional Association

7c. *Fraxinus latifolia* dominates or co-dominates with *Alnus rhombifolia* or *Umbellularia californica* in the tree overstory. Stands for this project were encountered and surveyed in the southern half of Sonoma County.

Fraxinus latifolia Alliance
Fraxinus latifolia Association
Fraxinus latifolia – Alnus rhombifolia Association

7d. *Salix lucida* ssp. *lasiandra* dominates in the overstory, sometimes with higher or similar cover by shrubs in the understory, such as *Rubus* spp. and *Salix lasiolepis*. Adjacent stands may be dominated by *Alnus* spp., *Quercus agrifolia* or conifers.

Salix lucida Alliance
Salix lucida ssp. *lasiandra* Association

8. A tree species of *Eucalyptus* dominates in planted or naturalized stands. Often found in groves, windbreaks, uplands, and along stream courses. Stands were observed, but not sampled for this project.

Introduced North American Mediterranean Woodland and Forest Macrogroup and Group

Eucalyptus (*globulus, camaldulensis*) Seminatural Alliance

---

[1] The *Alnus rubra* Alliance is placed in the Upland Vancouverian Mixed Woodland and Forest Group of the USNVC. It will likely be incorporated under the Vancouverian Riparian Deciduous Forest Group in the future as it has been for this project.

Class B. Shrubland Vegetation

Section I. Riparian or moist hillside settings with vegetation dominated or co-dominated by the following shrubs: *Frangula californica* (including all subspecies), *Morella californica*, *Rhododendron occidentale*, *Rubus armeniacus*, *R. spectabilis*, *Salix breweri*, *S. exigua*, *S. lasiolepis*, *S. melanopsis*, *S. sitchensis*, and/or *Sambucus nigra*.
*Note: if *Rubus ursinus* dominates, key to the *Gaultheria shallon – Rubus* (*ursinus*) Alliance in Section II below (step 5b3).

1. *Rubus armeniacus*, a nonnative from Europe, is strongly dominant in riparian sites, mesic clearings, disturbed areas, and stock ponds.

Vancouverian Lowland Grassland and Shrubland Macrogroup
Naturalized Nonnative Deciduous Scrub Group

Rubus armeniacus **Seminatural Alliance**
Rubus armeniacus Seminatural Association

2. *Morella californica*, *Rubus parviflorus*, *R. spectabilis* and/or *Salix sitchensis* dominate or co-dominate with *Rubus* spp.

Western Cordilleran Montane–Boreal Riparian Scrub Macrogroup
Vancouverian Coastal Riparian Scrub Group

2a. Vegetation dominated or characterized by *Morella californica*, *Rubus parviflorus*, and/or *Rubus spectabilis*. Stands may be small and are generally found close to the coast on moist or wet soils.

Morella californica – Rubus spectabilis **Provisional Alliance**
Morella californica – Rubus spp. Provisional Association
Rubus parviflorus Association
Rubus spectabilis Association

2b. *Salix sitchensis* dominates or co-dominates with *S. lasiolepis* along coastal or low-elevation streams, lagoons. A variety of subdominant trees and shrubs may be present, including *Acer*, *Alnus*, *Fraxinus*, *Salix*, and *Rubus*.

Salix sitchensis **Provisional Alliance**
Salix sitchensis Provisional Association

3. *Frangula californica*, *Rhododendron occidentale*, *Salix breweri*, *S. exigua*, *S. lasiolepis*, *S. melanopsis*, and/or *Sambucus nigra* dominant or co-dominant with *Baccharis pilularis* or *Rubus* spp.

Southwestern North American Riparian, Flooded and Swamp Forest Macrogroup
Southwestern North American Riparian/Wash Scrub Group

3a. *Frangula californica* and/or *Rhododendron occidentale* dominate or co-dominate with *Baccharis pilularis* or *Rubus*. Stands are found along springs, seeps, ravines, and hillslopes, often on sedimentary and serpentine substrates that retain water much of the year.

Frangula californica – Rhododendron occidentale **Provisional Alliance**
Frangula californica ssp. *californica* Provisional Association
Rhododendron occidentale – Frangula californica ssp. *tomentella* Provisional Association

3b. *Salix breweri* dominates along creeks and stream terraces, on serpentine-derived alluvium. Locally present along streams on serpentine in The Cedars area. Commonly found with other moisture loving plants, such as *Alnus rhombifolia*, *Baccharis salicifolia*, *Rubus*, and *Stachys albens*.

Salix breweri **Alliance**
Salix breweri Provisional Association

3c. *Salix exigua* or *Salix melanopsis* dominates along rivers and streams, or close to springs. They are often the first plants to colonize bars and cut banks, followed later by trees such as *Populus* and *Salix* spp.

Salix exigua **Alliance**
Salix exigua Association
Salix exigua – Salix melanopsis Association

3d. *Sambucus nigra* dominates in the shrub overstory, often preferring stream terraces, bottomlands, and localized areas in uplands, where there was past disturbance. One stand was encountered for this project, along a draw that was burned.

Sambucus nigra **Alliance**
Sambucus nigra Association

3e. *Salix lasiolepis* dominates or co-dominates with *Rubus* along stream banks and benches, slope seeps, and drainage stringers. If *S. sitchensis* is co-dominant, key to the *S. sitchensis* Alliance instead (step 2b). Emergent riparian trees are often present, such as *Acer, Alnus, Fraxinus, Salix,* and others.

Salix lasiolepis **Alliance**
Salix lasiolepis / Rubus spp. Association

Section II. Coastal scrub, dune/bluff, and disturbance-following vegetation dominated or co-dominated by drought-deciduous or seral (both deciduous and evergreen) shrubs. Includes *Artemisia californica, Baccharis pilularis, Ceanothus thyrsiflorus, Ericameria ericoides, Eriodictyon californicum, Eriogonum fasciculatum, Gaultheria shallon, Lupinus albifrons, L. arboreus, L. chamissonis, Rubus ursinus,* and *Toxicodendron diversilobum*. Resprouting, deep-rooted, sclerophyllous shrubs may at times be characteristic, but not dominant.

4. *Ericameria ericoides, Lupinus arboreus,* and/or *Lupinus chamissonis* are dominant, co-dominant, or characteristic (sometimes with as little as 5% cover) in the shrub overstory on coastal dunes or bluffs. A variety of herbs, including many of the following nonnatives, may be present with high cover in the understory: *Bromus diandrus, Carduus, Holcus, Rumex acetosella,* and *Vulpia bromoides*.

Vancouverian Coastal Dune and Bluff Macrogroup

California Coastal Evergreen Bluff and Dune Scrub Group

4a. *Lupinus arboreus* dominates or co-dominates with *Baccharis pilularis*, and may co-occur with high cover by *Vulpia bromoides, Festuca perennis, Bromus diandrus,* and other nonnative grasses.

Lupinus arboreus **Alliance and Seminatural Alliance**
Lupinus arboreus Association

4b. *Ericameria ericoides* and/or *Lupinus chamissonis* dominate as individuals or in combination with *Baccharis pilularis* or *Lupinus arboreus*.

Lupinus chamissonis – Ericameria ericoides **Alliance**
Lupinus chamissonis – Ericameria ericoides Association

5. Shrublands dominated or co-dominated by native, disturbance-following, naturalized, or planted species including *Artemisia californica, Cistus, Eriodictyon californicum, Eriogonum fasciculatum, Genista, Heterotheca oregana, Lupinus albifrons, Baccharis pilularis, Ceanothus thyrsiflorus, Gaultheria shallon, Rubus ursinus, Toxicodendron diversilobum,* and/or *Ulex europaeus*.

California Coastal Scrub Macrogroup

5a. *Eriodictyon californicum, Heterotheca oregana,* or *Lupinus albifrons* dominates in the overstory.

Central and South Coastal California Seral Scrub Group

5a1. *Eriodictyon californicum* or *Lupinus albifrons* dominates, often in stands that are open and/or display recent evidence of fire or other disturbance. The understory may be composed of mixed native and nonnative herbs, which sometimes have higher cover than the overstory shrubs.

> ***Eriodictyon californicum – Lupinus albifrons* Provisional Alliance**
> *Eriodictyon californicum* / Herbaceous Association
> *Lupinus albifrons* Association

5a2. *Heterotheca oregona*, a perennial herb that acts like a short-lived shrub, dominates herbaceous stands that have seasonal hydrologic disturbance. Found along sunny, rocky stream terraces, seasonally dry streambeds, sandbars in river drainages, and cobbled gravel bars in floodplains.

> ***Heterotheca* (*oregona, sessiliflora*) Provisional Alliance**
> *Heterotheca oregona* Provisional Association

5b. *Baccharis pilularis, Ceanothus incanus, C. thyrsiflorus, Gaultheria shallon, Rubus ursinus,* and/or *Toxicodendron diversilobum* dominate or co-dominate as shrubs. Shrubs are typically evergreen or winter-deciduous, not sclerophyllous or drought-deciduous species. Found along cool, coastal strips or on sheltered inland ravines and lower slopes, where species are tolerant of disturbance and tend to be overtopped and excluded by trees.

California North Coastal & Mesic Scrub Group

5b1. *Baccharis pilularis* dominates or co-dominates with *Frangula californica, Toxicodendron diversilobum,* or *Rubus* spp. in the shrub overstory. If *Calamagrostis nutkaensis* is co-dominant with *B. pilularis*, key to the *C. nutkaensis* Alliance (see Class C, step 9c3a). A variety of native and nonnative forbs and grasses may intermix in the herbaceous layer, sometimes with higher cover than *Baccharis* – including *Avena, Bromus, Danthonia, Deschampsia, Elymus glaucus, Festuca, Hypochaeris, Nassella pulchra,* and others.

> ***Baccharis pilularis* Alliance**
> *Baccharis pilularis – Frangula californica – Rubus* spp. Provisional Association
> *Baccharis pilularis – Toxicodendron diversilobum* Association
> *Baccharis pilularis* / Annual Grass – Herb Association
> *Baccharis pilularis* / *Danthonia californica* Association
> *Baccharis pilularis* / *Deschampsia cespitosa* Association
> *Baccharis pilularis* / *Nassella pulchra* Association
> *Baccharis pilularis* / Native Grass (Mixed) Association

5b2. *Ceanothus incanus* or *C. thyrsiflorus* dominates in the overstory shrub layer, often with moderately dense cover. *Diplacus aurantiacus, Heteromeles, Pseudotsuga menziesii, Quercus wislizeni,* and other species may intermix as subdominants in the shrub and tree layers. Stands of *C. incanus* are included in the *C. thyrsiflorus* Alliance since they are more limited in distribution and are ecologically similar to *C. thyrsiflorus*.

> ***Ceanothus thyrsiflorus* Alliance**
> *Ceanothus incanus* Provisional Association

5b3. *Gaultheria shallon* and/or *Rubus ursinus* dominate or co-dominate with *Anthoxanthum odoratum, Holcus lanatus,* or *Toxicodendron diversilobum* on hillslopes, rock outcrops, coastal bluffs, or flats. If *Arctostaphylos nummularia* is co-dominant with *Gaultheria*, key to the *Arctostaphylos* (*nummularia, sensitiva*) Alliance below (step 6).

> ***Gaultheria shallon – Rubus* (*ursinus*) Provisional Alliance**
> *Gaultheria shallon – Rubus* spp. Provisional Association
> *Rubus ursinus* Association

5b4. *Toxicodendron diversilobum* dominates, sometimes intermixing with subdominant *Baccharis pilularis* and *Rubus* spp. If *B. pilularis* is present and co-dominant, key to the *Baccharis pilularis* Alliance (step 5b1). For this project, stands were encountered close to the coast, although they are likely to occur inland as well.

Toxicodendron diversilobum **Alliance**

Toxicodendron diversilobum – Baccharis pilularis Provisional Association

5c. *Artemisia californica* dominates and may intermix with *Baccharis pilularis, Diplacus aurantiacus,* and others. One stand, which may represent the northernmost occurrence of *A. californica* in the state, was encountered during field reconnaissance along Highway 1, approximately two miles southeast of Fort Ross.

Central and South Coastal Californian Coastal Sage Scrub Group

Artemisia californica **Alliance** (no description provided)

5d. *Cistus, Eriogonum fasciculatum, Genista, Ulex,* or other Mediterranean shrubs not native to Sonoma County dominates in naturalized or planted stands. May be found invading disturbed areas, grasslands, or forest openings.

Naturalized Nonnative Mediterranean Scrub Group

5d1. *Genista monspessulana, Ulex europaeus,* or other broom species/hybrids dominate in the shrub overstory. Fire promotes broom invasions in woodland settings, however broom may invade coastal grasslands without fire.

Broom (*Cytisus scoparius* and Others) Seminatural Alliance

5d2. *Cistus, Eriogonum fasciculatum* or other naturalized/planted species dominates in the shrub overstory. *Eriogonum fasciculatum,* while native to other parts of California, does not occur naturally in Sonoma County. *E. fasciculatum* is often chosen for erosion control and slope stabilization projects because it grows relatively quickly, spreads well, and maintains a nice appearance year round. One stand was observed during field reconnaissance near Lake Sonoma, though other stands may be found elsewhere in the County. Planted stands do not fit under the *Eriogonum fasciculatum* Alliance, which is reserved for native vegetation.

Naturalized Nonnative Mediterranean Scrub Group (key to group level only)

Section III. Shrub vegetation dominated by evergreen sclerophyll-leaved species, including many that have developed growth strategies driven by a Mediterranean climate. Most of the core diagnostic species are endemic to California, including *Adenostoma, Arctostaphylos, Ceanothus cuneatus, C. oliganthus, Cercocarpus montanus, Quercus berberidifolia, Q. durata,* and shrubby *Q. wislizeni.*

California Chaparral Macrogroup

6. *Arctostaphylos nummularia* ssp. *nummularia* dominates or co-dominates with *Gaultheria shallon* or *Vaccinium ovatum* in maritime chaparral stands. *Arctostaphylos columbiana, Chrysolepis chrysophylla* var. *minor, Pinus muricata,* and *Pteridium aquilinum* are often present.

Californian Maritime Chaparral Group

Arctostaphylos (nummularia, sensitiva) **Alliance**

Arctostaphylos nummularia ssp. *nummularia* Provisional Association

7. *Cercocarpus montanus* and/or *Quercus berberidifolia* dominate or co-dominate with *Adenostoma fasciculatum.* Stands are mostly found inland from the coastal fog belt and are often composed of large shrubs occupying mesic sites such as north-facing slopes, concavities, and toeslopes with well-drained soils.

Californian Mesic Chaparral Group

7a. *Cercocarpus montanus* dominates or co-dominates with *Adenostoma fasciculatum. Diplacus aurantiacus* and *Toxicodendron diversilobum* are often present. Stands are frequently found on rocky, north-facing slopes, though they can occur on all aspects.

Cercocarpus montanus **Alliance**

Cercocarpus montanus – Adenostoma fasciculatum Association

7b. *Quercus berberidifolia* dominates or co-dominates with *Cercocarpus montanus*. Stands are found primarily on north-facing, steep slopes with well-drained soils. If *Adenostoma fasciculatum* is co-dominant with *Q. berberidifolia*, key to the mixed *Quercus berberidifolia-Adenostoma fasciculatum* Alliance directly below.

Quercus berberidifolia **Alliance**
Quercus berberidifolia Association
Quercus berberidifolia – Cercocarpus montanus Association

7c. *Quercus berberidifolia* and *Adenostoma fasciculatum* co-dominate and often occupy ecological interfaces between mesic sites that *Quercus* prefers and xeric sites that *Adenostoma* prefers. A variety of shrubs may intermix as subdominants.

Quercus berberidifolia – Adenostoma fasciculatum **Alliance**
Quercus berberidifolia – Adenostoma fasciculatum Association

8. *Arctostaphylos bakeri*, *Ceanothus jepsonii*, and/or *Quercus durata* dominate or co-dominate in shrub vegetation restricted to or adapted to ultramafic soils and substrates (e.g., serpentine, gabbro).

Californian Serpentine Chaparral Group

8a. *Arctostaphylos bakeri*, a serpentine endemic, dominates or co-dominates with *Quercus durata* in the shrub overstory, often on upper slopes, flats, and ridges. *Ceanothus jepsonii*, *Hesperocyparis sargentii*, *Heteromeles arbutifolia*, and *Melica torreyana* are commonly present.

Arctostaphylos (*bakeri, montana*) **Provisional Alliance**
Arctostaphylos bakeri Provisional Association

8b. *Quercus durata* dominates or co-dominates with *Adenostoma fasciculatum* or *Ceanothus jepsonii* on ultramafic soils. *Heteromeles arbutifolia* and/or *Umbellularia californica* are often present in stands.

Quercus durata **Alliance**
Quercus durata – Adenostoma fasciculatum Provisional Association
Quercus durata – Ceanothus jepsonii Provisional Association
Quercus durata – Heteromeles arbutifolia / Umbellularia californica Association

9. *Ceanothus oliganthus* and/or *Quercus wislizeni* var. *frutescens* dominate or co-dominate in the shrub overstory. These shrublands are more frost tolerant and typically found at higher, cooler, and more mesic sites than those in the California Xeric Chaparral Group.

Californian Pre-Montane Chaparral Group

9a. *Ceanothus oliganthus* dominates in shrublands that are often found in localized patches following fires. If *Quercus wislizeni* is co-dominant, key to the *Q. wislizeni* (shrub) Alliance directly below.

Ceanothus oliganthus **Alliance**
Ceanothus oliganthus Association

9b. Regenerating or shrubby *Quercus wislizeni* (var. *frutescens*) dominates or co-dominates with *Ceanothus oliganthus*. Stands that represent the possibly distinct *Q. wislizeni* var. *frutescens* and those with *Q. wislizeni* having shorter stature due to factors that limit height (e.g., fire) are included in this alliance. When *Q. wislizeni* dominates or co-dominates as an overstory tree, key to the *Q. wislizeni* (tree) Alliance. *Umbellularia californica* is often emergent, while a variety of thick- and soft-leaved shrubs intermix as subdominants.

Quercus wislizeni (shrub) **Alliance**
Quercus wislizeni var. *frutescens* Provisional Association
Quercus wislizeni – Ceanothus oliganthus Provisional Association

10. Sclerophyll (i.e., thick-leaved) shrublands dominated by one or more of the following taxa: *Adenostoma*, *Arctostaphylos canescens*, *A. glandulosa*, *A. manzanita*, *A. stanfordiana*, *A. viscida*, or *Ceanothus cuneatus*. Most stands occur on well-drained soils along exposures that are in full sun much of the growing season, including upper slopes, spur ridges, and convexities.

Californian Xeric Chaparral Group

10a. *Arctostaphylos canescens*, *A. manzanita* and/or *A. stanfordiana* dominate or co-dominate, sometimes with co-dominant *Adenostoma fasciculatum*. Found typically on volcanic, Franciscan, and greenstone substrates. One alliance is recognized for all three *Arctostaphylos* vegetation types, with associations specific to each species.

Arctostaphylos (*canescens, manzanita, stanfordiana*) **Provisional Alliance**
Arctostaphylos canescens Provisional Association
Arctostaphylos manzanita Provisional Association
Arctostaphylos stanfordiana Provisional Association

10b. *Arctostaphylos glandulosa* dominates or co-dominates with *Adenostoma fasciculatum* on convexities, outcrops, ridges, or slopes. Soils may be derived from serpentine or gabbro. Species commonly found as emergent trees or subdominant shrubs include *Arbutus menziesii*, *Arctostaphylos* spp., *Diplacus aurantiacus*, and *Heteromeles arbutifolia*.

***Arctostaphylos glandulosa* Alliance**[2]
Arctostaphylos glandulosa Association
Arctostaphylos glandulosa – *Adenostoma fasciculatum* Association

10c. *Arctostaphylos viscida* (e.g., *A. viscida* ssp. *pulchella*) dominates or co-dominates with *Ceanothus jepsonii* on serpentine substrates. *Ceanothus jepsonii* may occasionally exceed *A. viscida* in cover when present.

***Arctostaphylos viscida* Alliance**
Arctostaphylos viscida – *Ceanothus jepsonii* Provisional Association

10d. *Ceanothus cuneatus* dominates or co-dominates with *Adenostoma fasciculatum*, often on convexities with westerly exposures. A variety of shrubs may intermix, including *Arctostaphylos*, *Baccharis*, *Eriodictyon*, *Heteromeles*, *Quercus durata*, and others.

***Ceanothus cuneatus* Alliance**
Ceanothus cuneatus – *Adenostoma fasciculatum* Association

10e. *Adenostoma fasciculatum* dominates, often with subdominant shrubs such as *Arctostaphylos manzanita*, *A. stanfordiana*, or *Diplacus aurantiacus*. *Salvia sonomensis*, an understory shrub, may have higher cover than *Adenostoma*. If *A. fasciculatum* co-dominates with *Arctostaphylos* spp., *Ceanothus cuneatus*, *Cercocarpus montanus*, *Quercus berberidifolia*, or *Q. durata*, key to one of the latter alliances instead of *A. fasciculatum*.

***Adenostoma fasciculatum* Alliance**
Adenostoma fasciculatum Association
Adenostoma fasciculatum – *Arctostaphylos manzanita* Association
Adenostoma fasciculatum – *Arctostaphylos stanfordiana* / *Salvia sonomensis* Provisional Association
Adenostoma fasciculatum – *Diplacus aurantiacus* Association
Adenostoma fasciculatum Serpentine Association

---

[2] The *Arctostaphylos glandulosa* Alliance is placed in the Pre-Montane Chaparral Group of the USNVC. For this project, it fits better under the Xeric Chaparral Group because stands occupy relatively dry, southerly facing sites with shallow soils and are more similar ecologically to other xeric chaparral alliances of Sonoma County. Future versions of the USNVC may include an alliance of *A. glandulosa* under the Xeric Chaparral Group.

Class C. Herbaceous Vegetation

Section I. Vegetation of a) freshwater wetland or riparian settings with water or wet ground present temporarily, seasonally, or throughout the growing season, b) saline or alkaline lowlands where water accumulates in the winter, or c) tidal salt or brackish marshes with seasonal or ephemeral inundations. Includes herbaceous vegetation dominated, co-dominated, or characterized by: *Argentina, Azolla, Bidens, Bolboschoenus, Brasenia, Carex, Ceratophyllum, Distichlis, Eleocharis macrostachya, Grindelia stricta, Juncus arcticus, J. effusus, J. lescurii, J. patens, Lasthenia glaberrima, Lemna, Lepidium latifolium, Leymus triticoides, Ludwigia, Mimulus guttatus, Nuphar, Oenanthe, Persicaria, Pleuropogon, Sarcocornia (=Salicornia), Schoenoplectus, Scirpus, Spartina, Typha,* and/or *Xanthium.*

1. Freshwater stands dominated by aquatic, floating or submerged plants, including *Azolla, Brasenia, Ceratophyllum, Lemna, Ludwigia,* and/or *Nuphar.* Found along slow-moving streams, still ponds, lakes, or on ground surfaces after water levels have dropped.

Western North American Freshwater Aquatic Vegetation Macrogroup

1a. *Ludwigia hexapetala* or *L. peploides* dominates, creating mats in shallow water or overwet soil. Other aquatic plants such as *Azolla, Lemna, Polygonum,* and *Sparganium* may be present.

Naturalized Temperate Pacific Freshwater Vegetation Group

Ludwigia (hexapetala, peploides) **Provisional Seminatural Alliance**
 Ludwigia (hexapetala, peploides) Provisional Seminatural Association

1b. *Azolla filiculoides* or *Azolla mexicana* (=*A. microphylla*) dominates or characterizes stands on water or wet ground surfaces. If *Lemna* is co-dominant, key to this alliance.

Temperate Freshwater Floating Mat Group

Azolla (filiculoides, mexicana) **Alliance**

1c. *Brasenia, Ceratophyllum, Lemna,* or *Nuphar* dominates on water surfaces of streams, ponds, or lakes.

Temperate Pacific Freshwater Aquatic Bed Group

1c1. *Ceratophyllum demersum* dominates. One stand was encountered for this project, near the eastern border of Sonoma County in a dammed pond. Other stands are likely to occur in the county.

Ceratophyllum demersum **Provisional Alliance**
 Ceratophyllum demersum Western Provisional Association

1c2. *Brasenia schreberi* or *Nuphar lutea* dominates on the water surface. Algae and a variety of hydrophytes may intermix, including *Alisma, Carex, Hippuris vulgaris, Polygonum,* and *Oenanthe.*

Nuphar **spp. –** *Potamogeton* **spp. –** *Lemna* **spp. Freshwater Aquatic Provisional Alliance**
 Brasenia schreberi Provisional Association
 Nuphar lutea ssp. *polysepala* Provisional Association

2. Freshwater or brackish stands dominated by *Argentina, Carex pansa, C. obnupta, C. praegracilis, Juncus effusus, J. lescurii, J. patens, Oenanthe, Schoenoplectus, Scirpus microcarpus,* and/or *Typha,* where water is present throughout all or most of the growing season. Soils have high organic content and may be poorly aerated.

Western North American Freshwater Marsh Macrogroup

2a. *Schoenoplectus* and/or *Typha* dominate in the herbaceous layer. Stands are found along streams, ditches, shores, bars, and channels of river mouth estuaries; around ponds and lakes; and in sloughs, swamps, and freshwater to brackish marshes.

Arid West Freshwater Emergent Marsh Group

    2a1. *Schoenoplectus acutus* dominates or co-dominates with a species of *Typha*.

                      ***Schoenoplectus acutus* Alliance**
                        Schoenoplectus acutus Association

    2a2. *Schoenoplectus californicus* dominates or co-dominates with a species of *Typha*.

                        ***Schoenoplectus californicus* Alliance**
                        Schoenoplectus californicus Association

    2a3. *Typha angustifolia, T. domingensis,* and/or *T. latifolia* dominate in semipermanently flooded freshwater or brackish marshes. If *Schoenoplectus acutus* or *S. californicus* is co-dominant, key to the appropriate *Schoenoplectus* Alliance.

                      ***Typha* (*angustifolia, domingensis, latifolia*) Alliance**
                        Typha domingensis Association
                        Typha latifolia Association

2b. *Argentina egedii, Bolboschoenus maritimus, Carex nudata, C. obnupta, C. praegracilis, C. pansa, Distichlis spicata, Eleocharis macrostachya, Juncus effusus, J. lescurii, J. patens, J. occidentalis, J. phaeocephalus, Oenanthe,* and/or *Scirpus microcarpus* dominate or co-dominate in mesic or wetland settings. *Holcus, Hypochaeris, Leontodon, Rumex,* and *Vulpia bromoides* may intermix with similar cover. Stands may be found along seasonally flooded brackish marshes, coastal sand dunes, swales and plains, shallowly inundated woods, meadows, roadside ditches, mudflats, coastal swamps, lakeshores, marshes, and riverbanks.

Vancouverian Coastal/Tidal Marsh and Meadow Group

    2b1. *Argentina egedii* (=*A. anserina* or *Potentilla anserina* ssp. *pacifica*) dominates or co-dominates with *Bolboschoenus maritimus, Carex nudata, Distichlis spicata, Eleocharis macrostachya, Holcus lanatus, Juncus lescurii, Leontodon taraxacoides,* and *Rumex acetosella*. If *Oenanthe sarmentosa* is co-dominant, key to the *O. sarmentosa* Alliance below.

                        ***Argentina egedii* Alliance**
                        Argentina egedii Association

    2b2. *Carex praegracilis, C. pansa,* or *C. tumulicola* dominates or co-dominates with *Holcus lanatus* or *Lolium perenne*. Stands of *C. praegracilis* are not restricted to the coast. One stand was sampled near the eastern boundary of the county in a moist depression on a hillside.

                      ***Carex* (*pansa, praegracilis*) Provisional Alliance**
                      Carex praegracilis Provisional Association

    2b3. *Carex obnupta* dominates in the herbaceous layer in a variety of freshwater and brackish settings near the coast.

                            ***Carex obnupta* Alliance**
                          Carex obnupta Association

    2b4. *Juncus effusus, J. patens, J. occidentalis,* and/or *J. phaeocephalus* dominate individually or in combination near the coast or farther inland. Co-dominant species may include *Carex densa, Holcus lanatus, Hypochaeris radicata, Juncus bufonius,* and *Vulpia bromoides*.

                      ***Juncus* (*effusus, patens*) Provisional Alliance**
                        Juncus effusus Association
                        Juncus patens Provisional Association
                        Juncus patens – *Holcus lanatus* Provisional Association
                        Juncus patens – *Juncus occidentalis* Provisional Association
                        Juncus phaeocephalus Provisional Association

2b5. *Juncus lescurii* dominates or co-dominates with *Agrostis stolonifera*, *Argentina egedii*, *Eleocharis macrostachya*, or *Juncus phaeocephalus* in slightly brackish marshes or seeps near salt marshes.

Juncus lescurii **Alliance**
Juncus lescurii Association

2b6. *Oenanthe sarmentosa* dominates or co-dominates with *Argentina egedii* in freshwater to slightly brackish marshes.

Oenanthe sarmentosa **Alliance**
Oenanthe sarmentosa Association

2b7. *Scirpus microcarpus* dominates in marshes, roadside ditches, and along stream banks. Larger forbs such as *Conium maculatum*, *Oenanthe*, *Heracleum maximum*, and *Urtica dioica* may be present as subdominants.

Scirpus microcarpus **Alliance**
Scirpus microcarpus Association

3. Salt and brackish marshes dominated or co-dominated by *Bolboschoenus*, *Distichlis*, *Sarcocornia* (=*Salicornia*), and/or *Spartina*. May appear as sparsely vegetated mudflats at low tide, or during restoration (as along San Pablo Bay). Mudflats with trace amounts of cover by herbs are included here (see 3e).

North American Pacific Coastal Salt Marsh Macrogroup

Temperate Pacific Tidal Salt and Brackish Meadow Group

3a. *Bolboschoenus maritimus* dominates or co-dominates with *Sarcocornia* (=*Salicornia*) *pacifica*.

Bolboschoenus maritimus **Alliance**
Bolboschoenus maritimus Association
Bolboschoenus maritimus – *Sarcocornia pacifica* Association

3b. *Distichlis spicata* dominates or co-dominates with *Frankenia salina* and/or *Jaumea carnosa*. *Sarcocornia pacifica* may present as a subdominant.

Distichlis spicata **Alliance**
Distichlis spicata – *Frankenia salina* – *Jaumea carnosa* Association

3c. *Sarcocornia pacifica* dominates or co-dominates with *Jaumea carnosa*, *Distichlis spicata*, and/or *Lepidium latifolium*.

Sarcocornia pacifica (*Salicornia depressa*) **Alliance**
Sarcocornia pacifica Association
Sarcocornia pacifica – *Jaumea carnosa* – *Distichlis spicata* Association
Sarcocornia pacifica – *Lepidium latifolium* Association

3d. *Spartina foliosa* dominates on mudflats, banks, berms, and margins of bays and deltas.

Spartina foliosa **Alliance**
Spartina foliosa Association

3e. Mudflats or dry pond bottoms (sometimes in sites undergoing restoration) with trace amounts of cover by *Agrostis avenacea*, *Sarcocornia pacifica*, *Sesuvium*, and others. Cover by plants is so sparse and/or uneven that stands are not recognized by the USNVC.

Mudflat/Dry Pond Bottom Mapping Unit

4. Herbaceous stands dominated or characterized by *Eleocharis macrostachya*, *Grindelia stricta*, *Lasthenia glaberrima*, or *Pleuropogon californicus*. In the *Manual of California Vegetation* (Sawyer et al. 2009), these stands are recognized in a macrogroup associated with vernal pools, even though they do not always occur in vernal pool settings. Future versions of the hierarchy will likely split vernal pool and non–vernal pool stands into different alliances, groups, and macrogroups based on ecological and environmental differences. Many true vernal pool types occur in Sonoma County but are not treated in this report[3].

Western North America Vernal Pool Macrogroup

Californian Mixed Annual/Perennial Freshwater Vernal Pool / Swale Bottomland Group

4a. *Pleuropogon californicus* and/or *Lasthenia glaberrima* are present with high cover in the herbaceous layer. If *Eleocharis macrostachya* or *E. palustris* is present and co-dominant, key to this alliance instead of *Eleocharis*. Stands typically occur in vernal pools or vernally influenced marshes.

Lasthenia glaberrima **Alliance**
Lasthenia glaberrima – *Pleuropogon californicus* Association

4b. *Eleocharis macrostachya* dominates in the herbaceous layer along lakeshores, streambeds, swales, vernal pools, pastures, ditches, and ponds. If *Lasthenia glaberrima* or *Pleuropogon californicus* is present with high cover, key to the *L. glaberrima* Alliance above.

Eleocharis (*acicularis*, *macrostachya*) **Provisional Alliance**
Eleocharis macrostachya Association

4c. *Grindelia stricta* dominates or co-dominates with nonnative herbs such as *Raphanus sativus*, *Vulpia bromoides*, and *Bromus diandrus*. Stands may be found on slightly elevated or drier ground adjacent to coastal dunes, salt or alkaline marshes, or on bluffs, levees, and road margins.

Grindelia (*stricta*) **Provisional Alliance**
Grindelia stricta Provisional Association

5. Wetland herbaceous vegetation dominated or characterized by *Bidens frondosa*, *Carex barbarae*, *C. nudata*, *C. serratodens*, *Juncus arcticus*, *Lepidium latifolium*, *Leymus triticoides*, *Mimulus guttatus*, *Persicaria lapathifolia*, or *Xanthium strumarium*. Stands occupy settings where saturated soil or standing water throughout the growing season are key characteristics.

Western North America Wet Meadow and Low Shrub Carr Macrogroup

5a. Stands dominated or characterized by the species of *Carex*, *Juncus*, *Leymus*, or *Mimulus* mentioned above.

Californian Warm-Temperate Marsh/Seep Group

5a1. *Carex barbarae* dominates in seasonally or intermittently saturated wetlands.

Carex barbarae **Alliance**
Carex barbarae Association

5a2. *Carex nudata* dominates along rocky creeks and streams below the high water mark. If *Argentina egedii* is co-dominant, key to the *A. egedii* Alliance (see 2b1).

Carex nudata **Alliance**
Carex nudata Association

5a3. *Carex serratodens* dominates or co-dominates with *Agoseris heterophylla*, *Juncus arcticus*, or *Leymus triticoides*. Stands are often found on serpentine substrates.

Carex serratodens **Provisional Alliance**
Carex serratodens Provisional Association

---

[3] Vernal pool data collected from over 100 relevés in the Santa Rosa Plain from 2007–2009 have not been completely analyzed. The final classification and mapping will be treated under a separate vernal pool phase of the Sonoma County vegetation project. The vernal pool stands studied so far appear to fall largely within the *Lasthenia glaberrima* Alliance, but new associations may be defined and some samples may represent other alliances.

5a4. *Juncus arcticus* (var. *balticus* or *mexicanus*) dominates in freshwater, brackish, or alkaline settings. *Mentha pulegium, Poa pratensis,* and other hydrophytes may intermix as sub-dominants.

Juncus arcticus **(var.** *balticus, mexicanus***) Alliance**
Juncus arcticus (var. *balticus, mexicanus*) Association

5a5. *Leymus triticoides* dominates or co-dominates with *Briza maxima, Lolium perenne,* or other nonnative grasses or forbs. Stands are found on poorly drained floodplains, valley bottoms, and brackish marsh margins.

Leymus triticoides **Alliance**
Leymus triticoides Association
Leymus triticoides – Lolium perenne Association

5a6. *Mimulus guttatus* or another wetland *Mimulus* species dominates or co-dominates in the herbaceous layer with *Eleocharis, Juncus,* or *Lolium perenne*. Stands are found in moist or saturated settings along streams, ephemeral cascades, ditches, fens, seeps, and springs.

Mimulus **(***guttatus***) Alliance**
Mimulus guttatus Association

5b. Stands dominated or characterized by the nonnative or ruderal taxa mentioned above: *Bidens, Lepidium, Persicaria,* and/or *Xanthium*.

Naturalized Warm-Temperate Riparian and Wetland Group

5b1. *Lepidium latifolium* dominates in the herbaceous layer along intermittently and seasonally flooded freshwater and brackish marshes and riparian corridors. In alkaline or saline settings, *Distichlis spicata* is commonly present.

Lepidium latifolium **Seminatural Alliance**
Lepidium latifolium – Distichlis spicata Seminatural Association

5b2. *Bidens frondosa, Persicaria* spp., and/or *Xanthium* spp. dominate in marshes and regularly disturbed vernally wet ponds, fields, and stream terraces.

Persicaria lapathifolia – Xanthium strumarium **Provisional Alliance**
Bidens frondosa Provisional Association

Section II. Vegetation dominated or characterized by herbaceous species that occupy dry, seasonally moist, and usually well-drained sites that range from interior dry ridges and cliffs to ocean bluffs, dunes, and terraces with cooling summer fog and salty breezes. Stands are not wet or inundated as in Section I above. This group includes native and nonnative annual and perennial grasslands, seral herbaceous stands, dry cliff and canyon vegetation, and coastal dune/ bluff vegetation. Dominant, co-dominant, and characteristic taxa include: *Abronia, Agrostis gigantea, A. stolonifera, Allium falcifolium, Ambrosia, Ammophila, Anthoxanthum, Asclepias solanoana, Avena, Brachypodium, Brassica, Briza, Bromus, Calamagrostis, Carpobrotus, Centaurea, Cynosurus, Danthonia, Deschampsia, Elymus elymoides, E. glaucus, E. multisetus, Eriogonum cedrorum, E. luteolum, E. nudum, Erodium, Eryngium armatum, Eschscholzia, Festuca arundinacea, F. californica, F. idahoensis, Heterotheca, Holcus, Hordeum, Lasthenia californica, Leymus mollis, Lolium, Melica, Mesembryanthemum, Nassella, Phalaris, Plagiobothrys nothofulvus, Plantago erecta, Pteridium, Raphanus, Selaginella bigelovii, Streptanthus,* and/or *Vulpia*.

6. *Allium falcifolium, Asclepias solanoana, Eriogonum cedrorum, E. luteolum, E. nudum, Selaginella bigelovii,* and/or *Streptanthus morrisonii* characterize or dominate stands on exposed rock.

California Cliff, Scree, and Other Rock Vegetation Macrogroup

Central California Coast Ranges Cliff and Canyon Group

6a. *Selaginella bigelovii* dominates or characterizes small stands on rock outcrops, cliff faces, or skeletal soils over gently to steeply sloping, impervious substrates. Moss and lichen species often intermix.

*****Selaginella bigelovii* Alliance**

6b. Sparsely vegetated herbaceous stands (generally less than 2% absolute cover) characterized by *Allium falcifolium, Asclepias solanoana, Eriogonum cedrorum, E. luteolum, E. nudum,* and/or *Streptanthus morrisonii,* growing on steep serpentine barrens with exposed gravel and bedrock.

***Allium falcifolium – Eriogonum* spp. – *Streptanthus* spp. Provisional Alliance**
Eriogonum luteolum – Streptanthus morrisonii Provisional Association

7. *Eriogonum nudum* or *Heterotheca oregona* dominates or co-dominates with nonnative herbs in stands with recent or seasonal disturbance.

California Coastal Scrub Macrogroup

Central and South Coastal California Seral Scrub Group

7a. *Eriogonum nudum* dominates or co-dominates with *Bromus diandrus, Erodium botrys, Vulpia bromoides,* and others in herbaceous stands often occupying exposed convexities.

***Eriogonum* (*elongatum, nudum*) Provisional Alliance**
Eriogonum nudum Provisional Association

7b. *Heterotheca oregona,* a perennial herb that acts like a short-lived shrub, dominates herbaceous stands with seasonal hydrologic disturbance. Found along sunny, rocky stream terraces, seasonally dry streambeds, sandbars in river drainages, and cobbled gravel bars in floodplains.

***Heterotheca* (*oregona, sessiliflora*) Provisional Alliance**
Heterotheca oregona Provisional Association

8. Native and nonnative annual forb/grass vegetation AND native perennial grasslands growing within the California Mediterranean climate. Stands are generally found in relatively drier sites than those in the Western North American Temperate Grassland and Meadow Macrogroup, which is more common near the coast (see step 9). Includes vegetation characterized by, but not limited to, *Avena, Brassica, Bromus, Centaurea, Cynosurus, Elymus glaucus, Eschscholzia, Lasthenia californica, Lolium, Nassella, Melica, Plantago erecta, Pteridium aquilinum, Vulpia microstachys,* and *Plagiobothrys nothofulvus.*

California Annual and Perennial Grassland Macrogroup

8a. Herbaceous vegetation dominated, co-dominated, or characterized by native annual forbs and grasses such as *Eschscholzia, Lasthenia californica, Lupinus, Plagiobothrys, Plantago erecta,* and *Vulpia microstachys.* Commonly occurring taxa include *Avena, Bromus, Cryptantha, Geranium, Dichelostemma, Lolium,* and *Vulpia.* Stands are found on upland slopes, flats, and ridges.

California Annual Herb/Grass Group

8a1. *Eschscholzia californica, Lupinus bicolor,* and/or *L. nanus* dominate or co-dominate with a variety of native and nonnative forbs and grasses.

***Eschscholzia* (*californica*) – *Lupinus* (*nanus*) Provisional Alliance**
Bromus hordeaceus – Lupinus nanus – Trifolium spp. Association
Eschscholzia californica Association

8a2. *Plagiobothrys nothofulvus* dominates and intermixes with a variety of native and nonnative forbs and grasses.

***Plagiobothrys nothofulvus* Alliance**
Plagiobothrys nothofulvus – Daucus pusillus – Trifolium microcephalum Provisional Association

8a3. *Lasthenia californica, Erigeron glaucus, Calycadenia multiglandulosa, C. truncata, Hemizonia congesta, Lomatium, Lotus humistratus, Micropus californicus, Plantago erecta,* and/or *Vulpia microstachys* dominate individually or in combination in the herbaceous layer. *Lasthenia californica, Plantago erecta,* and/or *Vulpia microstachys* are often present, sometimes with sparse cover.

> ***Lasthenia californica – Plantago erecta – Vulpia microstachys* Alliance**
> *Erigeron glaucus – Lasthenia californica* Provisional Association
> *Hemizonia congesta – Lolium perenne* Provisional Association
> *Lotus humistratus – Plantago erecta – Lomatium* spp. Provisional Association
> *Micropus californicus* Provisional Association
> *Vulpia microstachys – Plantago erecta – Calycadenia* (*truncata, multiglandulosa*) Association

8b. *Bromus carinatus, Elymus glaucus, Melica californica, Nassella pulchra,* and/or *Pteridium aquilinum,* all native perennial grasses, are dominant or characteristic in stands, sometimes with equal or greater cover of nonnative herbs.

California Perennial Grassland Group

8b1. *Bromus carinatus, Elymus glaucus* and/or *Pteridium aquilinum* dominate or co-dominate near meadows, in forested openings, and on elevated flats. *Anagallis arvensis, Bromus hordeaceus, Geranium dissectum, Rumex acetosella,* and *Vulpia bromoides* are often present.

> ***Elymus glaucus – Bromus carinatus* Provisional Alliance**
> *Bromus carinatus* Provisional Association
> *Elymus glaucus* Association
> *Pteridium aquilinum* Provisional Association

8b2. *Melica californica* and/or *Nassella pulchra* are dominant, co-dominant, or characteristic in stands. *Achnatherum lemmonii, Avena, Bromus, Hemizonia congesta, Lolium perenne, Plantago erecta,* and/or *P. lanceolata* intermix as dominant, co-dominant, or characteristic taxa in associations of this alliance.

> ***Nassella* spp. – *Melica* spp. Provisional Alliance**
> *Melica californica* Provisional Association
> *Nassella pulchra* Association
> *Nassella pulchra – Achnatherum lemmonii* Provisional Association
> *Nassella pulchra – Avena* spp. – *Bromus* spp. Association
> *Nassella pulchra – Hemizonia congesta* Provisional Association
> *Nassella pulchra – Lolium perenne – Plantago erecta* Serpentine Provisional Association
> *Nassella pulchra – Melica californica* – Annual Grass Association
> *Nassella pulchra – Plantago lanceolata* Provisional Association

8c. Herbaceous vegetation strongly dominated by nonnative grasses and forbs such as *Avena, Brachypodium, Brassica, Briza, Bromus, Centaurea, Cynosurus, Danthonia pilosa, Erodium, Lolium, Nassella manicata,* and *Raphanus*. Native herbaceous species have insignificant cover in these stands, especially during the active growing season. Stands are found in foothills, rangelands, fallow fields, woodland openings, riparian areas, and disturbed settings.

Mediterranean California Naturalized Annual and Perennial Grassland Group

8c1. *Avena, Brachypodium, Briza, Bromus,* and/or *Erodium* dominate individually or in combination.

> ***Avena* spp. – *Bromus* spp. Provisional Seminatural Alliance**
> *Avena barbata* Seminatural Association
> *Brachypodium distachyon* Seminatural Association
> *Briza maxima* Provisional Seminatural Association
> *Bromus diandrus – Avena* spp. Seminatural Association
> *Bromus hordeaceus – Erodium botrys* Seminatural Association

8c2. *Brassica nigra, Raphanus sativus,* or another nonnative mustard dominates in the herbaceous layer, often in old or active agriculture lands.

> ***Brassica nigra* and Other Mustards Seminatural Alliance**
> *Brassica nigra* Seminatural Association
> *Raphanus sativus* Seminatural Association

8c3. *Centaurea solstitialis* or another nonnative species of *Centaurea* dominates herbaceous stands.

> ***Centaurea (solstitialis, melitensis)* Seminatural Alliance**
> *Centaurea solstitialis* Seminatural Association

8c4. *Cynosurus echinatus, Danthonia pilosa,* and/or *Nassella manicata* dominate or co-dominate in the herbaceous layer. *Anagallis, Avena, Lolium, Plantago lanceolata, Rumex,* and *Vulpia bromoides* are often present.

> ***Cynosurus echinatus* Seminatural Alliance**
> *Cynosurus echinatus – (Danthonia pilosa – Nassella manicata)* Provisional Seminatural Association

8c5. *Lolium perenne* dominates or co-dominates with *Avena barbata, Bromus hordeaceus, Hordeum marinum, H. murinum, Medicago, Trifolium subterraneum,* and other nonnatives in herbaceous stands. Often found on moist or poorly drained sites, on or off serpentine.

> ***Lolium perenne* Seminatural Alliance**
> *Lolium perenne* Seminatural Association

9. Herbaceous vegetation dominated, co-dominated, or characterized by native or nonnative perennial grasses. Stands are generally found in moister settings than those in the California Annual and Perennial Grassland Macrogroup (see step 8), and are often coastal. The grasses included are: *Agrostis gigantea, A. stolonifera, Anthoxanthum, Calamagrostis nutkaensis, Danthonia californica, Deschampsia cespitosa, Elymus elymoides, E. multisetus, Festuca arundinacea, F. idahoensis, Holcus, Hordeum brachyantherum* and/or *Phalaris aquatica.* Note: stands dominated by *Lolium perenne* key out in step 8 above.

9a. *Agrostis, Anthoxanthum, Festuca arundinacea, Holcus,* and/or *Phalaris* are dominant, co-dominant, or characteristic in herbaceous stands.

Western North American Temperate Grassland and Meadow Macrogroup

9a1. Nonnative, slightly mesic, disturbed pasturelands dominated or co-dominated by the following perennial grasses: *Agrostis gigantea, A. stolonifera, Anthoxanthum, Festuca arundinacea, Holcus,* and/or *Phalaris.* If native species are present and co-dominant, key to an alliance dominated or characterized by natives. Found in wet settings, including brackish marshes, meadows, stream terraces, wet pastures, agricultural wetlands, or tidal zones.

Vancouverian and Rocky Mountain Naturalized Perennial Grassland Group

9a1a. *Agrostis gigantea, A. stolonifera,* and/or *Festuca arundinacea* dominate or co-dominate in the herbaceous layer. The stands encountered for this project were dominated by *F. arundinacea,* though stands dominated by *Agrostis* may be present in Sonoma County.

> ***Agrostis (gigantea, stolonifera) – Festuca arundinacea* Seminatural Alliance**
> *Festuca arundinacea* Provisional Seminatural Association

9a1b. *Holcus lanatus* and/or *Anthoxanthum odoratum* dominate individually or in combination. Other co-dominants may include *Briza maxima, Pteridium aquilinum, Rumex acetosella,* and *Vulpia bromoides.*

> ***Holcus lanatus – Anthoxanthum odoratum* Seminatural Alliance**
> *Holcus lanatus* Seminatural Association
> *Holcus lanatus – Anthoxanthum odoratum* Seminatural Association

9a1c. *Phalaris aquatica* dominates in naturalized or planted stands. Other nonnative herbs, such as *Carduus pycnocephalus* may be present with similar cover.

Phalaris aquatica **Seminatural Alliance**
Phalaris aquatica Provisional Seminatural Association

9b. Native grasslands dominated, co-dominated, or characterized by the following perennial grasses: *Bromus carinatus, Elymus elymoides, E. glaucus, E. multisetus, Festuca californica, F. idahoensis,* or *Pteridium aquilinum*. May occur near the coast or inland.

Western Dry Upland Perennial Grassland Group

9b1. *Elymus elymoides* or *E. multisetus* dominates or co-dominates in stands on serpentine soils, often on southerly exposures. Stands of *Elymus multisetus* with *Eschscholzia californica* and/or *Plantago erecta* were encountered most often in the sites visited for this project; *Dichelostemma capitatum, Eriogonum nudum, Lotus humistratus,* and *Minuartia douglasii* were also commonly present.

Elymus (*elymoides, multisetus*) **Provisional Alliance**
Elymus multisetus – (*Eschscholzia californica* – *Plantago erecta*) Provisional Association

9b2. *Festuca idahoensis* dominates or co-dominates with *Danthonia californica* and/or *Elymus multisetus*. *Bromus carinatus, Elymus glaucus, Plantago erecta,* and a variety of native and non-native forbs and grasses may intermix as subdominants. Occasionally, the larger *Festuca californica* may replace *F. idahoensis* in somewhat shadier or less exposed sites.

Festuca idahoensis **Alliance**
Festuca californica Provisional Association
Festuca idahoensis – *Bromus carinatus* Association
Festuca idahoensis – *Danthonia californica* Provisional Association
Festuca idahoensis Ultramafic Provisional Association

9c. Native, mesic to moist, primarily coastal grasslands dominated, co-dominated, or characterized by *Calamagrostis nutkaensis, Deschampsia cespitosa, Danthonia californica, Eryngium armatum,* and/or *Hordeum brachyantherum*. *Baccharis pilularis, Briza maxima, Holcus lanatus, Nassella pulchra,* and/or *Vulpia bromoides* commonly intermix in stands. Found in a variety of settings, including dunes, bluffs, meadows, valley bottoms, alluvial slopes, terraces, meadows, and seasonally flooded areas with moderate salinity.

9c1. *Deschampsia cespitosa, Danthonia californica,* and/or *Eryngium armatum* dominate or co-dominate individually or in combination (if *Holcus lanatus* has the highest cover, but these three species have at least 10% combined cover, key to *Deschampsia*). Settings range from coastal dunes and bluffs to inland plains (e.g., Santa Rosa Plain) to montane meadows.

Western Cordilleran Montane–Boreal Wet Meadow Macrogroup

Western Cordilleran Montane–Boreal Mesic Wet Meadow Group

Deschampsia cespitosa **Alliance**
Deschampsia cespitosa – *Danthonia californica* Association
Deschampsia cespitosa – *Eryngium armatum* Provisional Association
Deschampsia cespitosa – *Holcus lanatus* Provisional Association

9c2. *Hordeum brachyantherum* dominates or co-dominates with *Bromus carinatus, Hypochaeris, Lolium perenne, Lotus corniculatus, Plantago erecta,* and *Trifolium subterraneum* in moist meadows, along stream terraces and coastal bluffs, and near seeps and springs.

Western Cordilleran Montane Shrubland and Grassland Macrogroup

Western Cordilleran Montane Moist Graminoid Meadow Group

Hordeum brachyantherum **Alliance**
Hordeum brachyantherum Association

9c3. *Calamagrostis nutkaensis* dominates or co-dominates with *Baccharis pilularis* OR stands are dominated or characterized by *Danthonia californica* with *Briza maxima*, *Nassella pulchra*, and/or *Vulpia bromoides*. Stands are found along valley bottoms, lower portions of alluvial slopes, terraces, floodplains, and ridges.

Vancouverian Lowland Grassland and Shrubland Macrogroup

Vancouverian Coastal Grassland Group

9c3a. *Calamagrostis nutkaensis* dominates or co-dominates with *Baccharis pilularis*. *Heracleum maximum*, *Holcus lanatus*, *Juncus patens*, and/or *Rubus ursinus* often intermix in stands.

Calamagrostis nutkaensis **Alliance**
Calamagrostis nutkaensis / Baccharis pilularis Association

9c3b. *Danthonia californica* dominates OR characterizes stands in combination with 1) *Nassella pulchra* or 2) *Briza maxima* and/or *Vulpia bromoides*. In the latter two cases, *Danthonia* and the other species share at least 15% relative cover in the herb layer, with other nonnative grasses and forbs sometimes having higher cover (e.g., *Cynosurus echinatus*, *Holcus lanatus*, and *Hypochaeris radicata*).

Danthonia californica **Alliance**
Danthonia californica – (Briza maxima – Vulpia bromoides) Provisional Association
Danthonia californica – Nassella pulchra Provisional Association

10. Coastal dune, bluff, meadow, and other vegetation dominated by herbaceous species such as *Abronia*, *Ambrosia*, *Ammophila*, *Carpobrotus*, *Leymus mollis*, and *Mesembryanthemum*.

10a. Native species, including *Abronia latifolia*, *Ambrosia chamissonis*, *Artemisia pycnocephala*, and/or *Leymus mollis* dominate or co-dominate on dunes or bluffs. Plants are adapted to salt spray, wind, and shifting sands and are thus capable of colonizing relatively unstable and sterile substrates.

Vancouverian Coastal Dune and Bluff Macrogroup

Vancouverian/Pacific Dune Mat Group

10a1. *Abronia latifolia*, *Ambrosia chamissonis*, and/or *Artemisia pycnocephala* dominate, sometimes with *Calystegia soldanella* or *Polygonum paronychia* occurring as associated species. *Cakile maritima*, *Ammophila arenaria*, *Camissonia cheiranthifolia*, and *Eriogonum latifolium* may be present.

Abronia latifolia – Ambrosia chamissonis **Alliance**
Ambrosia chamissonis Provisional Association
Artemisia pycnocephala – Calystegia soldanella Association
Artemisia pycnocephala – Polygonum paronychia Association

10a2. *Leymus mollis* dominates in the herbaceous layer. *Abronia*, *Artemisia pycnocephala*, *Cakile*, and other herbaceous species may be present as subdominants.

Leymus mollis **Alliance**
Leymus mollis – Abronia latifolia – (Cakile spp.) Association

10b. Nonnatives, including *Ammophila*, *Carpobrotus*, and/or *Mesembryanthemum* dominate on dunes, bluffs, or disturbed lands. Emergent shrubs such as *Baccharis pilularis* or *Lupinus arboreus* may be present.

California–Vancouverian Seminatural Littoral Scrub and Herb Vegetation Group 10b1.

Ammophila arenaria is strongly dominant in the herbaceous layer.

Ammophila arenaria **Seminatural Alliance**
Ammophila arenaria Seminatural Association

10b2. *Carpobrotus* and/or *Mesembryanthemum* dominate on bluffs, dunes, or disturbed lands, often forming impenetrable mats that prevent natives from establishing.

> ***Mesembryanthemum* spp. – *Carpobrotus* spp. Provisional Seminatural Alliance**
>
> > *Carpobrotus* (*edulis*) Provisional Seminatural Association

Chapter 8
Digital Elevation Models

Introduction

As introduced in chapter 2, three types of information can be derived from imagery: elevation models, feature maps, and thematic maps. This chapter reviews elevation models with a focus on digital elevation models (DEMs). The chapter introduces the types of DEMs and DEM-derived products and reviews the methods used to create them. It ends with a discussion on sources of DEM data. Chapter 9 furthers the discussion of DEMs by examining how DEMs and their derivatives can be used to support feature extraction and thematic mapping.

What Is a DEM?

Elevation models are created from point data that samples the x, y, and z coordinates of locations on the earth's surface. Two main types of elevation models exist: DEMs, which are raster datasets depicting the earth's topography as a regularly spaced grid, and triangular irregular networks, which connect irregularly spaced elevation points with triangular surfaces. In general, DEMs tend to be created from imagery, with triangular irregular networks created from survey data. DEMs are more commonly used and are, therefore, the focus of this chapter.

The earth's topography forms the natural foundation for working in three dimensions, where objects are placed above, on, or below the terrain surface. To work in 3D, the GIS analyst must have a model of the earth's topography. By representing topography, elevation models provide a 3D context for mapping and analysis and are an indispensable tool for the GIS analyst. For example, a GIS analyst who creates digital or hard-copy maps uses DEM derivatives like hillshades and contours for displaying topography. An analyst creating a vegetation map or habitat map relies on DEM derivatives such as slope and aspect to help predict vegetation and habit type. An analyst attempting to find the optimal location for a solar energy generation facility uses DEMs to find the amount of solar insolation that illuminates his landscape of interest.

Types of DEMs

There are many variations of DEMs. These include bare-earth DEMs, also known as digital terrain models (DTMs), highest hit DEMs, also known as digital surface models (DSMs), and digital height models (DHMs), also known as canopy height models (CHMs). Figure 8.1 provides a visual illustration of the different types of DEMs.

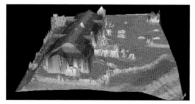
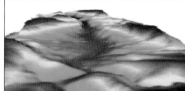

DSM: Digital surface model
Surface models depict topography and all objects on earth's surface, like trees and buildings. Lidar is typically used to create DSMs working with what is commonly referred to as first return data, as the elevation of the first returned laser pulse is used.

DTM: Digital terrain model
Terrain models, commonly referred to as bare earth, are void of things like buildings and trees. Use a DTM to create hillshades, determine slope of the topography or the aspect to the sun, calculate surface water flow, or set the base height of buildings and other features.

DHM: Digital height model
Less common but critical for 3D-enabling your GIS, height models are used to calculate height above ground for buildings, trees, and other features. Height models are created by calculating the difference between the terrain and surface models. Lidar is increasingly the way that DHMs are created, as in this profile of canopy heights in an old-growth Douglas fir forest.

Figure 8.1. Lidar-derived DEMs

DTMs

DTMs—also known as bare-earth digital elevation models—depict the elevation of the ground, typically in vertical units above or below sea level, which is typically represented by a raster value of 0. In figure 8.2, the blue line on the left-side image shows the location of the ground surface that the elevation values in the DTM will represent. The higher the spatial resolution of the image used to make the DTM, the more detailed the DTM.

DTM

A digital terrain model (DTM) depicts the elevation of the ground. For practical purposes a DTM is generally synonymous with a bare earth digital elevation model (DEM).

DSM

A digital surface model (DSM) represents the elevations of the surface trees, buildings, and other features projecting above the bare earth.

Figure 8.2. A conceptual illustration of a DTM and a DSM

Of notable importance to hydrologic modeling and mapping of water features and riparian areas is the *hydro-enforced DTM*. Hydro-enforcement of a DTM incorporates the true elevations of culverts, pipelines, and other buried passages for water into a DTM, creating a DTM suitable for modeling the flow of surface water. A hydro-enforced DTM is a prerequisite for accurate flood zone modeling, dam breach modeling, riparian zone modeling/mapping, and many other types of hydrologic modeling. A cousin of the hydro-enforced DEM is the *hydro-flattened DEM*. Hydro-flattened DEMs are not hydro-enforced (underground passages to water are not burned in), but surface water is "flattened" using hydrographic breaklines. Hydro-flattening improves the appearance of DEMs for display and cartography; it is required for lidar-derived DEMs that are to be included in the National Elevation Dataset (NED).

DSMs

DSMs, also known as digital surface models, represent the highest locations on the landscape. For open areas, the DSM will be equal to the DTM, because no feature is higher than the ground. In areas where buildings and vegetation exist, the pixel values of the DSM will represent the elevations of the vegetation and buildings. In figure 8.1, the red line on the right-side image shows the location of the surfaces that the pixel elevation values in the DSM will represent. Lidar-derived DSMs are also known as *first return* (or highest hit) elevation models, because the DSM elevation of a given location is equal to the highest elevation of the feature the laser pulse encounters at that location. For example, in a tree canopy, a single laser pulse may have a first return from the top of a tree, a second return from a branch lower on the tree, and a last return from the ground. The first return, or highest hit, is the return from the top of the tree.

DHMs

DHMs—also known as normalized digital surface models or CHMs—depict the *absolute* height of features above the ground, such as trees and buildings. DHMs represent the value of the DSM minus the value of the DTM, resulting in a measure of feature height. For example, a DSM pixel representing the top of a tree has a pixel value of 1,100 feet above sea level. The corresponding DTM pixel representing the area of ground of the bottom of the tree has a value of 1,000 feet above sea level. The difference between these two values—1,100 feet minus 1,000 feet—is 100 feet, representing the height of the tree at this location. DHMs are extremely useful in urban planning, forest management, and forest mapping because they provide accurate, high-resolution information about the height of forested stands and the heights of individual trees.

How DEMs Are Created

DEMs are created from imagery using one of two techniques: by using lidar point clouds or by using photogrammetry to extract points from optical or radar images taken from two or more perspectives. Lidar and interferometric synthetic aperture radar (IFSAR) are relatively new approaches to creating DEMs. Many DEMs are still produced using traditional photogrammetric approaches, which extract elevation information from imagery collected in stereo. In fact, very-high-resolution DEMs created from photogrammetric techniques are proliferating because of the recent boom in high-resolution aerial and UAS-collected imagery as well as new image processing techniques. Photogrammetric methods can generate elevation values for every photo pixel from the photos being collected, creating a very-high-resolution DEM.

DEMs Produced by Photogrammetry

DEMs produced by photogrammetry use imagery collected by either optical passive sensors or active radar sensors. The process of creating DEMs from remotely sensed stereo imagery—whether the imagery is collected in analog or digital format by an unmanned aerial vehicle (UAV), airplane, or satellite—hinges on the basic concept of parallax (McGlone, 2013). Parallax is the apparent displacement or difference in the position of an object viewed along two different lines of sight. Objects that are closer have a greater parallax than objects that are farther away. The concept of parallax can be used to determine elevations from remotely sensed imagery if the imagery is collected from two perspectives with overlap between them.

A simple example demonstrates the concept of parallax. With both eyes open, extend your index figure straight up in front of your face about half an arm's length in front of you. Using both eyes pick a distinct vertical object, and keeping your finger straight up, center your finger on the vertical object. Now, without moving your finger, close your left eye. Open both eyes again and then close your right eye. What happened? When closing one of your eyes, your finger appeared to move (this is apparent movement because you did not move your hand). Now, bring your finger closer to your face and repeat the experiment. What happened? You should have detected greater apparent motion. As stated in the definition, parallax is an apparent difference (not a real one), and nearer objects have greater parallax than farther ones. Moving your finger closer to your face increased the parallax. The parallax created by the images captured by our eyes is what allows humans to see in three dimensions.

The science and discipline of photogrammetry rely on the concept of parallax along with some mathematics—including a method called triangulation—to derive elevation data from stereo imagery (McGlone, 2013). Traditionally, photogrammetry was performed using a stereo pair of analog (film) aerial photos (i.e., two photos taken from two perspective views with at least a 50 percent overlap). More recently, the same processes have been applied to digital airborne and satellite imagery. The biggest advantage of using digital imagery is the ability to automate the photogrammetric process (i.e., remove the human analyst) and create higher resolution elevation models in a mostly automated workflow. The biggest disadvantage of creating elevation models using photogrammetry is that the process requires imagery for at least two points of view for every ground location. This becomes challenging in forested areas and dense urban areas.

Airborne platforms were the original source for photogrammetric DEMs and are still the workhorse for detailed photogrammetric terrain mapping today. The advent of UAVs in the past several years has only increased the use of airborne-derived photogrammetric terrain mapping.

The first widespread civilian effort to create DEMs from digital satellite imagery used the French SPOT (Satellite Pour l'Observation de la Terre) satellite launched into orbit in 1986. SPOT was a pointable satellite and therefore able to capture two perspective views. Until SPOT was launched, civilian earth-observing satellites were static in the vertical viewing position and lacked sufficient overlap to create parallax for a very large area. Now, several satellites including WorldViews 1–3, GeoEye-1, and Pleiades-1A and -1B are able to collect stereo imagery for creating DEMs over large areas.

Photogrammetric methods can also be used with IFSAR or dual-antenna radar imagery to collect surface elevation points. IFSAR image pairs for DEM generation are collected from slightly different positions, and the phase difference between the two images is analyzed to extract elevation information (McGlone, 2013). An example is the 2001 NASA Shuttle Radar Topography Mission (SRTM), which used dual-antenna radar to map surface elevations for the near-global scale from 56° south to 60° north at a 30-meter resolution.

DEMs Produced from Lidar

Lidar capitalizes on a synthesis of technologies including accurate global positioning systems (GPSs) and inertial measurement units to create a completely new way of deriving elevation data (Renslow, 2012). As discussed in chapter 3, airborne lidar sensors send out discrete pulses of laser light that bounce back and are recaptured by the aircraft's sensor. The durations of their paths are recorded and analyzed to extract elevation information. A single lidar pulse can have multiple returns indicating multiple objects such as treetops, branches, and the ground. This remote sensing technology has radically altered the collection of elevation data, resulting in significantly higher spatial resolution DEMs with very good vertical accuracy.

Lidar's laser pulses produce a highly accurate and dense "point cloud" of elevation points and—because lidar penetrates the tree canopy—DEMs produced from lidar data can very accurately depict both the ground (DTM) and the highest surface (DSM). In ArcGIS, DTMs, DSMs, and DHMs can be produced from the lidar point cloud by referencing the point clouds in LAS data format datasets or mosaic datasets. The various types of lidar-derived DEMs can then be derived by rasterizing the data using different filters. Other software packages offer solutions for creating the various flavors of lidar-derived DEMs from lidar point clouds.

DEM Derivatives

DEMs are the source of myriad derivative products—both raster and vector—with a wide range of applications for many disciplines. Below is a list of some of these useful DEM-derived products, many of which are discussed in more detail in chapter 9.

- Slope and Aspect. Slope rasters represent the rate of change of elevation (rise over run) for each pixel; aspect rasters represent the direction each pixel on the landscape is facing. These important DEM derivatives are very useful for thematic mapping, modeling, and planning. In ArcGIS, slope and aspect are easily created using raster functions or using the spatial analyst or 3D analyst extensions.
- Hillshade Rasters. These are "shaded relief" depictions of the landscape. Hillshades are great for visual reference when mapping features such as roads. New algorithms—such as the Esri multidirectional hillshade algorithm—provide more cartographic detail and a better depiction of hilly terrain than traditional hillshade algorithms. In ArcGIS, hillshades are easily created using raster functions or using the spatial analyst or 3D analyst extensions.
- Viewsheds. A viewshed represents the area that can be seen by a human from a selected point on the landscape. Viewshed analysis is a critical tool for architects, planners, and engineers. Viewsheds are produced in ArcGIS using the spatial analyst or 3D analyst extensions.
- Elevation Profiles. DEMs can be used to create elevation profiles. Elevation profiles graphically show the change of elevation for a route or a linear transect. Elevation profiles are commonly used for transportation planning, outdoor recreation, engineering, and hydrological analysis and flood modeling. Elevation profiles are produced in ArcGIS using the Elevation Profile Add-In and are built-in functionality for versions of ArcGIS Pro after 1.4.
- Elevation Contours. These are line vectors, with each line representing a constant elevation. Contours are widely used to depict elevation on hard-copy topographic maps, site plans, and engineering diagrams. In ArcGIS, contours are easily created using the spatial analyst or 3D analyst extensions.
- Volumetric Analysis. DEMs are used to determine the amount of material that is removed or added to an area that changes because of natural or man-made topographic modification. In ArcGIS, the cut and fill tools allow for comparisons of DEMs before and after a change, producing models that highlight the areas of erosion (or man-made material removal), and the areas of deposition (or

man-made filling). The cut and fill tools also provide the volume of material that eroded or was deposited in each area of the landscape.
- Solar Insolation Rasters. These quantify the amount of sunshine that illuminates a given area for a user-defined period. Solar insolation rasters are extremely useful for vegetation and habitat mapping, for planning, and for siting solar generation projects. Solar insolation is derived from DEMs in ArcGIS using Spatial Analyst.
- Hydrologic Derivatives. There are many DEM derivatives crucial for the study of hydrology and hydrologic processes including flow accumulation and proximity to water.
- Lidar-Derived Forest Canopy Metrics. Several lidar forest canopy metrics (e.g., tree height and canopy density) are very useful for understanding and measuring forest structure.

GIS analysts can produce the DEM derivatives discussed above using ArcGIS. Many of these derivatives can be directly rendered from the DEMs as part of image services served to web applications or accessible through geoprocessing services. For example, an elevation profile geoprocessing service could easily be deployed. The end user of the service would simply digitize a line on a map in a JavaScript app, in a web map, or in ArcGIS, and the service would process their input and return an elevation profile graphic.

Sources of Data for DEMs

DEMs are freely available in the USA through the National Elevation Dataset (NED). For the continental United States, 1/3-arcsecond (approximately 10-meter) and 1-arcsecond (approximately 30-meter) DEMs are available seamlessly. The USGS's 3D Elevation Program (3DEP) has resulted in an increasing availability of higher resolution, more accurate lidar-derived DEMs over ever-increasing areas of the country. These products include 1-meter bare-earth DEMs where lidar data are available that meets or exceeds USGS specifications (3DEP quality level 2).

Some states and counties have collected lidar data and provide access to the point cloud and/or derivatives like DEMs, building footprints, stream centerlines, and other lidar-derived features. Examples include the State of Indiana (http://gis.iu.edu/datasetInfo/statewide/in_2011.php) and Sonoma County, California (http://sonomavegmap.org/).

DEMs are available for download at the following national portals:
- The National Map (http://nationalmap.gov)—Provides access to the NED and to USGS's 3D Elevation Program Data
- NOAA's Digital Coast—Provides access to lidar data and lidar-derived DEMs for coastal areas of the United States
- USGS makes worldwide SRTM-derived DEMs available on EarthExplorer.

ArcGIS users have access to the World Elevation services through ArcGIS Online. This provides simple access to the best available DEMs for any given area. Since it's an image service, the raw pixel values representing elevation are available. The service has multiple functions configured, so users can display the service as raw elevation or as one of several derivatives including slope, aspect, or as a hillshade. Many of the other DEM derivatives mentioned in this chapter can be also accessed as analysis services in ArcGIS Online or as ArcGIS Server geoprocessing services.

Quality and Accuracy

The quality and accuracy of a DEM are the most important variables for determining the type and scale of mapping that the elevation data will support. For example, DEMs from high-resolution lidar datasets with many ground elevations collected per square meter are required for deriving accurate 1-foot contours used in engineering planning, while lower resolution datasets may support the derivation of 5- to 10-foot contours. When planning new collections of elevation data, it is very important to match the quality requirements of the data collection with the planned end uses of the elevation data being collected.

The accuracy of a DEM is an important consideration. To assess the accuracy of elevation data, highly accurate elevation ground survey points are collected across the project area. These checkpoints are used only for assessing the accuracy of the elevation data and not as an input in creating the DEM. Once the checkpoints are collected, the ground survey elevations at the locations of the checkpoints are compared to the elevations of the mapped ground locations (as measured by a lidar/IFSAR instrument or mapped photogrammetrically). The absolute values of the differences (deltas) between the surveyed elevations and mapped elevations represent the vertical errors at each surveyed location. See chapter 12 for a more in-depth discussion of positional accuracy assessment.

Quality and accuracy are often linked. For example, vertical accuracy of lidar data is highest in open terrain and decreases as the density of vegetation increases. Lidar accuracy decreases in vegetated areas for two reasons. First, as vegetation density increases fewer lidar pulses are able to penetrate all the way to the ground, decreasing the resolution of the

DTM. Second, if vegetation and ground returns are both present, ground returns can sometimes be confused with vegetation returns and vice versa. As a general rule, increasing the point density of a lidar collection will increase the accuracy of a lidar dataset. As a result, point density is the most critical consideration when planning a lidar collection.

Figure 8.3 illustrates the sparseness of lidar ground returns in areas of dense vegetation. The figure shows a lidar-derived bare-earth hillshade for an area in western Sonoma County, California. The area is very densely forested with very large, extremely tall, coast redwood trees (*Sequoia sempervirens*). In the hillshade, areas of sparse returns are evidenced by the large triangles where the hillshade algorithm had few ground returns to work with. The lidar collection in Sonoma County averaged eight pulses per square meter; had it been collected at a higher point density, ground returns from these very densely forested areas would have been less sparse.

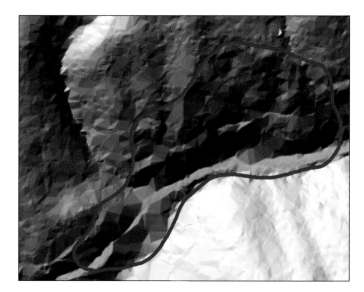

Figure 8.3. Sparse lidar ground returns in a dense forest

Summary—Practical Considerations

DEMs are among the most useful raster datasets available to a GIS analyst. They are used to visually depict topography in the form of slope, aspect, and shaded relief (discussed in chapter 9). They are used for thematic mapping to model or predict land cover, habitat suitability, etc. They are used to derive vector datasets such as contour lines and flood inundation areas. DEMs (along with solar information) are used to model the amount of solar radiation that a location receives. These are just a few use cases for DEMs—the list is long and crosses many GIS disciplines.

Chapter 9
Data Exploration: Tools for Linking Variation in the Imagery to Variation on the Ground

Introduction

Remotely sensed imagery can be used to create maps mainly because a very strong positive correlation exists between what occurs on the ground and what occurs on a remotely sensed image of the ground. Anything that detracts from this relationship can cause problems, errors, or both in using remotely sensed data to create a map (see chapter 6 for a discussion of these causes).

Linking variation in the imagery to variation on the ground is done through a process called data exploration or data mining. The goal is to discover the relationships between the imagery and the ground to most effectively extract information from the imagery and produce the best map possible. Now that you understand the fundamental concepts of imagery, it is time to look at the methods for extracting information from the imagery. These methods represent not only a series of historical developments over the last 75 years or so, but also a progression of improvements in computer processing, automation, and algorithm development.

This chapter reviews tools and techniques for linking variation in the imagery to variation on the ground. It begins by introducing the elements of an image that are used to make a map. The next section presents information about creating additional derivative bands that might aid in strengthening the link between the ground and the imagery. The third section presents tools that can be used to further evaluate the variability between the ground and the imagery. Finally, the last section discusses nonimagery geospatial datasets, which may be highly correlated with variation on the ground and can be used to augment the imagery.

Image Elements

All methods of using imagery to create a map rely on exploiting the confluence of evidence produced by the convergence of the elements of an image: its tone/color, shape, size, pattern, shadow, texture, location, context, height, and date (Spurr, 1960).

Table 9.1. Image elements

Element	Description
Tone/Color	The intensity of spectral response for the bands of a sensor
Shape	The form of the outline of an object
Size	The dimensions (length, width, area) of an object or relative measures such as larger than a house or smaller than a car
Pattern	The spatial arrangement of the objects including random, systematic, clustered, and/or regular
Shadow	Caused by an object interfering with the reflectance or radiance of energy
Texture	The arrangement or repetition of tone/color on an image; usually described as if the interpreter were touching the object and describing it as rough, smooth, mottled, uniform, striated, etc.
Location (x,y,z)	The coordinates of an object
Context	The neighbors of an object
Height	The measured height of an object or the relative height of an object with respect to another object
Date	The date(s) when an image is collected

Tone and Color

Tone and color are perhaps the most recognizable of the image elements. Tone is measured as the intensity of spectral response of each band. In a panchromatic image, if the object reflects or radiates lots of energy, the object will be white. If the object reflects or radiates little energy, the object will be black. Shades of gray represent the intermediate amounts of energy reflected or emitted by the object.

Color is derived from combinations of the various spectral bands of the imagery (e.g., true color, infrared). Color works the same way as tone, except that instead of using a single value to represent the shades of gray, color is expressed using three values because there are three primary colors. As introduced in chapter 3, the color that your eyes see depends on the combination of blue, green, and red light reflected by an object, because you see in only these three portions of the electromagnetic spectrum. Analog (film) photos use the subtractive primary colors (yellow, magenta, and cyan), while digital imagery is displayed on a computer monitor, which uses the additive primary colors (blue, green, and red). One of the most powerful aspects of remote sensing is the ability to sense energy in wavelengths of the electromagnetic spectrum beyond those your eyes can see. The nonvisible areas of the spectrum can be visualized on a computer screen or by using specialized films.

Both analog and digital remote sensors can detect energy in the near-infrared portion of the spectrum, creating photos/images that are called color infrared composites. These photos/images allow you to "see" in the near-infrared portion of the spectrum when we substitute the reflectance of the near-infrared light into one of the primary colors that humans can see. When you use film or a digital sensor to create a natural color photo/image, you use the same wavelengths that your eyes see (blue, green, and red) in order to mimic your eyes. However, to create a color–infrared composite, you must eliminate one of the visible wavelengths and substitute the near infrared. You do this by eliminating the blue wavelength and then producing a photo/image that shows the green, red, and near-infrared wavelengths displayed with the infrared band in red, the red band in green, and the green band in blue.

In this way, wavelengths beyond what your eyes can see can be made visible to you for use in both manual and digital interpretation. However, because manual interpretation is restricted to the colors humans can see, manual analysis is restricted to three bands at a time, while digital analysis allows you to mine all of the imagery bands simultaneously. Additionally, film sensors are limited to only the visible, ultraviolet, and near-infrared wavelengths, while digital sensors can obtain imagery in many other wavelengths.

Variation in tone allows you to discern other image elements such as shape, size, texture, and pattern. Figure 9.1 shows several combinations of different Landsat bands, chosen to highlight different types of features on the ground. You can see that color (a measure of spectral response in each wavelength sensed) is an extremely important element of imagery interpretation and analysis.

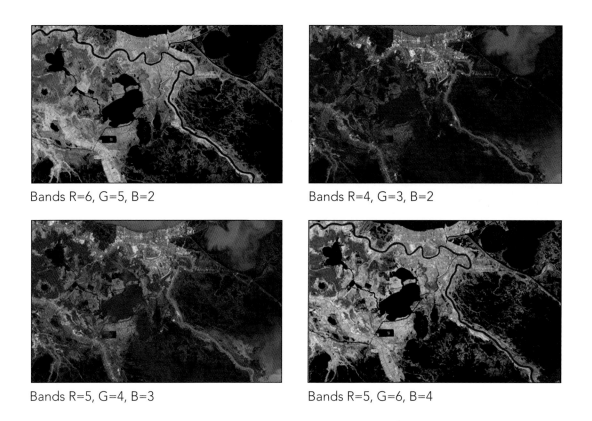

Figure 9.1. New Orleans, Louisiana, and the Mississippi River delta. Comparison of different combinations of Landsat 8 Operational Land Imager bands available as a dynamic service from Esri. (esri.url.com/IG91)

Shape

Shape refers to the form of the outline of an object and is one of the most powerful elements of interpretation, especially for objects that have a unique shape. For example, streams tend to be sinewy and utility lines straight. Three-dimensional shape is also important. Trees that are conical indicate conifers. Trees with more rounded crowns indicate hardwoods. It is common for objects built by humans to have regular shapes such as circles, squares, rectangles, etc., while objects in nature tend to be more amorphous. Pivot irrigation plots are round, buildings are often rectangular, and agricultural fields are usually angular. Perhaps the most used example of the power of shape for identifying objects created by humans are the pyramids in Egypt and the Pentagon near Washington, DC (figure 9.2).

Figure 9.2. The easily recognizable shape of the Pentagon in National Agricultural Imagery Program (NAIP) imagery

Size

Size is the extent of the object being identified. Size can mean the actual measurements of an object's dimensions or its size relative to other objects whose size is well known and fairly standardized (e.g., a car or a soccer field). Actual measurements can be made only if the scale of the photo or imagery is known. Size must, of course, be considered relative to the scale of the imagery.

Pattern

Pattern is derived from the spatial arrangement or configurations of objects. Human-created features tend to have regular patterns as well as shapes. For example, orchards and vineyards can be identified by the linear pattern of the trees or vines (figure 9.3). In addition to a regular pattern, objects may be spatially arranged in clusters, randomly, or in some other uniform or nonuniform way.

Figure 9.3. The distinctive pattern of vineyards on Sonoma County 6-inch infrared imagery

Shadow

Shadows exist when the ability of the sensor to capture reflectance or radiance of a visible feature on the ground is hindered by another feature. Hence, the object casts a shadow on the ground. This is significant because the illumination on the ground, in the shadow, is interfered with by the object casting the shadow. As a result, the spectral response for the shadowed feature is lower than that of a nonshadowed illuminated feature. Shadow is a useful element in interpretation because it can yield information about an object that is not apparent by looking at the object itself in an image. While most imagery is acquired with the constraint to minimize shadows by obtaining imagery while the sun is highest in the sky, objects that have height will still cast some shadow regardless of the sun angle. Shadows can hinder image interpretation and analysis because they obscure objects on the ground. However, the shape of a shadow can reveal the shape of the casting object. For example, figure 9.4 shows an image of oak trees. The species of the oak is indistinguishable from other oak species if you look only at the trees. But if you look at the shadows, you can see that the valley oak trees have spindly, complex shadows which let some light through, while the nearby live oak trees have round, compact shadows.

Figure 9.4. The unique shadow of valley oaks (*Quercus lobata*) versus those of coast live oaks (*Quercus agrifolia*)

The pattern of shadows can also provide information. For example, on a high-spatial-resolution image, a young field of corn will be uniform with few shadows, while a cornfield close to harvest will have a pattern of illumination and shadow caused by the individual corn plants. Likewise, on a moderate-spatial-resolution Landsat image, a regenerating stand of trees will be uniform with few shadows, while an old growth stand's illumination will be scattered and have dark pixels caused by the shadows of the large trees.

Texture

Texture is traditionally defined as the feel or appearance of the surface of an object. Is it soft or hard, smooth or rough? It is easiest to think about this element by imagining placing your hand on what you are seeing and then describing how it feels. Is it smooth or rough? Is it sharp or dull? Is it uniform or mottled? By imagining how objects in an image actually feel, you can better understand texture. For example, Christmas trees are conical in a regular pattern of rows. Now imagine taking your hand and pushing it down from above on the Christmas trees. You would cry out in pain because this would be analogous to pushing your hand down on a bunch of nails sticking up. The pointy conifer tops would hurt your hand as you pushed down on them. By contrast, pushing your hand down on a peach orchard would be much more pleasant. The orchard's deciduous fruit trees, with their billowy canopies, would feel soft to the touch.

However, in remote sensing you cannot "feel" the object you are attempting to map because you are distant from it and have only an image to look at or analyze. Instead, you must rely on changes in the image's appearance across space to assess texture. Texture is represented on an image by the repetition (or lack of repetition) and pattern of the tone, shadow, and color on the image. For example, figure 9.5 shows the textures of calm versus rough water in radar imagery. Calm water has little texture because the water is smooth, while rough water is highly textured because of the varying pattern caused by the wave crests and shadows. The variation in tone across space is used to determine image texture, which is usually measured as the standard deviation of the tone of a window of pixels, as shown in figure 9.6.

Texture is often used to distinguish urban from suburban areas. As seen in figure 9.7, suburban areas are highly textured because of the high tone variability resulting from the mix of houses, lawns, streets, and trees. Urban areas are uniformly bright or very dark (low variability = low texture) in all bands because they are made of stone, concrete, or steel, include large shadows, and usually have little vegetation. In contrast, vigorously growing agricultural crops tend to be uniformly bright in infrared bands.

Figure 9.5. The texture of calm versus rough water in radar imagery

Neighborhood sum (with equally weighted cells)

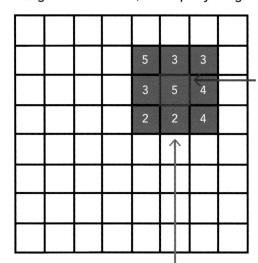

Processing cell.
The input value of the cell is 5. The output value is the standard deviation of the cell values of the 3 x 3 cell window surrounding the processing cell or, in this example, 1.30

Standard deviation is calculated as

$$\sigma = \sqrt{\frac{1}{N}\sum_{i=1}^{N}(\chi_i - \mu)^2}.$$

Figure 9.6. Many image texture algorithms replace the center cell value with the standard deviation of a window of adjacent pixels. This example uses a 3 × 3 rectangular pixel window, but other sizes and shapes can be used.

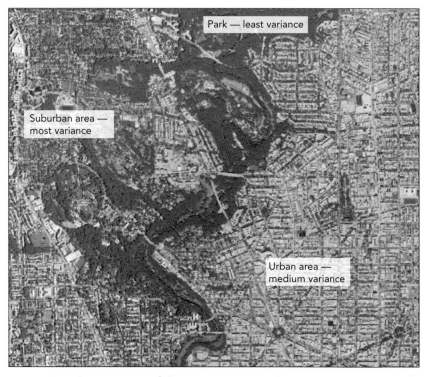

Figure 9.7. Comparison of the image textures (i.e., variance in spectral response across space) of parks and suburban and urban areas in NAIP 1-meter infrared imagery, Washington, DC

Location

Location is derived by registering the imagery to the ground so that each pixel has accurate x, y, and z coordinates. Historically, the location of an object of interest would be marked on a photo and the approximate location transferred to a map. However, given available GPS technology and photogrammetry tools available today, a much more precise location of any object on an image is now readily obtainable.

Location is useful because it allows the determination of relationships between features in the imagery and nonimagery variables (e.g., slope, aspect, distance to streams) that may affect the distribution of the feature or land cover/use classes to be mapped. Location is also helpful because it may contribute to knowledge specific to a certain area or region because certain species or cultural practices occur only in specific locations. For example, redwood trees grow in only limited areas from the central California coast to the southern Oregon coast, in the central Sierra Nevada Mountains, and in the Hubei province of China. So mapping redwoods in any other areas would be inaccurate. Various cultural practices may also occur in only certain places. For example, because the water table is so high in New Orleans, the cemeteries there consist of mausoleums to inter the dead above the ground.

Context

Context refers to the neighbors of an object. Context provides information about not only what surrounds the object of interest, but also the interaction between neighboring objects. It provides a clue to the identification of the object because a certain activity or phenomenon is typically encountered when certain objects are present together. For example, water surrounded by trees is more likely to be a lake than a swimming pool, which would more likely be surrounded by buildings. Also, if the lake is in the eastern US, those conifer trees around the lake are likely to be spruce or hemlocks because they thrive in wetter soils. A large building with ample parking adjacent to a golf course is most likely a country club. It is the combination of the golf course and the building together that helps identify the country club. Without the golf course, the building could have one of many different uses. While context has always existed in interpretation, more emphasis has been placed on it recently because GIS allows rigorous identification of what is surrounding an object. This spatial thinking has increased the use of context as a powerful element of image interpretation.

Height

Height is the distance between the highest and lowest points of an object. Like size, it can be absolute or relative. Relative height can be discerned using stereo imagery. Height can also be measured from stereo optical imagery or from lidar or radar imagery. Height is an important way to characterize an object and differentiate it from other types of objects. For example, object height is instrumental in distinguishing different vegetation classes, as shown in figure 9.8. While height has been an important element in image classification for decades, its use has increased greatly with the increasing availability of lidar data.

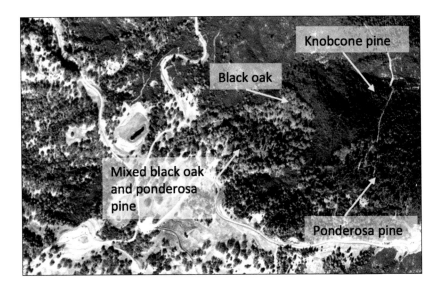

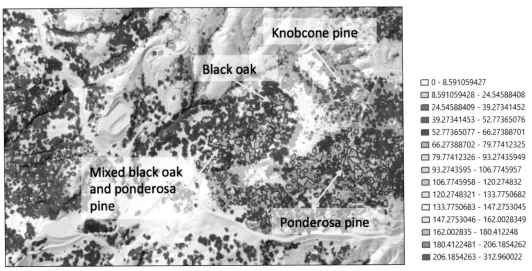

Figure 9.8. Vegetation height from lidar imagery used to distinguish different forest types

Date

The date when an image is taken can affect the tone of some objects, which can help in object identification. Historically, most maps have been created using a single date of imagery because the cost of imagery collection and acquisition was too high to allow for multiple date collects. Only very low spatial resolution imagery such as Advanced Very High Resolution Radiometer (AVHRR) data, which is free, could be used, but its poor spatial resolution was problematic. However, recently, the use of multitemporal image analysis has blossomed with NASA's launch of the Moderate Resolution Imaging Spectroradiometer (MODIS) imagery collected daily, USDA's provision of National Agricultural Imagery Program (NAIP) imagery every two to three years, the rapid adoption of drone imagery, the increasing constellation of high-resolution satellite imagery, the launch of the European Sentinel 2 A and B systems with 5 day revisit, and especially the USGS policy change in 2008 that made all Landsat imagery accessible on the web and free. Multitemporal imagery is useful in many thematic mapping projects and especially so for agricultural applications where crop changes occur frequently.

Figure 9.9 compares images of the same hardwood forest in California taken in the spring and the fall. In the spring image, all the trees are green and it is difficult to distinguish different oak species. But the fall image, taken when the deciduous leaves have turned color, allows for the identification of evergreen oaks versus deciduous oaks, and deciduous oak species from one another.

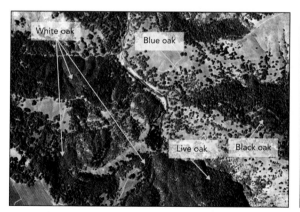
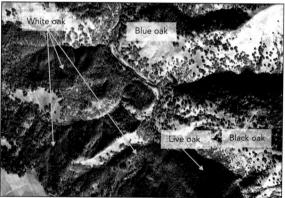

Figure 9.9. Comparison of spring versus fall imagery of forests in California. White oak, black oak, and live oak are indistinguishable in the spring imagery because they are all the same shade of green. In the fall imagery, the black oak is yellow, the white oak is brown, and the live oak is green. Blue oak looks similar to white oak in the fall imagery, but it is blue in the spring imagery compared to the white oak's green color.

Image dates can also be helpful in identifying crop types because some crop types are only grown at certain times of the year. For example, in the lower Colorado River region of the southwestern United States, crops can be grown all year round. However, a crop calendar for the region would tell you that lettuces are grown only in the winter when it is cooler, and corn is grown only in the hot summer months.

With the advent of freely available multitemporal imagery (e.g., MODIS, Sentinel, and Landsat), multitemporal analysis has become a useful tool in mapping. Looking at a time series of imagery can reveal patterns of change, as shown in figure 9.10, which shows Mount St. Helens before and after its May 1980 eruption. Multitemporal analysis is also commonly used to monitor crop health and predict crop yield. Many of the small satellite companies are targeting multitemporal analysis for predicting changing market conditions.

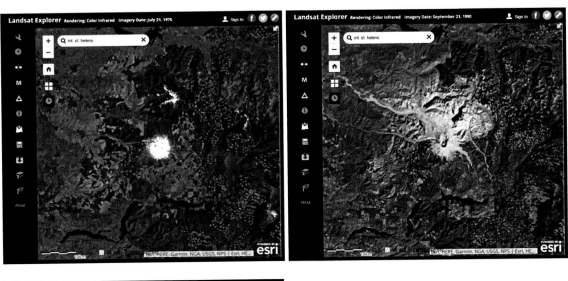

Figure 9.10. Multitemporal imagery of Mount St. Helens before (1975) and after (1990) the May 1980 eruption shown in Esri's Landsat Explorer dynamic image viewer. The image on the right is an image of the difference of NDVI values of the 1975 and 1990 images. Shades of green indicate vegetation regrowth in areas outside the blast zone. Shades of magenta indicate vegetation loss within the blast zone as well as forest harvesting outside of the zone. (esri.url.com/IG910)

Summary: The Confluence of Evidence

Whether it is manual or automated, the process of creating a map from imagery relies upon the *confluence of evidence*. While each element is influential in its own right, it is the combination of all the elements that gives remote sensing its power in mapping features, elevations, or land use/cover. As Spurr stated in 1948 (pg. 182), "Important though each incidental pictorial quality may be, it is the sum total of these qualities which gives an object its characteristic appearance." Raben also elaborated on this concept in 1960 (pg. 109) when describing photo interpretation: "Identifying objects in aerial photographs by direct recognition is a fairly simple process. Either the interpreter knows what the objects are because he has seen them before, or he does not know. . . . In order to identify the objects he has not seen before . . . the photo interpreter exploits the principle of convergence of evidence." Just as a detective compiles a series of clues that leads to a conclusion, the interpreter or automated algorithm chooses the image elements that are most predictive of the objects being mapped. The more elements that point to the same answer (i.e., confluence), the more confidence that the label is correct. For example, the important elements used to interpret a Christmas tree farm would be

1. Shape. The trees will be conical and may be viewed directly or seen from shadows cast by the trees.
2. Pattern. Unlike a natural forest, the trees will be in a regular pattern in lines and approximately equally spaced apart.
3. Color. The trees will be dark green when viewed in true color.
4. Texture. Christmas trees have pointy tops and are arranged in rows, usually with herbaceous vegetation between them which will cause the texture to be high.
5. Height. The trees should not be very tall (perhaps 3 to 12 feet).

From these five elements, you have sufficient evidence (i.e., confluence) that the object is a Christmas tree farm. Some other elements may also corroborate your conclusion. Location could also help because Christmas trees are grown only in certain regions of the world and only in specific areas within those regions. Context may also be useful because stands of Christmas trees often occur near farmhouses, so the proximity of a farmhouse would help validate that Christmas trees are present.

Derivative Bands

In addition to the original bands of imagery that are recorded by a sensor (e.g., blue, green, red, and near-infrared [NIR] for NAIP imagery or blue, green, red, NIR, middle infrared [MIR]1, MIR2, and thermal infrared [TIR] for Landsat Thematic Mapper), it is possible to calculate a number of additional bands from the original bands. The additional bands are called derivative bands because they are derived from the original or raw data. Derivative bands include simple ratios, transformations, or indices that are created to reveal or enhance the link between the variation in the image and the variation on the ground. Some of these derivative bands have been developed from theoretical knowledge of the physical and chemical properties of the objects being sensed. Others have been empirically derived from observation. In either situation, the goal is to tease more information from the imagery as it relates to what is happening on the ground to produce a more accurate map.

Ratios

One type of derivative band is a simple ratio. In this case, one original band is divided by another to create a new derivative band. For example, a common simple ratio is derived by dividing the infrared band by the red band (e.g., for Landsat TM that would be band 4/band 3). The reason that this derivative band is useful is that it provides more insight about vegetation. Because healthy vegetation absorbs red light for use in photosynthesis and reflects NIR light, by taking the ratio a single band can combine two original bands and provide additional information about vegetation health. This ratio is an example of using knowledge about the physical properties of vegetation to create a derivative band. The ratio of a red band to an NIR band (e.g., for Landsat TM that would be band 3/band 4) has empirically been found to help separate water from dense vegetation on steep slopes. In ArcGIS, ratio bands can typically be created using the math raster functions, but can also be created using map algebra inside the raster calculator or via Python/ModelBuilder.

Transformations

Transformations are more complex than simple ratios and result in the same number of derivative bands as there were original bands. However, the new derivative bands have been changed or transformed in some unique and valuable way. Two very common transformations are performed in digital image analysis. They are Principal Components Analysis (PCA) and the Tasseled-Cap Transformation (TCA).

Principal Components Analysis

PCA is performed from a statistical basis of data reduction by transforming the imagery into independent derivative bands in which the variance of the imagery is maximized into the first few principal components. PCA is a transformation used to remove the redundancy between the original bands and create independent or orthogonal transformed bands. Because the original image bands are so highly correlated, creating PCA bands has the advantage of creating new independent transformed bands and potentially reducing the number of bands needed. Band reduction occurs because when creating independent transformed bands, the majority of the variance in all the original bands is now represented by the first principal component. For Landsat TM imagery, it is common for up to 80 percent of the variance of the seven original bands to be represented in the first principal component band. Also, 90 percent to 95 percent of the original seven-band image variance is represented by the first three principal components. Historically, reducing the number of bands to analyze was important due to computer processing speed. Today, this is less of an issue but there is still power in using some of the principal components in the thematic mapping process. Figure 9.11 shows how the PCA works in a simple two- band case that is easy to visualize. On the left is a bispectral plot of band x versus band y. The digital number (DN) values for the imagery are simply plotted here. The hatch marks show the variance or range of values in both the x and y directions. Now look at the plot on the right, which shows the original bispectral plot, but also a plot that has been transformed. The direction of the x axis has been mathematically transformed to maximize the image variance. The y axis is orthogonal (at 90 degrees) to the x axis. These are the principal components. The first principal component is now transformed to represent the most variance possible in the imagery. This figure is only a graphic representation in two dimensions. In reality, PCA is done mathematically and in as many dimensions as there are bands. However, this figure allows one to easily understand the process.

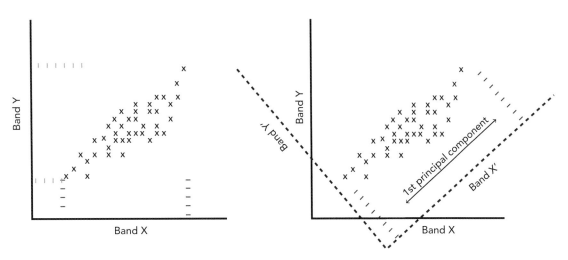

Figure 9.11. An example of principal components analysis graphically demonstrated for a two-band image

Tasseled-Cap Transformation

The tasseled-cap transformation (TCT) is an empirically based transformation of the imagery that has been observed to provide information about the brightness, greenness, and wetness of the imagery. It is not a data reduction, but instead provides additional information about the imagery. The process is similar to PCA in that the transform uses linear combinations of the original bands. However, the transform is sensor specific with the transform matrix being derived from the specific sensor (e.g., Landsat MSS, Landsat TM 5, SPOT-4, etc.) and the data from each image being processed through the transform matrix to provide the result. TCT is also called the Kauth–Thomas transform after the two scientists who first proposed this transform for Landsat Multispectral Scanner (MSS) imagery (Kauth and Thomas, 1976). Landsat MSS had four bands (G, R, NIR1, and NIR2), so the original transform had four derivative bands called soil brightness, vegetation greenness, and two others. Crist and Kauth (1986) expanded this transform for Landsat TM imagery to produce seven transformed bands. The first three transforms are the important ones, showing brightness, greenness, and wetness (figure 9.12). The transform is called tasseled-cap because a plot of brightness versus greenness reveals the shape of a cap with a tassel on the top. Many current projects use TCT to provide extra information about the linkage between variation on the imagery and the ground; they include the USGS National Land Cover Data project and the NOAA Coastal Change Analysis Program (C-CAP). Both of these programs use Landsat imagery to map large portions of the United States.

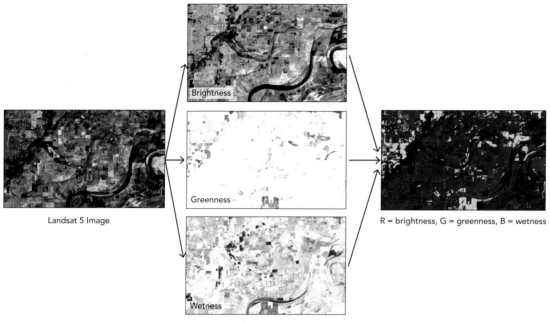

Figure 9.12. Example of a tassel cap transformation

In ArcGIS, the TCT is available as a raster function for Landsat TM/ETM, IKONOS, and QuickBird. (Raster functions are discussed in chapters 5 and 12). The TCT can be applied by directly using the raster function. Other transforms can be applied by using the spatial matrix raster function or using raster algebra in the raster calculator (Spatial Analyst). Raster algebra can also be developed and performed to execute transformations in ArcGIS using ModelBuilder or Python.

Indices

The third and final type of derivative band that can be created are indices. Perhaps the most common indices are vegetation indices. However, other indices provide information about other physical and chemical properties; for example, geologic applications. Vegetation indices reveal additional information about vegetation status, condition, and health. There are a large number of different variations of vegetation indices, all with slightly different purposes. All of these indices are some form of a ratio of original bands, sometimes with other factors or coefficients included. The most common vegetation index is the Normalized Difference Vegetation Index (NDVI). Figure 9.13 compares a Landsat 8 true color image to a NDVI image created using Esri's dynamic services. The equation for a NDVI is

$$NDVI = (NIR\ band - red\ band)/(NIR\ band + red\ band). \tag{1}$$

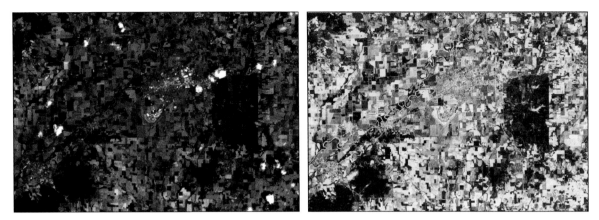

True color

Colorized NDVI created using Esri's dynamic Landsat services

Figure 9.13. Comparison of a Landsat 8 true color image to a colorized Normalized Difference Vegetation Index (NDVI) image of southern Indiana. Dark green is thick, vigorous vegetation; yellow represents fallow fields; and brown represents sparse vegetation or clouds.

Consideration of this equation reveals that it is a stronger version of the simple ratio of the NIR band divided by the red band described earlier. The benefit of normalizing an index is that it constrains the values it can attain from −1 to 1 instead of the 0-to-infinity values attained by a simple ratio. Again, this derivative band is very useful in determining vegetation health because healthy vegetation will reflect NIR light and absorb red light. By subtracting the red from the NIR in the top of this index and adding them together in the bottom, the ratio reveals a more definitive pattern of vegetation health. In addition, the ratio has the effect of normalizing the result for sun angle, topographic effects, and some atmospheric effects (Jensen, 2016).

Other vegetation indices include the following:
- The normalized difference moisture index, which uses the NIR and short-wave infrared (SWIR) bands to account for plant canopy moisture content.
- The soil-adjusted vegetation index (SAVI), which includes an additional factor to the NIR and red band ratio that accounts for the amount of soil visible as opposed to vegetation in the image.
- The enhanced vegetation index (EVI), which was originally developed for use with MODIS imagery and includes the use of the soil adjustment factor as in SAVI, with other coefficients modifying the red and blue bands used in the ratio with the NIR band. EVI is especially useful in areas of high biomass where NDVI gets saturated and does not work as well.

- The normalized burn ratio, which is used to map the severity of areas where the vegetation has burned. The index is a ratio of NIR and SWIR bands and is often used to evaluate pre- and postburn areas.

There are many other indices including more vegetation indices that exist and new ones are being developed every day.

Most image analysis software, including ArcGIS, allows users to program into the software whatever ratio, transform, or index they wish to use. In this way, the analyst can experiment with the latest ideas and results gained from a review of the current literature. In ArcGIS, NDVI is available as a mosaic dataset raster function (raster functions are discussed in chapters 5 and 12). NDVI can also be applied to a raster on-the-fly using the image analysis window in ArcGIS Desktop. All the indices discussed here, including custom indices, can be created in ArcGIS Online using the NDVI raster functions or raster algebra functions. Raster algebra is also accessible through the raster calculator in ArcGIS Desktop and ArcGIS Pro and via ModelBuilder and Python. Finally, NDVI is included as a defined function in many image services including the NAIP and Landsat 8 services available in ArcGIS Online.

Tools for Linking Variation in the Imagery to Variation on the Ground (Data Exploration)

The previous section of this chapter described many methods for creating derivative bands from the original imagery to potentially enhance the link between variations in the imagery and on the ground. These derivative bands can prove quite useful in improving the accuracy of the thematic map created when using these bands. However, the question remains, which of these original or derivative bands are the most predictive of the objects on the ground and will, therefore, result in the most accurate map? The following section discusses some tools that are available to conduct the data exploration necessary to evaluate these bands and select the best ones for the analysis. These tools include

- spectral pattern analysis,
- bispectral plots,
- feature space analysis, and
- divergence analysis.

Together, these tools provide insight to the analyst about selecting the bands for mapping that provide the best linkages between the variation in the imagery and the variation on the ground.

Spectral Pattern Analysis

Spectral pattern analysis (SPA) is a simple, yet powerful tool for evaluating which bands provide the most linkage between the imagery and the ground. To proceed, the objectives of the mapping, including a well-defined classification scheme, must be determined. Then, sample areas for each of the map classes (often called training sites or samples) must be acquired. SPA is an x,y plot of the mean value for each map class for each band in the imagery (original and derivative bands). The bands are on the x axis and the reflectance values (DNs) are on the y axis. Figure 3.1 (in chapter 3) shows a theoretical spectral pattern for some different land-cover types. Notice that the lines are curved, indicating that many wavelengths of electromagnetic energy were sampled. This type of result resembles that of chemical spectroscopy done in a laboratory or if hyperspectral imagery was acquired.

More commonly, the spectral pattern looks like figure 9.14. Note that the lines connecting the bands are now straight. This is an SPA of some Landsat TM imagery without any derivative bands. Because there are only six bands (the thermal band was excluded), the plot is not smooth; straight lines connect the samples. Looking at the SPA shows two boxes indicating the bands (4 and 7) that seem to provide the most separability between the different map classes. Therefore, these bands should definitely be used in the thematic mapping. Analyzing the spectral pattern shows the relationship between the map classes (the ground) and how they differ on the imagery. Some map classes will be easy to tell apart. Others may require creating derivative bands to aid in their discrimination. The SPA aids the analyst in determining the usefulness of each band in creating a thematic map of the area of interest. If additional derivative bands are created, they can easily be added to the SPA, which will then reveal whether those bands help separate the map classes and produce an accurate thematic map. In fact, in very simple classification schemes, SPA could be used to actually classify the land-cover types. For example, just looking at the NIR band could allow the analyst to separate water (very low NIR reflectance) from everything else in the imagery, resulting in a water versus nonwater map.

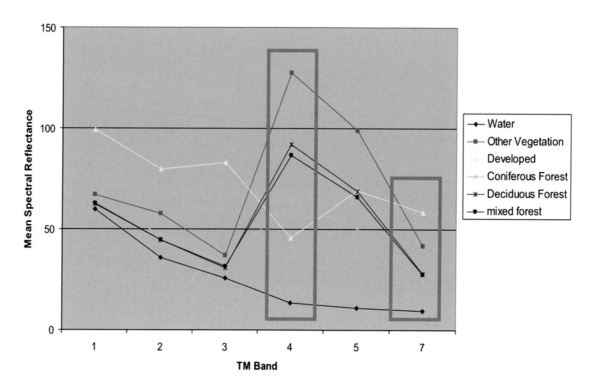

Figure 9.14. An example of spectral pattern analysis for a Landsat TM image

Incorporating multiple dates of imagery into the SPA can also help identify features. Figure 9.15 shows a simple example of using a year's worth of imagery and plotting the values of NDVI for each of three vegetation types: crops, pasture, and orchards. Instead of different bands on the x axis, there are days of the year. All three types have high NDVI values, but the pattern of these values differs throughout the year, allowing the analyst to be able to tell the vegetation types apart multitemporally.

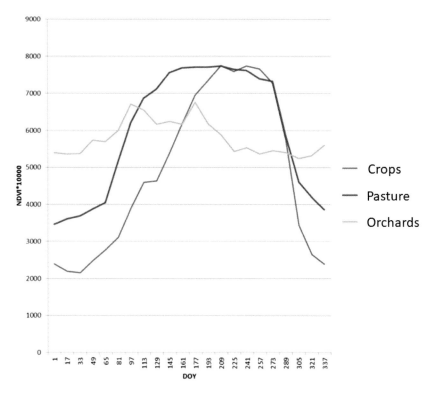

Figure 9.15. A multitemporal plot (esri.url.com/IG915)

This approach has become quite common and relies on a multitemporal stack or cube of imagery to produce effective results. In the last few years, this technique has been used extensively for many types of land-cover maps and is especially useful when mapping agricultural crops whose appearance heavily depends on the date of the year (phenology) (Wu et al. 2014). Applying many processors simultaneously to analyze this multitemporal image cube has greatly facilitated its use.

Bispectral Plots

Bispectral plots (BSPs) are another graphic tool for evaluating the linkage between the variation in the image and the ground. As the name suggests, two bands are plotted simultaneously, with one band on the x axis and the other on the y axis. Figure 9.16 is an example of a BSP. As in SPA, the reflectance or DN values are plotted for each map class. Careful inspection of the plot reveals interesting information about the map classes. This BSP is for Landsat TM imagery and plots band 4 (NIR) on the x axis and band 7 (a MIR band) on the y axis. Note that the water samples are near the 0,0 point on the plot. This result makes sense, because water absorbs both NIR and MIR energy. Also, note the circled map classes (mixed forest and deciduous). Their close proximity on the plot indicates that they may be difficult to tell apart in the classification process. Note that the conifers have lower NIR reflectance than the deciduous trees, so they can be separated in the classification process. It will be difficult to distinguish them only in areas where they form a mixed stand. The BSP allows the analyst to further explore the relationships in the data.

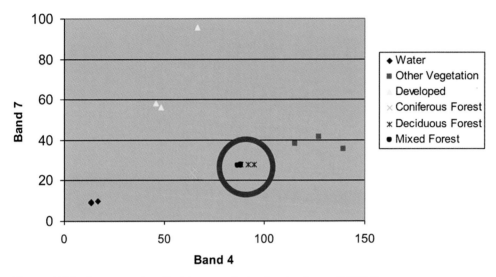

Figure 9.16. An example of a bispectral plot (esri.url.com/IG916)

While BSP analysis is quite useful, this type of analysis is not limited to just two dimensions. A trispectral plot is also possible and easy to visualize. Beyond 3D is also possible. Imagine an N-spectral plot. A computer can easily process this type of analysis in as many dimensions as needed. However, for human analysts, the BSP is the most common approach to explore the relationships between bands in the imagery.

Feature Space Analysis

Feature space analysis is a continuation of bispectral plots. Instead of a sample of data being plotted, the entire image dataset is used. Figure 9.17 shows a typical example of a feature space plot. Because the entire image is plotted here, additional insight into the information in the entire image can be obtained, and the degree of between-band correlation is revealed. The full dynamic range of the imagery is also revealed, showing areas where the image values are concentrated, but also showing the extremes.

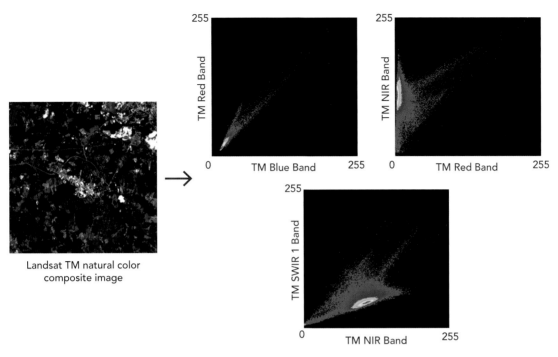

Figure 9.17. An example of a feature space plot for a Landsat TM image

Divergence Analysis

Divergence analysis (DA) is another method of determining which bands to use for creating the best thematic map for a given mapping project, and therefore exploring the linkage between the imagery and the ground. Unlike the three methods described above, DA is not a graphic technique. Rather, it is a statistical technique that is computed from the statistics (mean and covariance matrices) available from the map class samples (training samples) selected as part of the analysis. All bands used in selecting the map class samples, including any derivative bands, are included in the analysis. The analyst specifies the number of bands (n) to be used in the classification process, and the DA looks at every possible combination of those n bands and chooses the one combination that will provide the best separability of the map classes. Different equations/measures can be used to compute divergence. These include divergence, transformed divergence, Bhattacharyya distance, and Jeffreys–Matusita distance. Each of these different measures has advantages and disadvantages. Regardless of which divergence measure or measures are selected, the results of this analysis provide additional input into which bands of imagery (raw or derivative or both) will provide the best separability of map classes, and therefore the most accurate thematic map.

Introducing Other Sources of Geospatial Data to Capture Variation on the Ground

Imagery serves as a key data source for feature extraction and land use and land cover mapping. However, sometimes imagery doesn't capture all of the variation on the ground or there isn't a strong correlation between the ground and what is imaged by the sensor. In these cases, other sources of geospatial data may help characterize the landscape and improve map accuracy. Before GIS, photo interpreters usually had a topographic map on hand to aid in image classification. By looking back and forth between the imagery and the map, the interpreter could glean information on slope, aspect, elevation, and location. GIS now allows multiple coregistered datasets to be simultaneously available on a digital map during image interpretation. In addition, typical modern semiautomated mapping projects use machine learning (discussed in chapter 10) to predict map classes based on imagery and many other geospatial datasets (known collectively as independent variables). This section discusses the types and data sources of commonly used predictor variables that are not imagery or derived from imagery.

The most effective independent variables for mapping are empirical datasets that represent or measure some physical process or landscape characteristic, such as elevation, rainfall, or spectral reflectance (the remainder of this chapter discusses many more).

Data Sources

Elevation and Landscape

Elevation and its many derivatives (e.g., slope, aspect) are among the most critical geospatial data layers (second to imagery in their importance) for thematic mapping of land use, land cover, and vegetation. Elevation and its derivatives are often highly correlated with vegetation type and serve as an indispensable suite of predictor variables in land-cover mapping. Elevation data is publicly available for all areas of the United States as raster digital elevation models (DEMs) from the National Elevation Dataset at a pixel resolution of 10 meters. The creation of DEMs from imagery is discussed in chapter 8.

With the proliferation of lidar data during the past decade, many areas of the United States and Europe now have lidar-derived elevation data of much higher accuracy and resolution. Figure 9.18 illustrates the increased accuracy and resolution of lidar versus nonlidar elevation data. See chapter 4 for a discussion on publicly available portals for acquiring lidar data.

This section provides an overview of some of the more commonly used derivatives of elevation data, including slope and aspect, hillshades, solar insolation, and flow accumulation.

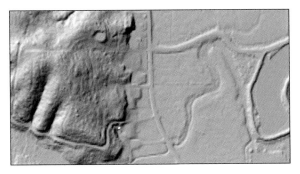
Lidar-Derived Shaded Relief

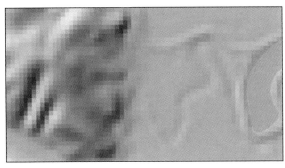
Shaded Relief Derived from Existing 10m USGS DEM

Figure 9.18. Comparison of elevation data created from a lidar versus a nonlidar source

Elevation

DEMs, in and of themselves and without any further processing, are very useful for visualization and analysis in thematic mapping projects and as predictor variables in machine learning, because elevation is often highly correlated with vegetation species occurrence.

Three categories of DEMs derive from lidar data: digital terrain models (DTMs), digital surface models (DSMs), and digital height models (DHMs). To recap information in chapter 8, DTMs depict the ground, DSMs depict the highest surface (the vegetation canopy and the tops of features such as buildings), and DHMs depict the heights of features such as trees. All these types of DEMs serve as key data sources for thematic mapping.

Figure 9.19 shows an example of a bare-earth DEM used to help find the extent of vernal pools in the Santa Rosa Plain of Sonoma County, California. Vernal pools are areas of wetland grasses that grow in topographic depressions over clayey soils where water pools during the rainy season. Imagery alone is helpful for mapping vernal pools, but the bare-earth DEM—by revealing subtle sinks in the landscape—improves an analyst's ability to identify and delineate these sensitive wetland habitats. Figure 9.19 shows the topographic depressions in the darker tones (delineated by light blue lines); the lighter tones represent areas of upland grasses between the vernal pools.

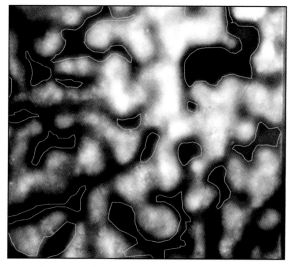 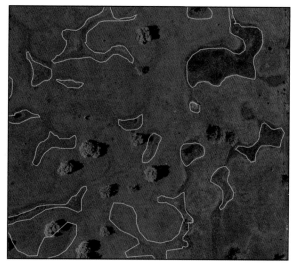

Vernal Pools (Blue Outlines) with Bare-Earth DEM   Vernal Pools (Blue Outlines) with 6-inch Imagery

Figure 9.19. Vernal pools in Sonoma County, CA, revealed using lidar derived DEMs. Vernal pools occur in areas with Mediterranean climate conditions. The pools are depressions in the landscape that temporarily fill with water during the rainy season, creating seasonal wetlands. During the dry season, the water in the pools evaporates as shown in the imagery on right side of the figure. Source: Sonoma County Agriculture Preservation and Open Space District

Slope and Aspect

Land cover is also often highly correlated with slope and aspect. Slope represents the rate of change of elevation (rise over run) for each DEM cell. Slope can be calculated using the ArcGIS Spatial Analyst Slope tool, which returns a raster with pixel values that represent the inclination of slope in either degrees or percent slope. Aspect identifies the downslope direction of the maximum rate of change in value from each cell to its neighbors. The ArcGIS Aspect tool returns a raster with pixel values representing the direction (typically in degrees) that the pixel faces. Flat pixels have pixel values of −1.

Slope and aspect rasters can be reclassified into ranges and combined into a single layer that represents both slope and aspect. For example, using the ArcGIS Reclass tool, slope could be reclassified from its integer values of percent slope into four classes: gentle, moderate, steep, and very steep. Aspect could be classified from its range of 0 to 360 degrees into eight directions: north, south, east, west, northeast, southeast, southwest, and northwest. The two reclassified rasters could be combined into a single hybrid of slope and aspect that is very useful for visualization and as a data source/predictor variable for thematic mapping. Figure 9.20 shows a slope–aspect raster.

Figure 9.20. Combined and reclassified slope and aspect

Hillshades

Hillshades show the landscape with various degrees of illumination and shadowing based on a user-defined sun location. With lidar's high positional accuracy and high resolution, lidar-derived hillshades are an excellent reference data source to assist in mapping features such as roads, archeological areas, earthquake faults, stream channels, and deformations caused by landslides or other mass wasting events. Hillshades are also very useful in manual image interpretation of vegetation types because they provide an instantaneous and easily understandable depiction of the relative slope, aspect, and elevation of any location.

Because lidar data penetrates the tree canopy, features under the canopy are revealed by the lidar hillshade, even though those features may be completely obscured by forest canopy. The upper image of figure 9.21 shows a high-resolution orthoimage with a road covered by the tree canopy, whereas in the lower image the road is clearly visible in the lidar-derived hillshade. The road is depicted by dashed yellow lines in both the imagery and the hillshade.

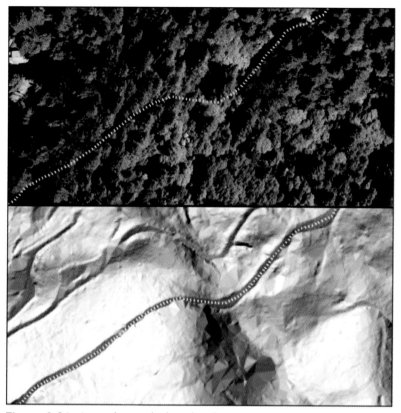

Figure 9.21. A road revealed under the tree canopy in a bare-earth, lidar-derived hillshade in Sonoma County, CA

Solar Insolation

Solar insolation represents the amount of solar radiation received by an area over some user-defined period. Solar insolation is calculated in ArcGIS using the Solar Radiation toolbox in Spatial Analyst. The Area Solar Radiation tool takes a bare-earth DEM as its input and returns a raster with pixel values that represent solar radiation over a time period.

Since the amount of available sunlight is correlated with vegetation type, solar radiation is a useful data source for thematic mapping and is often employed as a predictor variable in machine learning for land-cover and vegetation mapping.

Hydrology

Flow Accumulation

Flow accumulation represents the upstream catchment area for given pixel on the landscape. Flow accumulation is derived from a hydroenforced DEM. Hydroenforcement of a DEM imparts the true elevations of culverts, pipelines, and other buried passages for water into a DEM, creating a DEM suitable for modeling the flow of surface water.

In a flow accumulation raster, pixel values represent the number of pixels that flow into a given pixel, essentially providing a value for an upstream catchment area. Pixels with high flow accumulation represent areas of flow concentration (such as streams).

Flow accumulation is a prerequisite for creating many other layers that are important data sources for thematic mapping. Layers that require flow accumulation for their creation include stream centerlines (discussed below), vertical height above a river (discussed below), and other hydrologic terrain derivatives such as watershed boundaries. Flow accumulation rasters are created in ArcGIS by a series of Spatial Analyst functions that include Fill, Flow Direction, and Flow Accumulation.

Stream Centerlines

One of the uses of flow accumulation is to generate stream centerlines and flow networks. Stream centerlines and flow networks exist for the entire United States and are publicly available for download by way of the National Hydrography Dataset (NHD). NHD flowlines are typically derived from the 7.5-minute series of US Geological Service (USGS) topographic maps; as such, NHD features are used appropriately up to 1: 24,000 scale. One of the most useful elements of the NHD is that it is a geometric network, making it useful for all manner of hydrologic analysis and modeling.

Because of lidar's high positional accuracy, high spatial resolution, and ability to depict the ground surface even when it's obscured by canopy, lidar-derived stream centerlines are typically significantly more positionally accurate and spatially precise than traditional NHD centerlines.

Stream centerlines are useful as reference data sources for thematic mapping as well as for modeling the flow of surface water (if they exist as part of a geometric network). Figure 9.22 shows an example of lidar-derived stream centerlines (the blue lines in the figure) for an area of Sonoma County.

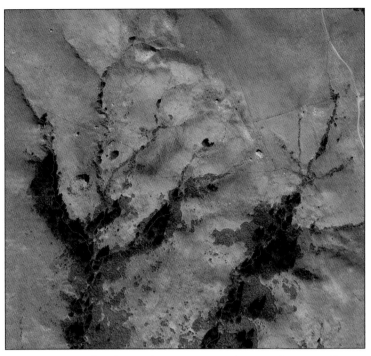

Figure 9.22. Lidar-derived stream centerlines in Sonoma County

Hydrologic Data Sources

Many terrain-based hydrologic variables and indices serve as useful data sources for thematic mapping. Such variables often rely on accurate flow accumulation models (discussed above), which are derived directly from preferably hydroenforced, lidar-derived, bare-earth DEMs.

Proximity to water, in both horizontal and vertical directions, is often highly correlated with vegetation type. Horizontal and vertical distances from water are useful data sources for vegetation mapping and are especially useful for accurately discriminating upland vegetation from wetland and riparian vegetation. Horizontal distance to water can be derived by creating a raster that represents the distance to the nearest water surface and/or a stream centerline. Vertical distance from water, also known as height above river (HAR) or height above channel, is a raster surface that represents the vertical distance above the nearest point on a stream (Dilts et al., 2010). A pixel in the channel (at the thalweg or centerline) would have a HAR of 0; a pixel on a side slope that was 15 meters above the thalweg vertically would have a HAR of 15 meters. Figure 9.23 shows a HAR from Sonoma County. Creating a HAR raster can be done using the Riparian Topography toolbox developed by Tom Dilts at the University of Nevada, Reno.

In addition to proximity to water, many other terrain-based hydrologic data sources can aid in thematic mapping. These include datasets that represent modeled floodplain extents and indices such as the topographic wetness index.

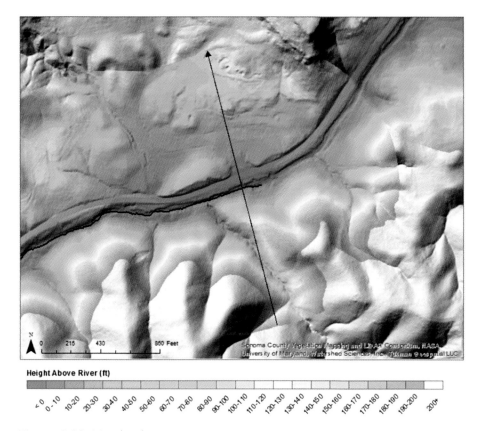

Figure 9.23. Height above river

Precipitation and Temperature

Precipitation and temperature are highly correlated with vegetation, very useful data sources in vegetation mapping, and often effective predictor variables in machine learning for vegetation mapping. In the United States, the National Center for Atmospheric Research (NCAR) provides publicly available nationwide precipitation data for many climatic metrics. Datasets include average annual precipitation and min/max/mean temperatures. NCAR distributes both long-term and short-term climate data in raster format. Note that the spatial resolution of many climate, precipitation, and temperature datasets is low and may not be applicable for thematic mapping projects.

A Case Study: Fog in Sonoma County

In areas of the country where marine fog influences climate, the amount of fog can be correlated with vegetation type and may serve as an important data source and predictor variable for vegetation. Sonoma County in Northern California has a Mediterranean climate with rain during the winter and rainless long summers. However, fog spilling off the relatively cold waters of the Pacific Ocean provides some summer moisture. In these areas, fog condenses on vegetation and drips onto the ground, providing measurable summer precipitation. Some areas, especially those adjacent to the coast, experience much more fog than the drier, more inland regions of the county. These fog-prone areas host vegetation that doesn't tolerate the heat and dryness of the less fog-prone areas, which support more xeric vegetation. The occurrence of fog is very localized; some very foggy areas occur very close to areas that are relatively fog free.

The authors of this book used fog as one of hundreds of predictor variables for a vegetation mapping project that resulted in a fine-scale vegetation and habitat map of Sonoma County. The fog data (Baldocchi and Waller, 2014) represented the average frequency of fog during the summer months of a ten-year period. The fog frequency information was obtained by classifying the land surface reflectance data from MODIS, which provides twice-a-day images of the United States. The resulting summer fog frequency raster proved to be highly correlated with vegetation type, and it improved vegetation map results, especially for fog-adapted species like the coast redwood (*Sequoia sempervirens*).

Soils

Soils data is available for the United States from the USDA Natural Resources Conservation Service (NRCS), typically at 1:12,000 scale. Soil type can be a useful data source for thematic mapping and is often highly correlated with vegetation and land cover. However, because soils are so difficult to map on a landscape scale, NRCS data often has limited utility for land-cover and vegetation mapping, especially at scales larger than 1:12,000. Because of the inherent difficulty of mapping soil types accurately (especially when the ground is obscured by vegetation), it is generally not advisable to rely heavily on soils data as a data source for thematic mapping. However, exceptions occur where specific soil types are accurately mapped and very highly correlated with vegetation. An example of this is serpentine-derived soils in California, which, because of their very low nutrient content and toxicity for some plants, host a unique array of specially adapted, slow-growing plant species.

Forest Canopy Metrics (Derived from Lidar)

As was discussed in chapter 4, one of the reasons that lidar is such a powerful tool for vegetation and land-cover mapping is that its multiple returns provide a 3D depiction of the forest canopy. The complex and data-heavy forest structure information of the raw point cloud can be distilled into many forest structure metrics, which are often highly correlated with vegetation type and are useful as data sources for thematic mapping, especially as independent variables in vegetation mapping projects.

The simplest forest structure metrics—canopy height and canopy cover—are the most common and simplest to derive from the point cloud. Other very useful lidar metrics include canopy volume profiles, which provide the percentage of lidar returns for a given area of land (e.g., a 20 × 20-meter raster cell) in each user-defined vertical stratum above the ground (e.g., 0–5 meters, 5–20 meters, 20–40 meters, 40+ meters).

Percentile height metrics, another more advanced type of lidar-derived forest structure metric, produce rasters with values that represent height at a user-selected percentile. For example, the 75th percentile height value for a 20 × 20-meter pixel would represent the height that 75 percent of the lidar returns in the vertical area above the pixel fell below.

Wildfire History

The presence or absence of wildfire on the landscape is an important data source for thematic mapping, especially in arid regions like the western United States. Many government agencies provide fire data. For example, the state of California's Fire Resource Assessment Program (FRAP) maintains an inventory of fire perimeters going back to the early twentieth century. Since the occurrence of fire is often correlated with vegetation type, this data layer is very useful as a predictor variable for vegetation mapping, as well as a reference map layer for thematic mapping in general. For each fire perimeter, FRAP's data contains an attribute for the year of the fire, which is very useful for classifying the perimeters into periods of occurrence. Figure 9.24 shows an area of the statewide fire history layer, classified into 20-year periods, for Sonoma County.

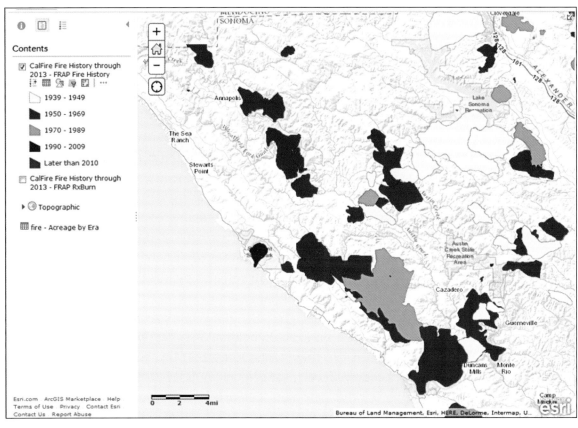

Figure 9.24. Sonoma County fire history in 20-year periods. http://frap.fire.ca.gov/data/frapgisdata-sw-fireperimeters_download. http://frap.fire.ca.gov/projects/fire_data/fire_perimeters_methods (esri.url.com/IG924)

Summary—Practical Considerations

This chapter first introduced the concept of image elements and how the confluence of evidence derived from a combination of elements allows you to convert imagery data into a map. Even though imagery products have greatly evolved over the last six decades, the elements have remained essentially the same as those listed by Spurr in the 1940s (Spurr, 1948).

You also learned the different methods for exploring imagery and discovering how the image varies (or does not vary) with the classification scheme. Many of these methods were first introduced during the severely disk- and memory-constrained 1970s and '80s to reduce the number of bands to be classified to only those discovered to be the most predictive of classes in the classification scheme. However, those techniques continue to be useful today because they clearly identify how the imagery varies with the classification scheme, and they allow for the detection of map classes that may be confused with one another.

Once possible map class confusion is detected, the analyst can bring in other ancillary datasets such as DEMs or soil maps, which may predict map class occurrence more effectively than the imagery. This chapter reviewed many of the most used ancillary datasets. Relying on them requires that they be positionally and thematically accurate, and of a scale similar to the map information being produced. It is critical that all data sources being considered for use in a mapping project be reviewed for their positional accuracy, even if the stated positional error of a dataset is very high. This is done most effectively in ArcGIS by using the Swipe tool on the Effects toolbar to visually compare the data source against positionally very accurate, high-resolution orthorectified imagery. Visual spot checks of the data source should occur across the entire geography of the project area, because the data source may have isolated or regionally specific positional accuracy problems that don't manifest in a check of just a single area.

In general, using existing thematic maps (e.g., soils or a historical map of land cover) to support the creation of a new thematic map should be done very cautiously. If a thematic dataset is to be relied upon to inform an image classification, it must be of known and acceptable thematic accuracy, because if a class is incorrectly labeled in the dataset, the error will likely be perpetuated into the new map. If thematic maps are to be used as source material or independent variables, they should have a stated minimum mapping unit as small or smaller than the map being made and should have map accuracies that exceed the target accuracy of the map being made. Additionally, the classification scheme used to create the dataset should be well understood.

Scale is another important consideration in the use of thematic data sources as independent variables or source material to compliment imagery in thematic mapping. If those sources are used, the data source map should have a maximum usable scale that equals or exceeds that of the map being made. If the data source thematic map has a maximum usable scale that is less than the map being made, its use could degrade the quality of the resulting map.

# Chapter 10
Image Classification

Introduction

At this point in the workflow, you will have acquired your imagery and ancillary data layers and examined them to ensure that they are accurately registered to the ground and relevant for your project. You have built a rigorous classification scheme and hopefully been able to test it in the field. You should understand and have controlled to the maximal extent possible any variation in the imagery or ancillary data that is not related to variation in your classification scheme. You will you have studied the variation in your imagery and performed data exploration to better understand it and how it varies with the objects you want to map on the ground. Now you are ready to extract information from the imagery—to make a map.

In chapter 9 you learned about the elements of imagery and how the confluence of those elements allows you to classify imagery; i.e., to extract features or create a thematic map. This chapter reviews the methods for image classification that constitute the processes for converting the imagery and associated ancillary data into map information. For this conversion, you will use either manual interpretation or combined manual interpretation and automated computer algorithms.

Both manual and automated image classification involve data exploration to establish relationships between the imagery, ancillary information, and features on the ground. In manual classification, data exploration is primarily heuristic, with the analyst gaining knowledge about the area to be mapped through visits, discussions with experts, and/or literature and document reviews. This knowledge is then used by the analyst to delineate and label the imagery using the confluence of evidence derived from the 10 image elements presented in chapter 9. In semiautomated classification, data exploration is accomplished both heuristically and quantitatively because the image is not only a picture examined by the analyst, but also a numerical dataset that can be rigorously analyzed along with coregistered ancillary datasets such as digital elevation models, soil maps, and hydrologic feature locations. Knowledge of what causes the map classes to vary on the ground is key regardless of the method used for turning the remotely sensed data into information, whether it be manual interpretation or some type of supervised or unsupervised semiautomated classification.

This chapter first examines the methods used in manual interpretation, and next focuses on semiautomated classification techniques such as unsupervised and supervised classification, as well as image segmentation, machine learning algorithms, and rulesets. The third section discusses map validation and editing procedures.

Basics of Manual Interpretation

Manual image interpretation is called photo interpretation or image interpretation, depending on whether it is performed on analog (film) aerial photographs or on digital imagery. Sometimes the interpreter has just a single photo or image to study, while at other times a stereo pair (i.e., a pair of images with two perspective views and at least 50 percent overlap) is used. Stereo viewing allows the interpreter to bring object height and a 3D perspective into the classification process.

Manual interpretation is an art and skill that improves with practice and requires a systematic, logical, and organized approach. It involves delineating and labeling objects of interest in a photo or image and deducing their significance. Like a detective or a doctor, the interpreter follows a series of clues provided by the 10 image elements (i.e., tone/color, shape, size, pattern, shadow, texture, location, context, height, date) that lead to a solution.

Generally, the interpretation should proceed from general to specific classes (i.e., using a hierarchical approach). For example, land-cover land-use maps usually are first interpreted for life-forms (e.g., trees, agriculture, urban) and then for specific classes within life-forms (e.g., white pine tree or red oak tree). In this way, a forest would first be delineated as separate from an urban area or croplands, and then the types of forests would be delineated and labeled within the forest life-form (e.g., deciduous versus evergreen, or oak versus redwood versus Douglas fir versus Sargent cypress stands of trees).

Readily identifiable objects are usually first delineated and labeled before unknown objects, and interpretation flows from known to unknown features. Through a process of elimination, easily identifiable objects such as water bodies, airports, or building footprints can be quickly mapped, allowing the interpreter to focus his or her intellect on distinguishing between less easily identifiable classes in the classification scheme such as forest tree species composition, wetland classes, or crop types.

While the mapping is proceeding, the interpreter must always keep in mind the quality of the imagery and ancillary data, and the conditions under which the imagery was collected. Time of day and season of year greatly affect the spectral qualities of the image across the landscape. Because many trees and shrubs are deciduous, a fall image looks very different from a spring image in the temperate regions of the world. For example, it is very common for imagery acquired for infrastructure mapping to be collected when deciduous trees have lost their leaves. As a result, the same imagery would be of limited use for differentiating between deciduous tree species. Similarly, an image collected under clouds will be very different from one collected on a cloud-free day. There is nothing quite as frustrating as having identified a strong spectral response in an image for a particular map class, only to have the response become unreliable when the mapping moves to images captured on a different day or time of day.

The tasks of manual interpretation are fairly straightforward, as shown in figure 10.1. First, the interpreter reviews the imagery and gathers ancillary information such as topographic maps, soil maps, field samples, and field notes. Information on management history, agricultural practices (e.g., crop calendars), and military and political history also provide important clues for object identification.

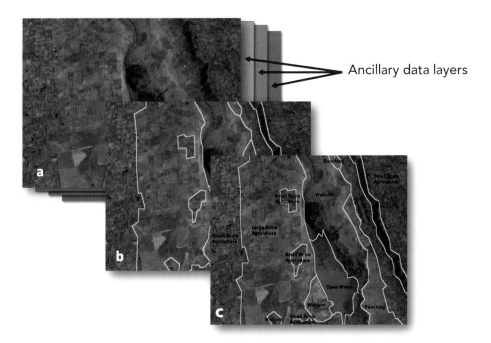

Figure 10.1. Tasks of manual interpretation of infrared Landsat imagery in Africa. First, the image is examined and ancillary data is compiled (a). Next, objects are delineated based on the classification scheme (b). Finally, the objects are labeled using the rules of the classification scheme (c).

Manual interpretation was once an entirely analog process, but over the last 15 years it has been greatly assisted by the integration and display of multiple coregistered images and ancillary data layers in a GIS. Figure 10.2 shows an ArcGIS MXD created for manual interpretation of vegetation types in Sonoma County, California. Layers in the MXD include:

- A variety of imagery sources from multiple dates (National Agricultural Imagery Program [NAIP] 1-m 4-band from 2009, NAIP 1-m 4-band from 2013, 1-foot resolution airborne 4-band imagery collected in 2011, and 6-inch resolution airborne 4-band imagery collected in 2013). The 6-inch true color imagery is the uppermost layer in figure 10.2.
- The data and labels of the sample vegetation plots collected to create the classification scheme.
- Vegetation labels and photographs for locations collected during multiple field trips (green and magenta dots in figure 10.2).
- Roads.
- Streams and height above streams (shown as yellow and dark blue in figure 10.2).
- Wildfire history.
- Various soils layers (shown as light blue and beige polygons in figure 10.2).
- USGS topographic maps.
- Previously created vegetation maps.
- Several lidar products including vegetation height, a digital elevation model (DEM), and bare-earth hillshade.

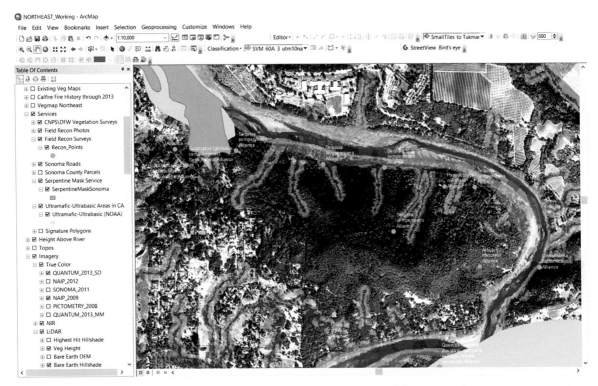

Figure 10.2. Using ArcGIS to integrate the various data types useful in manual image interpretation.
Source: Sonoma County Agriculture Preservation and Open Space District

Next, the interpreter delineates polygonal outlines around features (objects) on the imagery based on the rules of the classification scheme and the size of the minimum mapping unit (MMU). Once the outlines have been drawn, each object is labeled by relying on the indications developed from the ten image elements (i.e. tone/color, shape, size, pattern, shadow, texture, location, context, height, date) and the rules of the classification scheme. As the process proceeds, the analyst develops heuristic rules relating to variations in the 10 image elements to map classes. For example:

- Water is dark in all bands, relatively flat and smooth.
- Annual crops are in row patterns, bright in the infrared and green bands, less than 2 meters high, and usually occur on gentle or flat slopes.
- American urban areas are bright in most bands with a square pattern of flat streets and buildings over 2 meters high.
- Forests are less bright than crops in the infrared and green bands, occur on all aspects and slopes, and have trees more than 3 meters high.

Often an interpretation key, such as the one displayed in figure 10.3, is developed to aid the interpreter in labeling objects by presenting descriptions and pictures of how different classes typically appear on the imagery. Figure 10.4 exemplifies how the confluence of the ten image elements can be used to distinguish four simple life form classes.

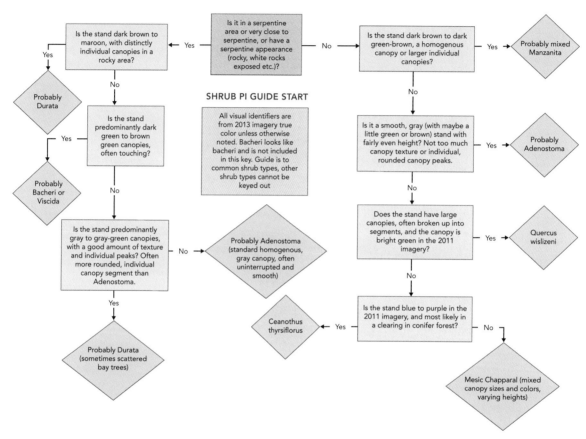

Figure 10.3. Manual interpretation key for shrub types in Sonoma County, California

Suburban
Color/tone: highly variable in all bands
Shape: rectangular houses, straight streets
Size: small buildings with small yards
Pattern: rectangular blocks
Texture: highly variable
Location: near an urban area
Context: near highways and schools
Height: highly variable
Date: not significant

Water
Color/tone: dark in all bands
Shape: simple
Size: variable
Pattern: none
Texture: usually smooth
Location: zero slope
Context: variable
Height: none
Date: important in capturing tidal and seasonal fluctuations

Vineyard
Color/tone: high in the infrared and green bands in the spring and summer
Shape: mostly rectangular, simple
Size: large fields
Pattern: planted rows
Texture: uniformly variable
Location: flat or gently sloped
Context: occur in all contexts
Height: <2 meters
Date: leaf out in spring, lose leaves in fall

Forest
Color/tone: variable from species to species but generally high in infrared and green in the spring and summer
Shape: complex
Size: variable
Pattern: variable
Texture: smooth in dense stands, highly varied in sparse stands
Location: all slopes and aspects
Context: occur in all contexts
Height: >3 meters
Date: important in distinguishing evergreen from deciduous species

Figure 10.4. Four life-form classes identified by the confluence of the 10 image elements

Often, the interpreter employs stereo viewing equipment to enhance the interpretation. The ability to view the imagery in stereo or 3D provides the interpreter with not only the ability to incorporate height into the process but also a more realistic view that further aids in the interpretation. Stereo viewing is based on the concept of looking at two different perspectives of the same object simultaneously. One eye views one perspective while the other eye views the second perspective, and then the brain fuses the images together to create the 3D effect (many have experienced this same effect wearing special glasses to view a 3D movie at a theater). While it is possible to view in stereo with some practice without any special device, using an instrument called a stereoscope can greatly aid the process. The simplest stereoscopes simply aid the eyes in looking straight down (one at image one and the other at image two) instead of converging on a single focus as your eyes normally do. More advanced stereoscopes use mirrors that make it even easier for one eye to look at one image while the other looks at the second image (remember, the images are of the same area, just taken from different perspectives). Finally, the most sophisticated and expensive stereoscopes use not only mirrors but also optics to enlarge the images, providing the most effective interpretation.

The accuracy of the interpretation depends on the quality of the imagery and ancillary data, the equipment used to perform the interpretation, the detail and difficulty of the classification scheme, and, most importantly, the skill of the interpreter. It is up to the interpreter to simultaneously consider the ten elements of the imagery within the context of the classification scheme. This can be a difficult task requiring patience, focus, and knowledge of the phenomena as well as the geography being mapped. Spurr's (1960) admonishments regarding the qualities of the interpreter are still valid today: "Both visual and mental acuity are required. Equally essential are extensive training and field experience in the specialty involved. Geological photo interpretation must be done by the geologist; soils interpretation by the soil scientist, and forest photo interpretation by the forester" (Stephen Spurr, *Photogrammetry and Photo-Interpretation* [New York: The Ronald Press Company, 1960], 234.

It is beyond the scope of this book to investigate the specifics of manual image interpretation techniques for each of the different specialties such as infrastructure, situational analysis, land form, wetlands, archaeology, agriculture, pollution, water resources, forestry, geology, snow and ice, or soils mapping. This information can be found in several textbooks including:

- Lillesand, T.; R. Keifer; and J. Chipman. 2015. *Remote Sensing and Image Interpretation*. 7th ed. New York: John Wiley & Sons.
- Paine, D., and J. Kiser. 2012. *Aerial Photography and Image Interpretation*. 3rd ed. New York: John Wiley & Sons.
- Phillipson, W. R. ed. 1996. *Manual of Photographic Interpretation*. 2nd ed. Bethesda, MD: American Society of Photogrammetry and Remote Sensing.

Manual interpretation has been and continues to be the most-used method for land-use/cover classification and feature extraction. However, manual interpretation can be extremely time consuming, costly, and inconsistent. No matter how rigorous the classification scheme, each interpreter may delineate and label the same area differently depending on their expertise, training, and mood. Even the same individual's interpretations can vary from day to day, and within a day. As a result, for the last 40 years, efforts have been made to automate part or all of the mapping process.

Semiautomated Image Classification

Introduction

The ultimate goal of semiautomated image classification is to make the mapping process faster, less expensive, more consistent, and more accurate. Like manual interpretation, semiautomated image classification relies on some or all of the same ten image elements used in manual interpretation. What differentiates semiautomated classification from manual interpretation is that computer algorithms rather than humans determine the confluence of evidence. Semiautomated classification involves data exploration to establish relationships between the imagery, ancillary data, and features on the ground. Computer algorithms are then used to classify the imagery based on those relationships. If the classes in the classification scheme are strongly correlated with the imagery and ancillary data used, much of the work can be successfully automated, producing results as good as or better than manual interpretation at a much lower cost, especially for large project areas. However, manual interpretation is always still needed to evaluate the output of the computer algorithms and to edit errors, hence the term *semiautomated* instead of *automated* image classification.

Semiautomated methods classify either pixels or objects, which are pixels grouped into spatially adjacent clusters of similar color/tone and texture (figure 10.5). Then, the pixels or objects are classified. Per-pixel classifiers group individual pixels into classes that are spectrally similar. They rely almost entirely on each pixel's color/tone, date, and texture to label each pixel. Object-based mapping first groups spectrally similar and spatially adjacent pixels into objects, and then classifies the objects using manual rulesets, unsupervised or supervised algorithms, or a combination of both.

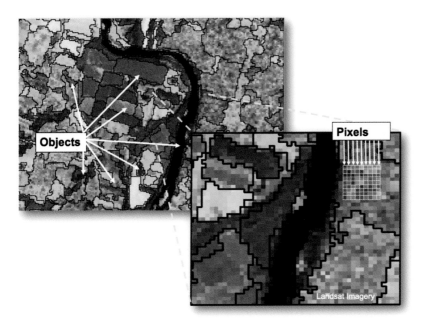

Figure 10.5. Pixels versus objects. Objects are created from spatially adjacent pixels of similar color/tone and texture.

This section begins with a brief discussion of how semiautomated image classification has evolved over the last 40 years. The next subsection discusses the concepts of image object creation, also called image segmentation. The third subsection details the multiple approaches to the semiautomated labeling of either pixels or segments. The section concludes with a discussion of practical concepts for consideration when deciding which image classification method to use.

History

Semiautomated image classification started when per-pixel classifiers developed in the 1970s to classify digital multispectral satellite imagery. Initially, there were only two major per-pixel approaches to traditional semiautomated image classification: supervised and unsupervised. The supervised approach mimics manual interpretation, while the unsupervised approach uses statistical clustering. There are advantages and disadvantages to each approach, which led to many analysts finding ways to combine them into hybrid approaches. Because of the per-pixel classifier's reliance on at most three of the ten image elements (color/tone, texture, and date), the accuracies of early per-pixel classification of thematic maps were often low. Users were also not used to having the landscapes or features depicted in pixels rather than as polygons or features. The result was that many map users rejected semiautomated image classification in favor of long-accepted manual interpretation techniques.

In the late 1980s and early 1990s, the incorporation of location into semiautomated image classification became possible with the introduction of GIS technology and the implementation of GPS technologies and algorithms that allowed for precise image registration. Because it became possible to accurately coregister the imagery and the ancillary layers to the ground and to one another, it also became possible to incorporate the element of location into semiautomated classification. In this way, the accuracy of per-pixel classifiers often improved through the use of location-based rulesets, which relate a pixel's location characteristics to its probable label. For example, both water bodies and forests on northeast facing mountain slopes can have very low spectral responses in all bands. As a result, it is not uncommon to have forest pixels misclassified as water. But still-water bodies are flat and do not occur on slopes. If the imagery is accurately registered to the ground and to a DEM, a ruleset can be developed that checks to make sure that only pixels with zero slope are classified as still-water, which distinguishes the water pixels from the forested slopes.

As computers improved in processing speed, and software gained sophistication, machine learning algorithms such as Random Forests and Support Vector Machines have been applied to image classification. While traditional supervised and unsupervised classifiers are restricted to classifying only continuous data, machine learning algorithms mine the imagery and ancillary datasets to establish relationships between the map classes and both continuous and categorical independent variables, resulting in the development of more complex and robust classification rules.

Almost simultaneously with the advent of digital high-spatial-resolution imagery in the late 1990s, new technologies were developed to segment imagery into spectrally and texturally homogeneous areas (objects) and classify these segments rather than each pixel. Pixels represent arbitrary delineations of rectangles on the ground. Segments delineate meaningful variation across the landscape, which can then be tied to the classification scheme. While powerful in the classification of moderate-resolution imagery (e.g., Landsat, Sentinel), object-oriented classification is pivotal for semiautomated classification of high-resolution airborne or satellite imagery because of the mixture of shadow and illuminated pixels within objects, and the common need to group pixels together to map vegetation classes (e.g., forests) instead of individual features (e.g., trees). Object-oriented classification also allows for the inclusion of the image elements of area, shape, pattern, and context into semiautomated classification. At the turn of the last century, lidar technologies became operational for topographic mapping. Suddenly, high-density DEMs and digital surface models (DSMs) and their products became readily available, allowing semiautomated approaches to incorporate height and to finally be able to use all 10 of the image elements available for manual interpretation.

Image Segmentation

Image segmentation is the general term for the automated delineation of image objects from groups of spatially adjacent pixels (as shown in figure 10.6). The goal of segmentation is to create objects with more between-object spectral and textural variability than within-object variability. Like manual delineation, image segmentation divides an image into relatively homogeneous, characteristically significant, and spatially unique segments or objects. Measures of the 10 image elements can then be calculated for each object and used in object-based image classification.

Figure 10.6. Example of image segmentation of six-inch high-resolution aerial imagery into objects that are outlined in yellow. Source: Sonoma County Agriculture Preservation and Open Space District

Some segmentation software focuses on mapping the entire landscape, while others concentrate on extracting features such as buildings, tanks, or airplanes. In any of these algorithms, the analyst specifies a series of parameters that determine how the objects are generated. It is possible to create large objects that incorporate very general vegetation/land-cover types or very small objects that may divide a specific cover type into multiple segments created from manual interpretation. Usually, more objects are produced than required by the classification scheme. After the objects have been classified, boundaries between adjacent objects with the same map label are dissolved, resulting in map polygons or features that represent different map classes as defined by the classification scheme.

Most segmentation for remote sensing is based on either edge detection or growing regions of spectral homogeneity from some starting point called a seed. Usually, optical images, infrared bands, or a normalized difference vegetation index (NDVI) are used as the input images along with a measure of image texture derived from the imagery. Texture is important to include so that segments can be created that delineate highly heterogeneous areas, such as a sparse stand of trees or a suburban area, from homogeneous areas (e.g., crops, water bodies). The amount of heterogeneity tolerated within an image segment is an adjustable parameter in the segmentation software. Some algorithms also include parameters that control how complex the boundary of the image segment can be.

The power of the segmentation process is twofold. First, because polygons have been the standard map unit for decades, map users tend to be more comfortable with polygon-based maps than pixel-based maps. Second, object attributes can include measures of image elements that do not exist for arbitrary rectangular pixels. For example, an object's area is easily calculated, as is the complexity of its border (i.e., its shape), and its neighboring objects can be evaluated to create a measure of context.

It is critically important that segments be carefully created to ensure that the variation delineated between the segments is correlated with the variation in the classification scheme. It is very easy to create segments in remote sensing software packages, but it is very difficult to create good segments. Often, the objects are too small with many adjacent objects representing the same map class. As a result, some of the small objects will be misclassified, the map will be noisy, and postclassification editing to correct the errors will be tedious. Conversely, overlarge objects will not capture the variation of the classification scheme, and the large segments will need to be cut in postclassification editing, also a tedious task. An important overriding consideration will always be the size of the classification scheme's MMU. Multiple iterations of segmentation should be run on the imagery to determine the best segmentation parameters, and large-area projects should be separated into regions, each run with separate segmentation parameters that best delineate that region's variation.

Furthermore, segmentation is most effective when it is combined in a stepwise manner with classification (i.e., a hierarchical approach). For example, a ruleset might segment an image, classify the objects into life-form classes (e.g., shrub, grass, hardwood forest, conifer forest, and mixed forest), merge adjacent objects with the same life-form class, and resegment within the merged broad classes using segmentation parameters tailored specifically to the life-form class.

Classification Algorithms

The process of classification groups data into classes with similar characteristics. Classification algorithms that are used to automatically classify pixels or objects are highly varied and range from simple, heuristically developed rulesets, to sophisticated statistical techniques, to relatively new and complex machine learning methods. Figure 10.7 diagrams the semiautomated classification algorithms most commonly used today. They fall into three broad categories: clustering or unsupervised classification, supervised classification, and combined approaches.

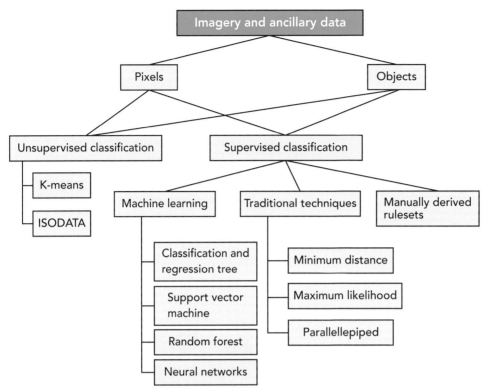

Figure 10.7. Diagram of image classification algorithms

Unsupervised Classification

Unsupervised classification algorithms currently exist primarily as per-pixel classifiers. However, ArcGIS does have an unsupervised object classifier. Unsupervised classification uses statistical clustering algorithms to sample the values of the pixels of an image, and then group similar sample pixels into clusters, as shown in figure 10.8. All cluster calculations are performed on the cell values in multivariate attribute space and are not based on any spatial characteristics. The clustered pixels do not need to be, and rarely are, spatially adjacent. Each cluster is statistically separate from the other clusters. How the imagery is sampled and how the samples are clustered depends on the unsupervised classification algorithms and parameters chosen by the analyst.

Once the clusters have been developed, the values of the remaining pixels in the image are then statistically compared to the summary statistics of the clusters, and each pixel is placed into the cluster that it most resembles, as shown in figure 10.9. The multivariate statistics calculated from the sample's input bands will determine into which cluster each pixel is placed.

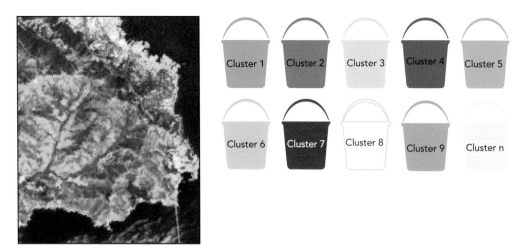

Figure 10.8. Unsupervised classification of southern Marin County, California. First, the image is sampled and then the sample pixels are iteratively grouped into clusters that are statistically similar.

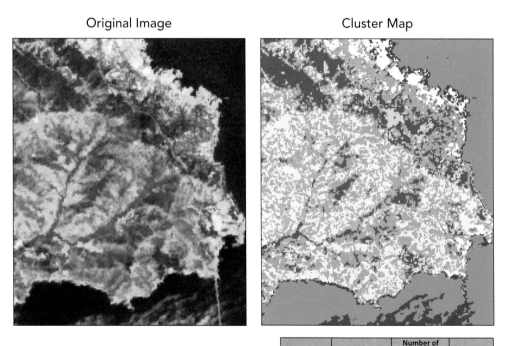

Figure 10.9. The rest of the image's pixels are classified into the clusters that they most resemble based on the multivariate statistics of each cluster.

Color	Name	Number of Pixels	Area sqm
	Cluster 1	6920	4325000
	Cluster 2	5530	3456250
	Cluster 3	6757	4223125
	Cluster 4	1798	1123750
	Cluster 5	2738	1711250
	Cluster 6	1	625
	Cluster 7	644	402500
	Cluster 8	1125	703125
	Cluster 9	11669	7293125
	Cluster 10	2209	1380625
	Cluster 11	651	406875
	Cluster 12	4673	2920625
	Cluster 13	910	568750
	Cluster 14	1217	760625
	Cluster 15	1599	999375

Chapter 10 : Image Classification

There are many unsupervised algorithms, but all of them require continuous digital data as an input. Different unsupervised classification algorithms will probably produce a different set of clusters, but the clusters would still be spectrally unique as defined by the unsupervised algorithm. The most commonly used algorithms are K-means and the iterative self-organizing data analysis technique (ISODATA). In K-means, the analyst specifies the number of clusters desired. The image is sampled, and each pixel is placed into the cluster closest (in Euclidian distance) to the multivariate mean vector, which is established as an arbitrary starting place of the clustering process. Once all of the sample pixels have been placed in clusters, new cluster means are calculated and the sample pixels are reclassified into clusters again based on the shortest distance to the new means. The means of the new clusters are recalculated again, and the previous step is repeated. The iteration process reiterates, updating the mean values until the user-defined number of iterations is reached or until less than 2 percent of the pixels change from one cluster to another.

The ISODATA clustering algorithm is similar to K-means, except that the number of clusters is allowed to vary from iteration to iteration based on analyst-determined measures. If the distance between two clusters falls below a specified minimum, then the clusters are merged. Conversely, if a cluster's variance exceeds a specified maximum, then the cluster is divided in two. As with K-means, this process is iterated until there is no significant change in cluster statistics or when a specified maximum number of clusters is reached.

The outputs of an unsupervised classification are the spectral statistics for each cluster and a map with each pixel labeled as the cluster to which it is statistically most closely associated. It is then the job of the analyst to determine the map class label for each cluster (as shown in figure 10.10) to create a map as defined by the classification scheme (figure 10.11). Labeling each cluster can be an arduous, time-consuming, and subjective process. Some clusters will clearly represent a single map class and can be easily labeled. Other clusters will include a combination of multiple map classes and can be very difficult to label. Notice that cluster 7 occurs in urban areas and along the coastline in the southwest portion of figure 10.11, where beaches but no urban structures exist. This result points to two problems. First, our classification scheme is missing a bare land class. Second, cluster 7 is informationally confused and will need further analysis to remove the confusion. An approach known as "cluster busting," in which clusters that represent multiple map classes are separated and iteratively run through additional clustering algorithms to separate them into additional spectrally unique clusters can be very effective in teasing more informational classes out of the imagery (Jensen 2016).

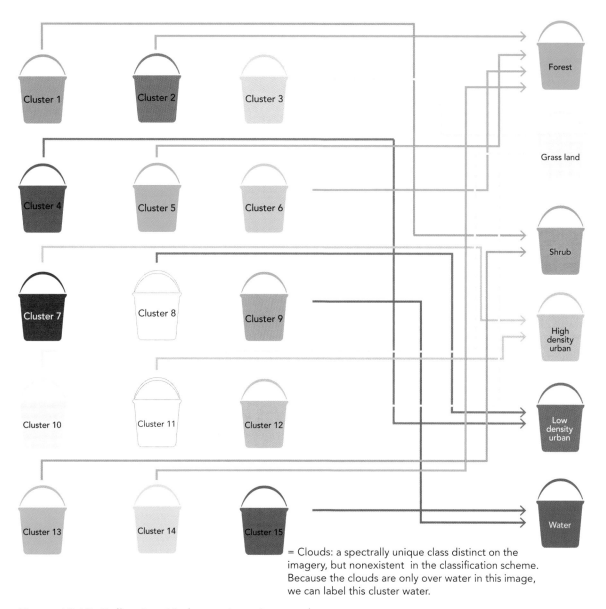

Figure 10.10. Collapsing 15 clusters into six map classes

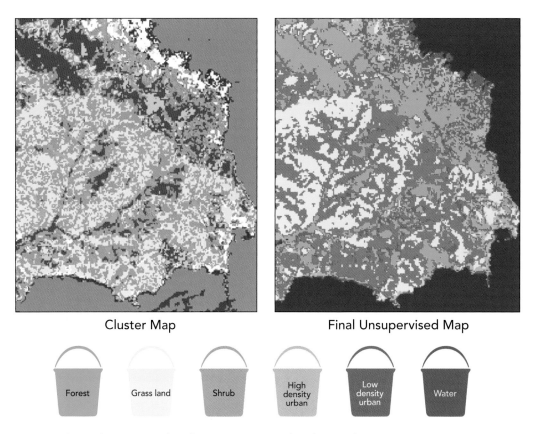

Figure 10.11. Converting the cluster map into a land-cover/use map

    The analyst decides how many clusters the unsupervised classification should produce, and knowledge of the study area is essential to produce an accurate map. The number of clusters must be greater than or equal to the number of map classes, and most often the number of clusters exceed the number of map classes. The number of clusters varies with the complexity of a project's classification scheme; in general, the more classes to be mapped, the more clusters that should be generated. The analyst must work iteratively with the number of clusters in the unsupervised classification to achieve the best separation of the map classes of interest.

    The power of unsupervised classification is that it clearly separates out the spectral information present in the image. Every cluster is statistically different from all the other clusters and is unique. Additionally, unique clusters unknown to the analyst can be discovered and mapped into a map class. For example, in figure 10.10 cluster 15 was found to represent clouds over water, which are spectrally unique but not a map class. Because the clouds only occur over water in this example, cluster 15 can be relabeled as water.

    The challenge in unsupervised classification is to determine the informational class of each cluster. While you know that the clusters are spectrally unique, you do not know which map class each cluster represents. If each cluster represents one and only one map class, an accurate map can be produced. However, there is no guarantee that a spectrally unique cluster will represent only one informational class, and therefore the clusters may be informationally confused, which will cause errors on the map.

Supervised Classification

Supervised classification relies on rules to decide into which map class a pixel or an object will be placed. The rules relate the dependent variable—the map classes—to the independent variables—the imagery and the ancillary data. As in manual interpretation, the rules can be developed heuristically by the analyst based on her knowledge of the project's classification scheme, imagery, and landscape. The rules can also be automatically and more rigorously generated from sets of samples, called training sites, which are collected across the project area for each map class.

The process of supervised classification mimics that of photo interpretation. Just as the photo interpreter must be "trained" to recognize characteristics of objects in an image, so too must the computer be "trained." The image interpreter uses the elements of image interpretation and a confluence of evidence to recognize features, cover, and land-use types. Similarly, in semiautomated classification, the analyst either builds a model of heuristic rulesets or uses statistics and data exploration or mining techniques to generate the rules. Supervised classification relies on the analyst's knowledge of the area to be mapped, which allows him to "supervise" or "train" the image classification.

As shown in figure 10.7, there are three types of supervised classification algorithms: manually derived rulesets, traditional techniques, and machine learning. All of the three types of techniques can be used to classify either pixels or segments. Image classification with rulesets is the simplest form of supervised classification, and it most closely mimics manual interpretation. Traditional techniques were the first classifiers developed to classify multispectral imagery and are still used extensively today. Machine learning classifiers are relatively new, but very powerful for the classification of complex classification schemes and distinguishing subtle differences between map classes. This section first discusses the requirements of good training samples for supervised classification, and then reviews the basics of each of the supervised classification techniques.

Training Sites

All supervised classification methods require examining sample areas of the imagery where the map class (according to the classification scheme) is known. The samples are referred to as training sites, and the rigor of their examination can range from informal viewing of the imagery to meticulous mathematical analysis of the values of the independent variables of all the samples. Training data is absolutely required for supervised classification, while strong knowledge of the ground, even if not specific sample sites, is required to label the results of the unsupervised classification. The rigor required depends upon the complexity of the classification scheme and how closely correlated the variation in the scheme is with the variation in the imagery and ancillary data. Simple classification schemes with high correlations between the imagery, ancillary data, and the map classes require little analysis. Complex schemes seeking to tease apart map classes that appear to be indistinguishable on the ground or imagery require in-depth analysis.

Great care must be taken to efficiently and effectively generate appropriate training sites for each map class. Poor selection of training sites can result in producing an inaccurate map because the classifier will be "poorly trained," resulting in the inability to properly identify each map class. The more spectrally different the map classes are, the easier it is to select good training sites. For example, it is relatively easy to conduct a supervised classification for a project containing only four distinct land-cover types: water, forest, developed, and other vegetation. These land-cover types are very easy to tell apart on the imagery. It would not take much training for an image interpreter to distinguish them, nor would it be difficult to select good training sites for a supervised classification.

However, as the classes to be mapped become more spectrally similar, it is increasingly important that subtle differences between the map classes are represented in the training sites. The number of pixels in a training site, the number of training sites needed per map class, and the method of delineating the exact training site are all important issues that the analyst must consider. Because statistics are generated from each training site, a sufficient number of pixels must be selected to obtain a viable training site. A good guideline is to have approximately $10 \times n$, where n is the number of image bands. In other words, if 7 bands of imagery are being used in the classification analysis process, then there should be approximately 10 times 7 or 70 pixels in each training area. A very small number of pixels is usually insufficient to produce representative statistics, and too many pixels can result in a training site with large variances that can potentially result in its class being erroneously overmapped.

The number of training sites per map class greatly depends on the complexity of the classification scheme. Simple classification schemes require relatively few training sites. Complex schemes with many classes require a very large number of training sites to effectively represent each map class's range of spectral and independent-variable conditions, especially in large project areas. Some map classes are inherently less variable than others, quite homogeneous, and easy to map. Some are more heterogeneous and require more effort to properly "train" the classifier. Some map classes turn out to be homogeneous in their spectral heterogeneity (e.g., suburban areas, sparse stands of trees), which means that some measure of texture will be an important independent variable.

For example, water is a rather homogeneous class, and typically only a few (5 to 10) good training sites are needed to separate water from the other map classes. Of course, if mapping water conditions (turbidity or salinity, etc.) is the objective, then training sites for each of these conditions are necessary. Conversely, urban or developed areas are inherently quite variable. A suburban environment is really a mosaic of buildings, trees, grass, roads, sidewalks, etc. in a unique pattern depending on population density. Therefore, to accurately represent this mosaic, many more training sites are required to adequately capture the spectral variation in this map class.

The class-by-class distribution of training sites usually mirrors the distribution of the classes on the landscape. Classes that occur commonly across the landscape have a large number of training sites, and classes that don't occur often have a smaller number of training sites.

There are many methods for delineating training sites. If pixels are to be classified, the simplest method is simply to use a box or rectangle to select a group of pixels. While simple, a rectangle leaves the analyst very little flexibility about including or excluding certain pixels, because everything in the box/rectangle is included and everything else is excluded. The next method is also user based, but allows the analyst to draw a polygon rather than just a rectangle, as shown in figure 10.12. The power of this method is that it is still quite simple, yet provides more discretion to the analyst to select pixels that should be part of the training site while excluding others. Both the rectangle and polygon methods rely exclusively on the analyst to select appropriate training sites. A final method, based on a region-growing algorithm, can also be used to select training sites. In this technique, the analyst selects a starting point or "seed" for the training site to begin and sets some parameters to define how the site will grow. Then the algorithm "grows" the training site by adding pixels to it based on the defined criteria. If the adjoining pixels are similar enough to the site as defined by the selection criteria, they are added to the training site. When all adjoining pixels have been found to be effectively different from the training site, the site is complete and no further pixels are added. This method has the advantage of using statistical similarity to determine whether a pixel is part of the training site. However, the criteria for this selection must still be set by the analyst, so the method is not purely objective.

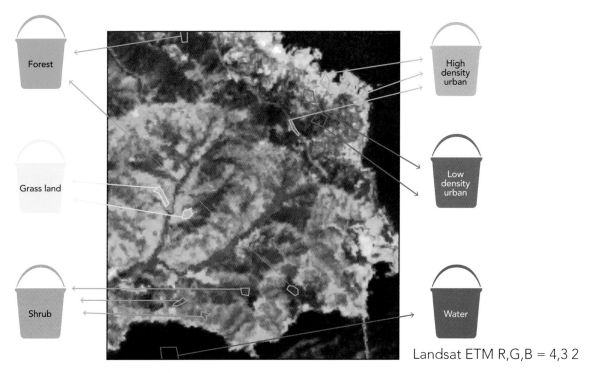

Figure 10.12. Collection of training sites from Landsat imagery of southern Marin County, California

If objects rather than pixels are to be classified, then objects can be used as training sites as long as the image segmentation detail matches the detail of the classification scheme. Too-small objects do not adequately capture the variation of the map class. Overlarge objects containing more than one map class result in too much variation within the training site, which causes map errors. Figure 10.13 shows a portion of Sonoma County, California, that has been segmented. The field-verified training objects are outlined in yellow with the remaining objects outlined in white.

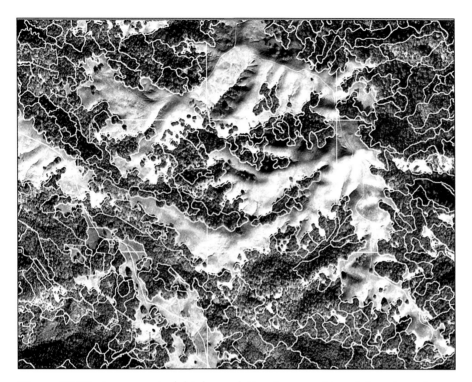

Figure 10.13. A segmented, high-resolution image with training sites shown as selected field-verified training sample objects outlined in yellow. Non-training site segments are outlined in white.

Training-site quality is the most critical consideration for producing high-quality and accurate maps using supervised classification. Selecting good training sites requires strict adherence to the project classification scheme, which must be composed of labels and rules, mutually exclusive, and totally exhaustive (see chapter 7 for a more thorough discussion of the requirements of a robust classification scheme). The scheme also needs to be easy to apply in the field so that sample collection is accurate and cost effective, and map classes must also be easily and accurately identifiable in the field and on the imagery.

The whole process of training-site sample collection provides the analyst with an understanding of both the causes of map class variability and the relationship between map class variability and variability of the imagery and ancillary data layers. While training-site samples can be collected solely from remotely sensed imagery, they are best collected by visiting and labeling sites on the ground with the imagery and ancillary datasets in hand.

Guidelines for Training-Site Sample Selection

Collecting training samples that fully capture the variation within and between map classes is a key factor in supervised classification, and the following guidelines should be followed:

- Informational homogeneity—Each sample must represent one and only one map class.
- Completeness—An adequate number of samples must be collected for all map classes, and the samples should capture the range of independent-variable conditions for a given map class. Samples must also capture all the significant spectral variance in the imagery across the project area. One of the best ways to ensure the complete capture of spectral variance is to run an unsupervised classification of the project area and check that all unsupervised clusters are represented by training samples. Spectrally unique clusters that do not relate to map classes may indicate a condition on the ground or in the imagery that is not captured in the classification scheme.
- Spectral homogeneity—The samples should have more spectral variability between classes than within classes. In other words, the spectral per-band statistics of one map class's training samples should be more similar to one another than they are to those of other map classes. If the spectral statistics of different map classes overlap, then there probably will be confusion between those map classes on the map.
- Minimum size—Samples should be larger than the MMU established for the project.
- Project-wide distribution—For each map class, sampling should be performed so as to collect samples that represent the class's distribution throughout the project area.
- Accurate—Each training sample must be accurately and correctly labeled. If there is ambiguity about a sample's label, then the ambiguity must be noted and captured as part of the training sample's data.
- Cost effective—Training sample collection can be time consuming and expensive, and therefore must be well planned and implemented. If samples are to be collected in the field, then special care must be taken to ensure cost-effective and safe access to sample locations.
- Consistent—Personnel collecting training sites must be well trained in the classification scheme, and data entry forms should be used to impose consistency in data collection. Minimum data collected should include the map class label for the sample site, the name of the collector, date, location, and comments. Additional information depends on the detail of the classification scheme. Today, this

data is captured most effectively and efficiently in the field through the use of an application designed for a smartphone, tablet, or laptop computer, which allows the data collector to take advantage of viewing imagery and ancillary data of the site as well as error checking, pull-down menus, and other time-saving measures (figure 10.A).

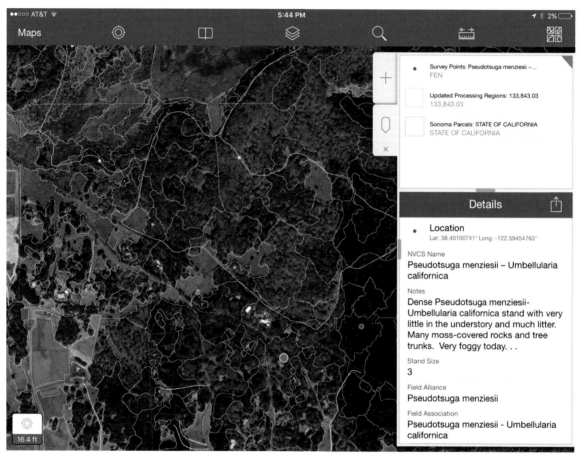

Figure 10.A. Example of using Collector for ArcGIS app for iPad field data collection in Sonoma County, California. The data collector is used to quickly label points in the field with a field-verified thematic map class. This configuration of Collector also includes many datasets for navigation and reference, including high-resolution imagery, parcels, USGS 7.5-minute topographic maps, and other data layers. Point data collection is facilitated with pull-down menus, and videos and photos are easily captured. (esriurl.com/IG10A)

The importance of thoroughly reviewing training-site data cannot be overstated. Manual review of training sites using reference data such as high-resolution orthophotography always reveals mislabeled training sites and data entry errors. Removing those training sites from the classification greatly increases classification accuracy.

Manually Derived Rulesets

The first type of supervised mapping technique we will discuss is manually derived rulesets (see figure 10.7). Manually derived rulesets can be used to classify either pixels or objects. The analyst builds the rules after studying the imagery, ancillary data, and landscape. The analyst may or may not collect training samples and perform data exploration on the samples to understand how map classes vary with the imagery and ancillary data. Instead of keeping the rules in the back of their head as done in most manual interpretation, or developing an interpretation key such as the one shown in figure 10.3, the analyst codes the rules and builds a model that relates the different map classes to the independent variables (imagery and ancillary data) available to the project. The resulting model provides consistency in the image classification, but can also oversimplify the landscape. Heuristically built rulesets are best used with simple classification schemes in areas where the map classes are highly correlated with the imagery and ancillary data. For example, you can create a rule that defines persistent smooth water as follows:

- Color/tone: low in all bands
- Shape: not indicative, because water exists in a variety of shapes
- Size: not indicative, because water exists in a variety of sizes
- Pattern: none
- Texture: low in all bands
- Location: zero slope, flat aspect, elevation variable
- Context: not indicative, because water can occur in many contexts
- Surface Height: zero
- Date: multiple dates needed to map persistent water to account for tidal changes and the impact of rainy versus wet seasons

Case Study—Rulesets for Chesapeake Bay Regional Watershed Mapping

In 2015, the Chesapeake Bay Program (CBP), a regional partnership that leads and directs the restoration and protection of the Chesapeake Bay and its watershed, commissioned an upgrade to their watershed landscape data from the Chesapeake

Conservancy, working with the University of Vermont and WorldView Solutions, to create a one-meter resolution dataset that covers all counties that intersect the Chesapeake Bay watershed boundary.

The Conservancy's methodology for the project relied on rule-based workflows to produce a map and dataset with 12 categories: water, emergent wetlands, tree canopy, shrub land, low vegetation, barren, structures, impervious surfaces, impervious roads, tree canopy over structures, tree canopy over impervious surfaces, and tree canopy over impervious roads. Figure 10.B shows three example rulesets developed for the project.

Data used in the project included leaf-on 2013 one-meter resolution; four-band NAIP imagery; height information derived from the most recently available lidar datasets; and most recently available, high-resolution, leaf-off imagery (sized between 15 centimeters to 1 meter per pixel). If available, county, state, and federal planimetric datasets were also used to enhance the classifications of features such as structures and roads.

Because it was adaptable to diverse landscapes, a rule-based classification methodology was selected by the Conservancy to classify Maryland, New York, West Virginia, and Washington, DC. The multistate area was broken into regions composed of 8 to 12 mosaicked NAIP images, each the size of USGS quarter quadrangles. NDVI and DSMs were calculated and the mosaicked images were segmented. Analysts evaluated the segment characteristics and formulated between one and eight rules for each land-cover type to create a mutually exclusive classification. Rulesets for each region were edited from previous efforts or written from scratch, depending on the location and how well previous rulesets transferred. Some classes had qualities that persisted throughout the watershed; for example, generally tree canopy segments were designated as trees if they were above two meters in height and had a comparatively high mean NDVI value. Once the data was categorized to the satisfaction of the analyst, the results were manually edited and local planimetric data was incorporated. The datasets were reviewed for consistency with local experts in each watershed county. Accuracy assessments were conducted, followed by additional editing, before the land cover was incorporated into CBP's Chesapeake Bay water-quality assessments. Overall accuracies averaged 90 percent.

```xml
<classes name="All classes">

<class name="Forest" threshold="0.88">
   <attribute band="0" name="Texture_Mean" operation="lt" tolerance="5" value="130.00000"/>
   <attribute band="0" name="Spectral_Mean" operation="between" tolerance="5" value="38.00000, 130.00000"/>
   <attribute band="4" name="Spectral_Mean" operation="between" tolerance="5" value="4.00000, 219.93280"/>
   <attribute band="5" name="Spectral_Mean" operation="between" tolerance="5" value="0.00000, 0.43681"/>
  </rule>
</class>

<class name="Shrubland" threshold="0.88">
   <attribute band="0" name="Texture_Mean" operation="lt" tolerance="5" value="130.00000"/>
   <attribute band="0" name="Spectral_Mean" operation="lt" tolerance="5" value="130.00000"/>
   <attribute band="4" name="Spectral_Mean" operation="between" tolerance="5" value="1.00000, 4.00000"/>
   <attribute band="5" name="Spectral_Mean" operation="between" tolerance="5" value="0.00000, 0.43681"/>
  </rule>
</class>

<class name="Herbaceous Vegetation" threshold="0.94">
   <attribute band="0" name="Spectral_Mean" operation="between" tolerance="5" value="95.00000, 210.00000"/>
   <attribute band="5" name="Spectral_Mean" operation="between" tolerance="5" value="-0.12000, 0.20000"/>
   <attribute band="5" name="Texture_Range" operation="lt" tolerance="5" value="0.14000"/>
   <attribute band="3" name="Spectral_Mean" operation="between" tolerance="5" value="125.00000, 200.00000"/>
   <attribute band="5" name="Texture_Mean" operation="between" tolerance="5" value="-0.08000, 0.20000"/>
   <attribute band="1" name="Texture_Variance" operation="between" tolerance="5" value="15.00000, 405.00000"/>
   <attribute band="3" name="Texture_Variance" operation="lt" tolerance="5" value="200.00000"/>
   <attribute band="4" name="Spectral_Mean" operation="lt" tolerance="5" value="2.00000"/>
  </rule>
</class>

<class name="Water" threshold="0.85">
   <attribute algorithm="binary" band="5" name="Spectral_Mean" operation="lt" tolerance="5" value="-0.15000"/>
              <attribute band="3" name="Spectral_Mean" operation="between" tolerance="5" value="21.01359, 82.00000"/>
              <attribute band="0" name="Spectral_Mean" operation="between" tolerance="5" value="68.00000, 138.00000"/>
   <attribute band="7" name="Spectral_Mean" operation="between" tolerance="5" value="0.06000, 0.20000"/>
              <attribute band="5" name="Texture_Mean" operation="lt" tolerance="5" value="-0.15000"/>
              <attribute band="0" name="Texture_Variance" operation="between" tolerance="5" value="0.00627, 175.00000"/>
```

Figure 10.B. Example manually derived ruleset for one of the Chesapeake watershed regions. Source: Chesapeake Conservancy

Traditional Techniques

The next type of supervised mapping technique we will discuss is traditional image-based supervised classification (see figure 10.7). Traditional techniques use statistics to develop rules for classifying *continuous* data such as imagery bands or indices, but sometimes also DEMs, DSMs, and products derived from them (e.g., slope, aspect). First, training sites are digitized. Statistics derived from the values of the pixels composing the training sites are calculated including minimum, maximum, mean, and standard deviation for each continuous dataset as well as covariances and correlations between bands (figure 10.14). In traditional supervised classification, each pixel in the imagery is compared to each training sample's spectral statistics for each map class. Each pixel is then assigned a map class based on the pixel's spectral similarity to each training site as determined by the supervised classification algorithm employed by the analyst.

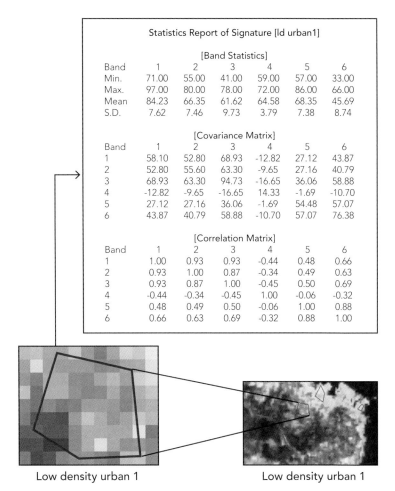

Figure 10.14. Multivariate statistics for one training site

A variety of rules/algorithms can be used to label all the unknown pixels in the image based on the training statistics. The most common of these rules in increasing order of complexity are 1) minimum distance, 2) parallelepiped, and 3) maximum likelihood.

Figure 10.15 shows a graphic representation of the common rules/algorithms that are used for labeling the unknown pixels in the image based on the knowledge gained from the training statistics. Part a of this figure simply shows a bispectral plot in which the pixels from the training sites are plotted. This example shows only two dimensions (a bispectral plot), but the computer can process this same type of analysis in as many dimensions as there are bands in the imagery. A quick look at Part a of the figure shows thematic mapper (TM) band 4 (NIR) on the x axis and TM band 3 (red) on the y axis. The locations of the pixels by vegetation/land-cover type confirms that the training data is valid. For example, the water pixels all show low-NIR and red reflectance, as expected. The coniferous trees are lower in NIR reflectance than the deciduous trees, as also expected. If some pixels were in the wrong place in this graph, this would be reason to investigate whether the training site was properly collected or labeled.

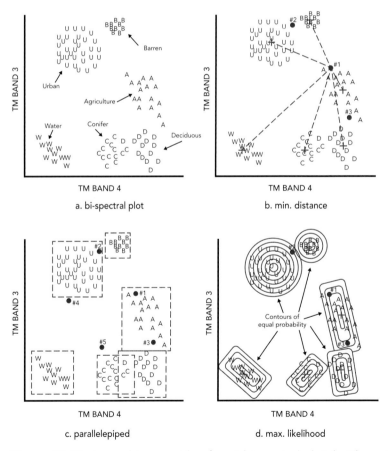

Figure 10.15. A graphic example of: a. A bispectral plot, b. The minimum distance algorithm, c. The parallelepiped algorithm, and d. The maximum likelihood algorithm.

296   Chapter 10 : Image Classification

Figure 10.15b shows how the minimum distance rule works. The distance from an unknown pixel to the center of each training site is measured, and whichever distance is closest determines the label for that pixel. For example, unknown pixel #1 in the figure would be labeled as A (Agriculture) because it is the closest to the center of the A training site. Calculation of this minimum distance is simple and fast. However, there can be some labeling issues as demonstrated in classifying unknown pixel #2. Its minimum distance is to the center of B, and so pixel #2 would be labeled as B (Barren). However, careful examination of the graph indicates that it might be more accurate to label #2 as U (Urban). Note that Urban is a much more variable map class than is Barren, and therefore the pixels in the Urban training site are much more spread out than those in the Barren training site. Upon visual inspection, the analyst would be much more likely to label #2 as Urban. However, using the minimum distance algorithm, the pixel must be labeled barren. The minimum distance algorithm does not account for variance in any way. Therefore, while this rule is simple and quick to compute, it may result in inaccurate labeling of unknown pixels by ignoring the variation in the map classes.

The parallelepiped (or box) rule/algorithm attempts to solve some of this problem. Computing variances is computationally intensive, so the parallelepiped algorithm uses a surrogate for variance: minimum and maximum. A parallelepiped is drawn around each training-site plot using the minimum and maximum values for each band (figure 10.15c). Only unknown pixels that fall within the parallelepipeds are labeled. For example, unknown pixel #1 falls inside the A box and is labeled Agriculture. Now, unknown pixel #2 is labeled Urban because it falls within that box. However, the issue with this algorithm is that pixels falling outside of the boxes (see pixels #4 and #5) are left unclassified. This algorithm is still quite efficient and indirectly incorporates a measure of variation. The obvious shortcoming is that depending on the quality of the training data, some or many unknown pixels can be left unclassified.

The third common rule/algorithm to label unknown pixels in a supervised classification using the training sites collected is the maximum likelihood algorithm. This method is demonstrated in figure 10.15d. In this algorithm, the mean and variance are computed for each training area, resulting in a probability that any unknown pixel is part of each map class. The unknown pixel is then labeled with the map class that has the highest probability. This method can result in very accurate labeling when using effective training data. However, the method does require computations of the variance statistic and the assumption that the data is normally distributed to determine the probabilities for each unknown pixel. Figure 10.16 shows the Marin County area classified using a maximum likelihood classifier.

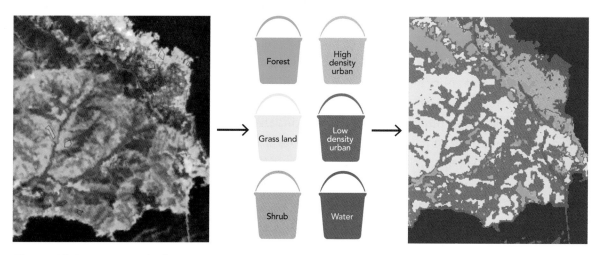

Figure 10.16. Map results from using a maximum likelihood classifier

Overall, the power of supervised classification is that the analyst "trains" the computer using well-chosen training sites that represent the information required in that mapping project. Supervised training samples are informationally unique. Each is delineated by the analyst to represent only one known and defined map class, making each training sample informationally unique. However, there is no guarantee that training samples will be spectrally unique. Map classes may be spectrally similar even though they are informationally distinct. If the training sites represent distinguishable spectral information, then the analyst can produce an accurate map. If some of the map classes are not spectrally separable, then the thematic map will not distinguish them and the map will have errors. This is the opposite situation that is faced in unsupervised classification, where the clusters are spectrally unique but possibly informationally confused.

Analysis can determine the potential spectral confusion between training samples of different map class types by using the data exploration techniques described in chapter 9. While you cannot ensure spectral uniqueness, you can do your best to reduce spectral confusion by choosing your imagery wisely (to maximize spectral variation between classes) and delineating your training sites with a minimum of within-site spectral variation.

Machine Learning Techniques

The final type of supervised mapping technique we will discuss is machine learning (see figure 10.7). Machine learning is an integral and growing component of data science. It is used across many disciplines including medicine, engineering, ecology, and information technology. Over the past two decades, machine learning has provided the tools behind many technological breakthroughs including speech recognition, image pattern recognition, and self-driving cars.

For the purposes of deriving information from imagery and geospatial data, machine learning mines the spectral information in the imagery and ancillary independent variables (such as slope and aspect) to find linkages between the independent variables and the map classes. These linkages are exploited by the machine learning algorithms and used to develop rules or decision trees that predict vegetation type across nonsampled areas.

There are many desktop applications for machine learning. The most powerful and widely used is the statistical package R, which is freely available. Beyond its core functionalities, R's huge ecosystem of users has developed packages (more than 6,000 as of this writing) that implement all commonly used machine learning algorithms. R is well integrated with Esri products, so that machine learning algorithms in R can directly read from and write to spatial and tabular datasets developed in ArcGIS. In ArcGIS Desktop, several common machine learning algorithms are available with the Spatial Analyst extension, making it easy to perform machine learning directly from within ArcMap and ArcGIS Professional.

During the past several years, there has been a growing adoption of web services for data science and machine learning. Web-services-based machine learning provides access to vast computing power, which can speed up machine learning tasks for large datasets by orders of magnitude. Web-services-based machine learning also enables publishing machine learning models via APIs, allowing them to be used repeatedly across the Internet. At this time, there are a growing number of options for performing machine learning using web services. These machine-learning-as-a-service platforms include Microsoft Azure Machine Learning Studio, Amazon Machine Learning, IBM's Watson Analytics, Google Prediction API, BigML, and DataRobot.

There are hundreds of machine learning algorithms. This chapter discusses a few of the commonly implemented algorithms for land-cover, land-use and vegetation mapping: Classification and Regression Tree (CART), Random Forests, Support Vector Machines (SVMs), and Artificial Neural Networks (ANNs).

CART, Random Forests, and SVMs are very effective for thematic mapping because they can accommodate a wide range of independent variables, including both categorical variables (such as geology or soils) and continuous variables (such as spectral reflectance or elevation). In addition, unlike other classifiers such as the maximum likelihood classifier, these machine learning algorithms don't require any assumptions concerning the distribution of the independent variables. These three algorithms are excellent tools for identifying both simple and complex relationships between variables that traditional techniques might not uncover.

CART

CART analysis (Breiman et al., 1984) is a nonparametric algorithm that uses a set of field-verified training data to develop a hierarchical decision tree. This decision tree is created using a binary partitioning algorithm to successively split a multidimensional cloud of explanatory data into increasingly homogeneous subsets. Once the final tree is generated, it can then be used to label all the unknown areas on the map. CART analysis has been widely used in the last two decades both for pixel-based and object-based image classification. The largest federal pixel-based land-cover mapping programs—the Gap Analysis Program and the National Land Cover Database—have relied heavily on CART as a core technology (**https://gapanalysis.usgs.gov/**).

Inputs for CART are independent variables for each training site (such as spectral information from high-resolution imagery and Landsat imagery, image texture, slope, aspect, geologic parent material, and other ancillary data). CART "mines" the sample data and builds rules that are if-then statements in hierarchical "trees" that condition the prediction of vegetation classes.

Case Study—a CART Decision Tree for Classifying a Forested Stand as Deciduous or Evergreen

This example is a simple illustration of CART. For this example, the authors used 2,551 training sites from field work in Sonoma County labeled simply as deciduous and evergreen forest. CART was run on these sites in R with over 150 independent variables using "rpart." For presenting a simple tree for the purpose of illustration, CART was run with a very-high-complexity parameter, producing a simple, highly pruned tree that used only the most powerful independent variables for discriminating deciduous from evergreen forest. The independent variables that were used are as follows:

B5_DIFF—The difference in Landsat 8 TM reflectance in band 5 (NIR) (spring minus winter reflectance)

NDVI—Normalized Difference Vegetation Index from late fall 2013 high-resolution orthophotography. In this image, deciduous vegetation—except in riparian areas—had mostly lost its leaves.

HAR—Lidar-derived Vertical Height Above River (vertical distance from the nearest stream channel)—see chapter 8 for details on this lidar derivative. Units in feet.

DIST_STR—Horizontal distance from the nearest stream channel. Units in feet.

In figure 10.C, the figures under the green and red "leaves" of the CART decision tree show the probability of being evergreen (left number) and the percentage of the total number of observations in the leaf. The top-level split used the Landsat 8 band 5 difference between winter and spring—this value is much higher for deciduous vegetation than for evergreen vegetation—hence the preponderance of evergreen observations on the right branch of the tree and deciduous observations on the left branch of the tree.

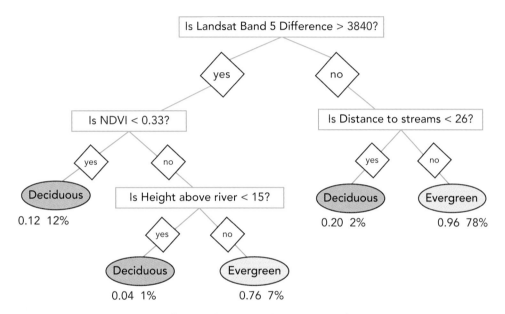

Figure 10.C. An example of a simple CART decision tree for predicting evergreen versus deciduous vegetation in Sonoma County, California

---

Random Forests

Random Forests represents an evolution of CART and was developed by Leo Breiman and Adele Cutler in the mid-2000s (Breiman, 2001). Breiman and Cutler's algorithm for Random Forests is implemented in R in the "randomForest" package.

Random Forests is likely the most widely used machine learning algorithm in thematic mapping at present. AmericaView, a nationwide partnership of remote-sensing scientists, performed a rigorous and extensive review of machine learning algorithms for land-cover mapping. They used data from various sources and across many geographies and found that Random Forests is the most effective machine learning algorithm when compared to other popular algorithms for land-cover mapping (Lawrence and Moran, 2015).

Whereas CART creates a single decision tree, Random Forests creates many trees (a "forest" of trees). To apply the model, an observation is run down all of the trees, and the most common prediction (the modal prediction from all the trees) is used as the final prediction. As such, Random Forests is in and of itself an ensemble classifier. Breiman's concept was that a bunch of "weak learners"—the individual trees—could be combined into a "strong learner"—the ensemble prediction.

Randomness is an important component of the Random Forest algorithm. For each tree in the "forest," only a random selection of the training data is run down the tree (two-thirds of the samples, sampled with replacement). The remaining one-third of training observations are "out of bag" and used for assessing classifier accuracy and for assessing independent-variable importance. Note that out-of-bag accuracy estimates generally overrepresent classification accuracy. For a valid evaluation of accuracy, test sites should be set aside and used as an independent source for validation (see Chapter 12).

In addition to the randomness in selecting the training data run down each tree in the forest, Random Forests imparts randomness in the way it deals with independent variables. Instead of using all independent variables at each node (decision point) in the tree, a random subset of variables is used—the number used is defined in the "mtry" parameter, discussed below. From this subset of randomly selected variables at each node, the one that optimizes the split is used.

There are several configurable parameters in Random Forests. One is the number of trees in the forest—the "ntree" parameter. Larger numbers of trees produce more stable models and are preferable, but computation time will increase as the number of trees increases. A safe starting point that should in most cases provide very stable models is 999 trees and—with today's computing power— should execute reasonably fast for all but the largest datasets. Another configurable attribute of Random Forests is the "mtry" parameter. This parameter determines the number of randomly selected independent variables that Random Forests will use at each tree node. The recommended value for this parameter is the square root of the number of independent variables. There is conflicting evidence in the literature as to the degree of influence that the "mtry" parameter has on classification results (Cutler et al., 2007; Strobl et al., 2007).

In addition to the classification itself, Random Forests produces a number of other outputs. The importance matrix provides useful information about the importance of each of the independent variables that are used to train the classification. This matrix shows each independent variable as rows and each map class as columns. By sorting this importance matrix in descending order for a given map class, one can see the list of independent-variable importance in descending order. Figure 10.17 shows a small subset of a Random Forest importance matrix for a vegetation classification in Sonoma County, California. The most important independent variables for *Quercus garryana* (Oregon white oak)—a deciduous oak—are shown and include a green index (i.e., [(mean green) – (mean red)]/ [(mean green) + (mean red)]), various measures of NDVI, and three Landsat TM 8 band 5 (NIR) spectral difference images that represent the difference in NIR reflectance between spring months (March, April, and May) and winter (February). Because *Quercus garryana* is deciduous, Random Forests found these band-difference images between leaf-on periods and leaf-off periods key to classifying these stands. In figure 10.17, the values in the right-hand column represent independent-variable "importance" (they are listed in descending order). Specifically, importance is the mean decrease in prediction accuracy (i.e., the mean decrease in the percentage of out-of-bag observations classified correctly) that would result from excluding the independent variable. The mean decrease in prediction accuracy is reported on a 0 to 1 scale (with 1 representing 100 percent).

Independent Variable Abbreviation	Independent Variable Description	Quercus garryana Alliance
MN_HINDVI	% of canopy w/ high NDVI in '13 orthos (not including non-veg areas)	0.080
MN_GREENDX	Green index (Green-Red)/(Green + Red), 2013 orthos	0.075
MN_B5DF_32	Mean Landsat 8 band 5 difference, March minus Feb	0.060
MN_B5DF_42	Mean Landsat 8 band 5 difference, April minus Feb	0.057
MN_LONDVI	% of canopy w/ low NDVI in '13 orthos (not including non-veg areas)	0.055
MN_NDVI	Mean NDVI, 2013 orthos	0.048
MN_NDVI_RA	Ratio of NDVIs between 2011 and 2013 orthos	0.036
MN_B5DF_52	Mean Landsat 8 band 5 difference, May minus Feb	0.033
MN_SOLARRA	Mean solar radiation	0.030
MN_SLOPE	Mean slope from lidar-derived bare-earth DEM	0.028
MN_BRIGHT	Mean 2013 ortho brightness index (from Ecognition)	0.027
MN_TM_NDVI	Mean Landsat 8 NDVI from 5/25/13	0.024
MN_BARE	Mean ground elevations from lidar-derived bare-earth DEM	0.023
MN_TM_GN	Mean Landsat 8 tasseled cap greeness from 5/25/13	0.022
MN_Wtr1AbAr_AV	Mean AVIRIS leaf water absorption index	0.022
MN_P90_30F	Mean lidar 90th percentile height from lascanopy	0.020
SD_P10_30F	Standard deviation lidar 10th percentile height from lascanopy	0.018
MN_STD_30F	Standard deviation lidar height from lascanopy (all returns)	0.018
MN_PRECIP	Mean average annual precipitation	0.017

Figure 10.17. A sample of a Random Forest importance matrix from Sonoma County, California

In addition to the importance matrix, Random Forests outputs a proximity matrix. The proximity matrix shows the number of times that a training observation is found in the same terminal node as another training observation. Lastly, Random Forests provides a list of the overall most important independent variables, as well as overall out-of-bag accuracies for each map class. In ArcGIS, Random Forests is available as part of the Spatial Analyst extension and part of the Segmentation and Classification toolset. Esri uses the OpenCV implementation of Random Forests.

Support Vector Machines

SVMs are widely used in machine learning, and in the past decade have become popular for thematic mapping using remotely sensed data and imagery (e.g., Mountrakis et. al, 2011). Like Random Forests, SVMs are nonparametric supervised classifiers. SVMs project the training data in a nonlinear manner into a feature space with a higher dimension than the input data's feature space. This projection is done using a kernel function and results in a linearly separable dataset (Congalton, 2010).

SVMs perform very well as a machine learning algorithm for thematic mapping and have been widely adopted. Recent literature shows that thematic map accuracy for maps produced using SVMs is close to that, and often exceeds, the accuracy of maps produced using Random Forests (Zhang et al., 2015; Waske et al., 2009; Ballanti et al., 2016).

Most implementations of SVMs provide a parameter for adjusting class weights. Setting this parameter to true (allowing the algorithm to adjust weights) often improves classifier performance. If the number of training samples per class is unbalanced (some classes have many training samples and others have few, for example), then SVMs can produce suboptimal results that are biased toward the common classes and underrepresent the prediction of the uncommon classes (He and Ma, 2012). Adjusting for class weights obviates this problem.

Another SVM parameter is the kernel, which is the function used to create decision spaces. For thematic mapping, the radial basis function (RBF) kernel is recommended. When using the RBF as the kernel, C (cost) and gamma parameters are also required. Cost defines the cost of misclassification of the training data—a small value for C sets a low cost of misclassification and allows a wider cushion between classes; a high value for C sets a high cost for misclassification, creating tighter margins between classes. Gamma controls how far the influence of a single training site reaches when the decision boundary is determined by the algorithm. Several tools, such as the "tune.svm" R package, can be used to select optimal C and Gamma values for a particular classification. In ArcGIS, SVM is available as part of the Spatial Analyst extension and part of the segmentation and Classification toolset. Esri uses the LibSVM implementation of Random Forests.

Artificial Neural Networks

ANNs were developed to mimic central nervous system and brain function. Though ANNs have recently been widely used in image and voice recognition, they have not seen broad adoption for thematic mapping using remotely sensed datasets.

Like other supervised approaches, ANNs are developed from a set of training data (samples with known dependent and independent variables), and the resulting model is used to make predictions for cases where no training data exists. Nodes are the most atomic unit of ANNs, analogous to neurons in a living organism. ANNs organize nodes into a connected system that is organized into three layers—the input layer, the hidden layers, and the output layer. The input layer is essentially the input data (akin to the independent variables in Random Forests or SVMs), the hidden layers are the network of interconnected nodes that embody the weighted rules of the model, and the output layer is the model's answer. An input to a neural network traverses the interconnected nodes guided by decision rules "learned" from the training data, eventually passing through the hidden layers to the output layer, which produces a prediction.

Because they are prone to overfitting and can be very complex (Congalton, 2010), ANNs aren't widely used for thematic mapping at this time. However—as evidenced in their broad adoption in pattern recognition and "deep learning"—ANNs are a powerful and promising technique.

Machine Learning Best Practices

Independent-variable selection is an important consideration for producing maps. Independent variables depicting elements that are correlated with the variation of the features being mapped should be used in machine learning. These include all types of imagery (aerial, satellite, hyperspectral) and especially NIR imagery, data layers that depict elevation and topography, and data layers that represent all manner of biogeochemical processes (climate, hydrology, geology, etc.) that play into the distribution of the thematic features being mapped. The final section of chapter 9 discusses many of the image and nonimage datasets that serve as useful independent variables.

The following are some of the independent variables that are effective for fine-scale land-cover and vegetation mapping:
- Spectral bands and indices (e.g., NDVI) from high-resolution orthophotography
- Spectral bands and indices derived from multitemporal Landsat imagery
- Landsat band-different images (spring minus winter, and summer minus winter)
- Landsat tasseled-cap transformations (brightness, greenness, and wetness)
- Hyperspectral band indices
- Canopy volume profiles derived from the lidar point cloud
- Canopy height strata metrics derived from the lidar point cloud
- Lidar-derived slope and aspect
- Lidar-derived elevation

- Lidar-derived canopy height
- Lidar-derived landscape metrics
- Fire history
- Shape indices that characterize stand shape (for object-oriented machine learning approaches)
- MODIS-derived fog/cloud frequency
- Average annual precipitation and other climate data
- Annualized solar radiation
- Height above river and horizontal distance from stream centerlines

CART, SVMs, and Random Forests perform well with a large number of independent variables and effectively winnow out the variables that aren't important. When selecting independent variables, it is important to consider their scale, positional accuracy, and thematic accuracy. These considerations are discussed in the final section of chapter 9.

Combined Approaches

With multiple algorithms available to classify imagery and ancillary data and with increasing computer power and capacity, it was inevitable that analysts would start to combine algorithms to improve classification accuracy. One of the first innovations was developed by Chuvieco and Congalton (1988) to combine the power and minimize the challenges of traditional per-pixel supervised and unsupervised classifications by developing a hybrid approach.

Figure 10.18 presents a comparison between the per-pixel supervised and unsupervised six-map-class classification of a study area in Marin County, California. Even a quick comparison of these two thematic maps reveals some differences that occur between classification approaches. The supervised approach, based on training sites, tends to produce a thematic map that is blocky, in which the map areas are more contiguous and larger. The unsupervised approach, based on spectral clustering, tends to produce a thematic map that is much more speckled, in which the map areas are more broken up and smaller. This is because the supervised approach begins with thematic information (the map classes) and derives statistics (from the training sites) that represent those map classes. On the other hand, the unsupervised approach begins with spectral uniqueness, resulting in finer discrimination of the pixels, which then must be labeled into map classes by the analyst.

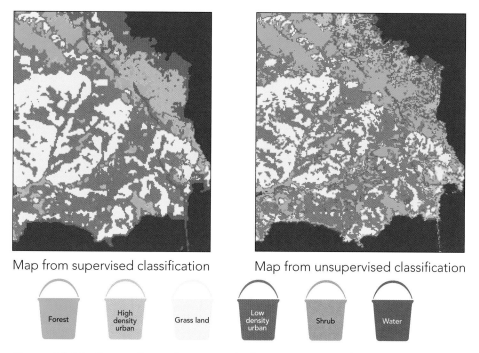

Figure 10.18. Comparison of map results from a supervised versus an unsupervised per-pixel classification of southern Marin County, California

The hybrid classification approach combines the spectral signatures or supervised training samples and unsupervised clusters to create an optimal set of spectral signatures. In this approach, informationally unique training samples are collected, and an unsupervised clustering algorithm is run on the imagery to produce spectrally unique clusters. The spectral statistics for both the training samples and clusters are then run through a Euclidian distance-clustering algorithm that creates a dendrogram, which shows how close the means of clusters and training samples are to one another in Euclidian multidimensional space, as shown in figure 10.19.

Because the results of a statistical clustering are to group all the clusters together into one final cluster, the analyst must determine the cut-off point for the clustering to stop. As can be seen in figure 10.19, a measure of Euclidean distance is used to determine whether a cluster should be merged. Clusters that group first (toward the left in the graph) are the most similar. Therefore, the power of this method is threefold:

1. Unsupervised clusters are merged with supervised training sites, thereby labeling the unsupervised cluster to the appropriate map class in a very objective manner while still ensuring excellent spectral separability. For example, figure 10.19 shows that unsupervised cluster 9 and supervised training site Water are spectrally very close to each other, and that cluster 9 represents water. The same can be said of clusters 1, 6, and 13, and Shrub.

2. Unsupervised clusters that do not merge readily with any supervised training sites indicate that there may be more spectral information in the imagery than is currently being represented by the map classes. Cluster 15 is such an example, as it merges very late in the statistical clustering process. This cluster should be examined closely to see what potential information it contains. As you already know, this cluster represents clouds over water—an area where no training sites were collected.
3. Supervised training sites of one map class merge with supervised training sites of another map class. This situation reveals that these two map classes are spectrally similar, and more work must be performed by the analyst to make sure that they can and should be distinguished. Figure 10.20 shows that Forest and Low-density urban are merged. In this case, this clustering makes sense, because low-density urban areas tend to have lots of trees, and the remote sensing device is above, sensing the tree canopies that are hiding the houses. It may be that other geospatial data layers beyond just spectral information are needed to separate out these two map classes. In other cases, the supervised training sites that are merged together do not make sense. In this case, the analyst has just learned that there is spectral confusion between the two map classes and that further data exploration is required before these classes can be accurately mapped.

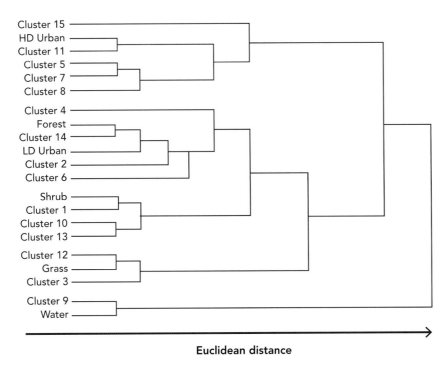

Figure 10.19. A dendrogram showing the results of a hybrid classification that clusters supervised training sites and unsupervised clusters together

The result is spectrally and informationally unique training statistics as well as the identification of informationally unknown but spectrally unique unsupervised clusters. Hybrid classification approaches are meant to take the best of traditional supervised and unsupervised classification and put them together to create the most accurate thematic map possible. Before the advent of faster computer power and advanced machine learning approaches, combining the benefits of traditional supervised and unsupervised classification was the best way to generate an accurate thematic map. However, the approach still has merit, especially because ArcGIS uniquely includes an unsupervised classifier for objects, making it possible to use this approach in object-oriented classification.

Recently, analysts have also combined machine learning techniques. As discussed above, Random Forests, SVMs, and CART can be powerful tools for creating maps. There are scores of other algorithms not discussed here that can also produce very good results for thematic mapping. In a given project, one algorithm may accurately map a map class in a given environmental setting, and another algorithm may map the same area poorly. But for another map class, the second algorithm may do a much better job than the first.

Because most machine learning algorithms run quickly (most of the work is in collecting and performing quality control on the training data), there is potential benefit in running more than one algorithm and combining the results in such a way that the final classification draws the best results from each algorithm—for example, by using the result from the algorithm that has the highest confidence in its prediction. This ensemble approach of running two or more machine learning algorithms and combining their results provides the opportunity for further increasing map accuracy beyond what can be obtained by using a single machine learning algorithm. For example, in the Sonoma County vegetation mapping project, both SVM and Random Forests were run against the training data and assessed for accuracy, allowing the analysts to choose the algorithm that best mapped each class.

Map Validation and Editing

Once the imagery and ancillary data have been classified, the map must be validated to discover and correct errors before it is formally assessed for accuracy. (See chapter 12 for an in-depth discussion of accuracy assessment). Errors can be corrected through reclassification or editing, depending on the level and type of errors perceived. Systematic errors are often corrected using rulesets, but also by the reclassification of confused classes. Random appearing errors require manual editing. This section discusses some of the commonly used methods of map validation.

If there are enough training sites, a very effective approach to validation is to split the training data into a pool that is used for developing the classification and a pool that is used for testing. The testing pool that is withheld from the classification process remains independent and can be used to independently assess classification accuracy during the process of classification. Figure 10.20 illustrates this workflow—in this example, 80 percent of the training samples (by class) are randomly selected to be used for model training; the remaining 20 percent of training samples are set aside as testing sites to be used to assess the accuracy of the classifiers. Once the classification is complete (using the training samples), the classification rules produced by the classification algorithms are run on the withheld testing sites. By reporting resulting accuracy using error matrices (see chapter 12), an independent assessment of classification for each choice of parameters or classification technique is produced. The process of running machine learning to create a classification on training data, testing the classification on the withheld testing data, and then changing parameters and repeating the process is the key to refining a machine learning based map. By iterating through this process many times, one can optimize the combination of classification parameters and algorithms, choosing the combination that produces the most accurate results. Of course, once the withheld sites have been used to refine the classification parameters and algorithms, they are no longer independent and cannot be used in formal accuracy assessment.

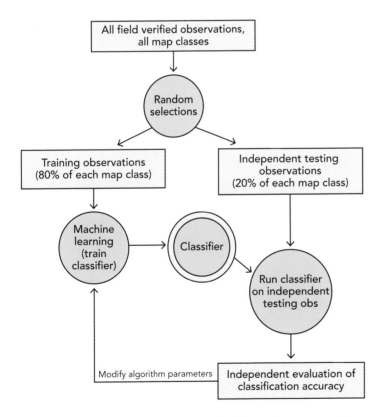

Figure 10.20. The workflow for withholding observations from modeling for refining classification algorithms and parameters

Even with fine-tuning the classification and algorithms using the techniques above, there will still be errors on the map that need to be manually edited before formal accuracy assessment. Editing needs can be slight for a simple map or require months of effort for a complex map with small MMUs.

Discovery of errors requires field trips if possible and a careful examination of the entire map. Comparison to historical maps is often useful, taking into account that the historical maps' scale, classification scheme, and registration may be substantially different from the newly created map. Consultation and draft map review with local experts are also recommended, not only for their knowledge of the project area but also to build local ownership of the map. Fortunately, at the end of the project the analyst will know considerably more about the project area than when the project was initiated, and the editing will be well informed.

Summary—Practical Considerations

This chapter reviews the various methods and techniques for converting image and ancillary data into map information. You have learned that even the most effective computer classifiers will require manual interpretation for the identification of training sites or clusters, and for map validation. Therefore, it is critical that the analyst is well-trained and fully understands the causes of map class variation across the landscape being mapped.

Both manual and semiautomated techniques require large fixed costs to initiate. In general, small project areas of less than 10,000 acres are mapped more cost effectively with manual techniques. Semiautomated techniques become cost effective with larger projects, where the marginal cost of mapping each additional acre decreases.

The more the map classes highly correlate with the imagery and ancillary data, the less expensive the creation of the map. Projects with many map classes are usually more expensive than projects with fewer classes. It is always tempting, but rarely cost effective, to add additional map classes during the map-making process. The more classes, the more likely for confusion between classes, which decreases the accuracy of the map.

While per-pixel classifiers have been the mainstay of semiautomated classification for decades, they are giving way to image segmentation and object-oriented classifiers as high- and very-high-spatial-resolution imagery becomes increasingly available and accessible.

Chapter 11
Change Analysis

Introduction

Imagery and GIS can be used to identify and map objects and resources at one point in time, as detailed in chapter 10, or to monitor how objects and resources change over time. As technology advances and populations increase, our planet is changing more rapidly than ever before. Changes in land cover and vegetation occur naturally through time and as a result of disastrous events such as fires, hurricanes, tornadoes, earthquakes, and tsunamis. They also occur as a result of human activities such as farming, logging, fishing, building, and wars. Resources are becoming scarcer, evidenced by rising land values and increasing conflicts over land use and ownership. To efficiently mitigate and respond to disasters and effectively manage and sustain our scarce resources, you need timely and accurate information about where resources occur and how they are changing over time. One of the most valuable uses of imagery is for change detection, because imagery provides an unbiased view of the landscape that is clearly understandable and interpretable by humans. As more and more imagery becomes available, your ability to analyze changes all over the world will continue to improve. This chapter focuses on the requirements for using GIS and imagery to monitor change.

Change detection is the analysis of information about a location over two or more points in time. The goals of change detection are many and include

1. detecting change,
2. measuring the extent and magnitude of change,
3. updating existing maps or GIS layers to incorporate change,
4. identifying the causes of change, and
5. assessing the impacts of change on environmental, economic, and political conditions.

Imagery and GIS support all these goals and are the primary sources of information for the first three worldwide.

Like any image classification, change detection requires
1. development of a classification scheme that clearly defines what type of change will be detected and mapped,
2. capturing the variation between the dates of imagery or maps that is related to the change of interest, and
3. controlling all nonchange variation between the various dates of imagery or maps to be compared in the change detection analysis.

This chapter reviews GIS and remote sensing methods for monitoring change and presents the pros and cons of each method. First, the different ways change is characterized are introduced. How you decide to characterize change affects the development of your classification scheme. The second section reviews the common methods used to perform change analysis. The chapter ends with a discussion of the practical considerations in performing a change analysis so that all nonchange variation is controlled.

Characterization of Change

Mapping and monitoring change can be very simple or very complex. Sometimes, it is only necessary to quickly identify areas that have changed. For example, you might want to see the areas in a rural–urban interface zone that were burned in a recent fire, or the location of houses destroyed by a recent tornado. In other cases, it is important to know more specific change information by map category. For example, it may be important for the commodities market to know that 25 percent of the area in Kansas that was corn last year is planted in wheat this year or for a city zoning board to understand the amount of forest land in the unincorporated area adjacent to the city that has been converted to development in the last five years.

In all cases, it is important that the change mapped be carefully characterized so that all map users clearly understand how change has been defined. Consider the example of mapping the change caused by a wildfire. The insurance adjuster may be interested only in damage to the structures within the fire boundary, while the ecologist may be interested in the intensity of the fire throughout the burned area, which will affect the soil and in turn the ability of the area to regenerate vegetation. Both are interested in change caused by the wildfire, but the type of change they want to identify is different. As with any mapping project, the type of change to be mapped needs to be defined by classification rules that result in totally exhaustive and mutually exclusive map classes (see chapter 7 for a discussion of the characteristics of a robust classification scheme).

There are four different types of change classification schemes: change/no change, magnitude of change, to/from map classes, and causes of change.

- Change/no change. This simplest change analysis provides a map of areas that have changed. Figure 11.1 shows Esri's swipe tool, which allows the user to quickly identify areas flooded during the 2016 floods in Louisiana. Images captured before and after the flood are registered to one another and displayed simultaneously on top of each other. By moving the slider back and forth and comparing the images you can see the extent of the flood at the time of the second image. A good example of an operational program that maps change continually can be found on the US Forest Service Active Fire Mapping Program website (**https://fsapps.nwcg.gov/afm/**), which uses low-spatial, high-temporal resolution MODIS and Visible Infrared Imaging Radiometer Suite imagery to detect and monitor wildfires daily in the United States and Canada.

Figure 11.1. Using the ArcGIS swipe tool to compare images before and after flooding in Louisiana in 2016 (esriurl.com/IF111)

- Magnitude of change. Often understanding that a change has occurred is not enough. Instead, you want to map the magnitude of change. An investor in crop futures might want to understand the magnitude of difference in crops in a region as they sprout, mature, and are harvested. A forester might want to measure how the percentage of tree cover/hectares has been decreased by clear-cuts, selective harvests, or insect defoliation. Mapping the magnitude of change requires more sophisticated methods than those used to map change.
- Before and after map classes. With additional analysis, a change map can be created that maps the areas of no change, and the before and after map classes of areas that have changed. Analysis of change for to/from map categories is significantly

more work and more difficult that producing just change/no change or magnitude of change analysis, because each map class before and after the change must be identified. NOAA's Coastal Change Analysis Program (C-CAP) is a good example of an operational program that provides to/from map class change analysis from Landsat imagery throughout the coastal areas of the United States. The program produces an online Land Cover Atlas (https://coast.noaa.gov/digitalcoast/tools/lca), which can be used to identify where and what type of change has occurred (figure 11.2).

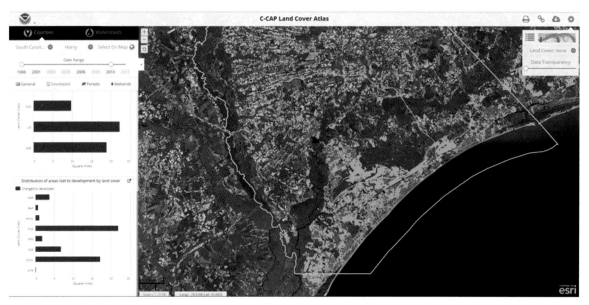

Figure 11.2. NOAA's C-CAP Land Cover Atlas illustrating in bright green the areas near Myrtle Beach, South Carolina, that have changed from the map classes agriculture, barren, wetland, forest, grass, and scrub to the developed map classes between 1996 and 2010. The charts at the left show the acreages of changed areas by map class. https://coast.noaa.gov/ccapatlas/ Source: NOAA

- Causes of change. The most difficult change mapping involves the identification of the causes of change. This can be easy when the change is easily identifiable and spectrally significant (e.g., deforestation from clear-cuts or the impacts of many natural disasters such as floods, wildfires, and tornadoes) and difficult when the change is subtle, or there is very little information about the change (e.g., illegal selective logging in the tropics, or marijuana cultivation under a closed canopy of trees).

Methods of Change Detection

Three types of methods can be used for change detection: 1) map-to-map comparisons, 2) image-to-image comparisons, and 3) image-to-map comparisons.

Multitemporal Map-to-Map Comparisons

Multitemporal map-to-map comparisons evaluate changes in map categories over time. Map-to-map comparisons are made possible through GIS technology, which allows for geospatial analysis of two or more digital maps that have been registered to one another. Because satellite and digital airborne imagery are relatively new, map-to-map comparison is often the only method available when attempting to complete retrospective change detection for the years before digital imagery became commonplace.

In multitemporal map analysis, two or more maps are compared to one another in a GIS. Three types of multitemporal map comparisons are possible: comparing two historical maps to one another, comparing a historical map to a newly created map, and comparing newly created maps from different dates of imagery. Table 11.1 summarizes the pros and cons of several types of multitemporal map analysis.

The success of map-to-map analysis centers on the ability to control differences between the maps that are not caused by change. Many maps are derived from remotely sensed imagery and are, therefore, one step removed from the change that appears on the imagery. As a result, multitemporal map comparison is easily confounded by differing classification schemes and registration issues, varying scales, and the methods used to create the maps being compared. Knowledge of these factors and taking every step to control them is necessary for effective map-to-map comparison. Historical maps are often of questionable registration, and are rarely accompanied by a robust classification scheme or any documentation concerning how the maps were made. Therefore, the probability that non-change issues will confound the analysis is high. On the other hand, new maps created from historical imagery require judgments about what existed on the landscape before a change occurred. Because you cannot travel back in time to see the landscape as it existed, you must make an educated guess. And if that guess is incorrect, you will introduce map error into the change analysis.

Table 11.1. Comparison of methods for multitemporal map comparisons

Types of Multitemporal Map Comparisons	Pros	Cons
Comparing Historical Maps	Often the only method available especially for analyses including dates before the 1970s. Relies on maps created during historical times, which tend to be more reliable than trying to create new maps from historical imagery.	Expensive to digitize the maps if they are not already digital. Historical maps are of questionable registration, accuracy, and classification schemes. As a result, there is a high potential for the combination of classification scheme differences, registration differences, method differences, and map error to result in an analysis that reflects the differences between maps, and not actual change.
Comparing Historical Maps to a New Map	Because at least one map is newly created, it is possible to control some of the nonchange variation better than when comparing historical maps. For this to be successful, the creator of the new map must fully understand the classification scheme and methods used to create the historical map.	High potential for the combination of classification scheme differences, registration differences, method differences, and map error to result in an analysis that reflects only map differences and not actual change.
Comparing New Maps from New and Historical Imagery	Classification schemes and methods can be forced to be identical, minimizing any confusion caused by classification scheme or method differences. Coregistration of the imagery captured since the 1970s is fairly straightforward.	Registration of pre 1970s historical imagery is often difficult if not impossible. Often cannot reliably identify past land-use/cover classes on historical imagery.

Multitemporal Image Analysis

Multitemporal image analysis compares differences in spectral responses of multitemporal imagery captured over the same area. The underlying assumption is that areas where map classes have remained constant will be represented by little or no spectral change, and that change will cause spectral differences. The change between crops and buildings or trees and no trees is evident when you are on the ground, looking at the landscape. These types of changes are also apparent in satellite or airborne imagery. When an agricultural field changes from corn to broccoli, or is converted to a housing development, or when a wetland is drained, the resulting spectral response captured by the remote sensing instrument also changes. Image analysts exploit the changes in image spectral response to detect and monitor change.

Multitemporal image analysis for change has become increasingly popular with the advent of accurately registered, inexpensive, and accessible multitemporal and multispectral imagery worldwide. Registration is key, because if the images are not accurately coregistered, the multitemporal image analysis will result in identifying image misregistration as change. Multispectral is also key, because the amount of information in a multispectral image makes it possible to use classification algorithms to identify and label multiple classes of change and no change. While it is possible to do change analysis using the near-infrared band from different sensors (e.g., Sentinel versus Landsat), considerable attention needs to be paid to minimizing any spectral differences caused by the different sensor systems. The bands used must also be inspected carefully for any atmospheric variances between the dates of the imagery. Slight atmospheric differences can be removed through a variety of techniques. However, clouds and cloud shadows in either date will seriously confuse the analysis.

Several methods are commonly used in multitemporal image analysis to detect change: manual interpretation of multitemporal imagery, image differencing, unsupervised classification, supervised classification, and continuous change analysis.

Manual Interpretation of Change from Imagery

Manually inspecting multitemporal images is the most prevalent type of change analysis. Change is identified through manual image interpretation and captured using on-screen digitizing. Often, the change is so dramatic that it can be identified using only one date of imagery, such as the one in figure 11.3, which shows an analysis of the impact of the 2004 tidal wave on Banda Aceh, Indonesia.

Figure 11.3. Single date change detection of the impacts of the 2004 tidal wave on Banda Aceh, Indonesia, as seen by DigitalGlobe's QuickBird satellite. Source: DigitalGlobe

Typically, images of two different dates are examined side by side or using a slider, as shown in figure 11.1. Figure 11.4 shows another tool for manual change analysis that combines two bands of different dates of Landsat into a *change image* that shows urban development in western Las Vegas, Nevada. Rather than the analyst swiping across the screen to identify change, the multitemporal change image helps to focus the analyst's attention on the areas that have changed. The change image is created by placing a band from one date (1993) in the red and blue channel of the computer display, and the band from another date (1998) in the green channel. Areas in shades of gray have similar band reflectance in both years. They are in shades of gray because all three channels of the display are reading the same spectral reflectance for the unchanged pixels. In the 1998 image, magenta (blue + red = magenta) areas are darker and green areas are brighter. It is up to the analyst to identify how those changes in reflectance relate to changes on the ground. Figure 11.5 shows how a change image works mathematically.

Figure 11.4 uses the near-infrared bands of two Landsat images. If you understand the vegetation of Las Vegas and how vegetation reflects near-infrared energy, you can turn the change image into a map of change. Las Vegas vegetation is mostly either sparse desert vegetation or irrigated landscapes. The soil in desert areas is light colored and bright. When land is cleared for new developments, it changes from desert vegetation to bright soil, thus the reflectance is higher in the near infrared and green on the change image. After the development is complete, the bright cleared soil becomes revegetated with irrigation or paved with black asphalt. Near-infrared reflectance then decreases, and the areas show up as magenta on the change image. In some areas, the change is from desert vegetation to irrigated vegetation.

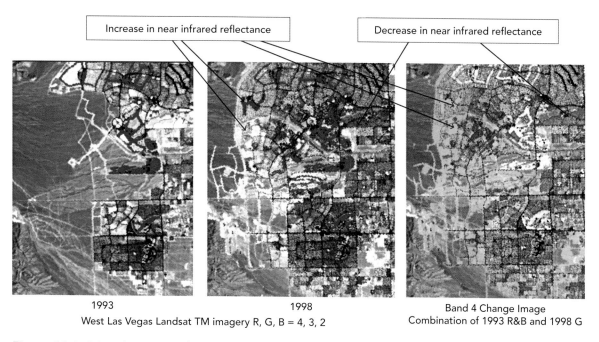

Figure 11.4. A Landsat near-infrared change image highlighting urban development between 1993 and 1998 in the west Las Vegas area (esriurl.com/IG114)

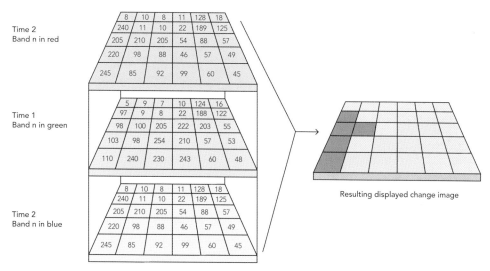

Figure 11.5. Example pixels from a change image. Most areas in the image will be gray, indicating no change, because each pixel has relatively the same value in each band. Areas that show up green were brighter in Time 1 than in Time 2. Areas that show up in magenta were brighter in Time 2 than in Time 1.

It is also possible to use an index (e.g., a normalized difference vegetation index [NDVI], tassel-cap greenness) instead of a specific band to create the change image. Indices have the advantage of normalizing many of the atmospheric effects. Also, the values are easier to deal with, because the range of the input bands will be from −1 to 1, rather than a huge range that would result from band values from images with dynamic ranges greater than eight bits.

Image Differencing

Image differencing involves subtracting the spectral values or index values of the image dates from each other for each pixel (figure 11.6). In areas of no change, the resulting values of the subtraction should be close to zero. Difference values in changed areas will show either an increase or a decrease in spectral reflectance. The differences in spectral reflectance can then be related to change on the ground. Esri's Landsat Explorer (figure 11.7) is an online tool that allows users to compare images and perform image differencing on Landsat 8 and Global Land Survey archive imagery anywhere in the world.

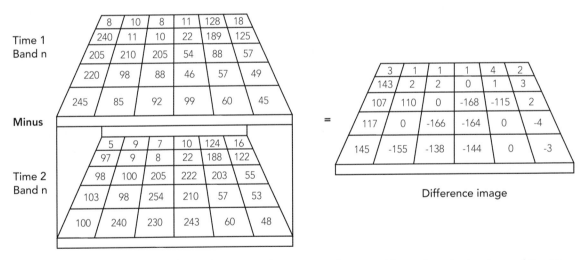

Figure 11.6. The pixel values from the Time 1 band are subtracted from the pixel values of the Time 2 band to result in a new image.

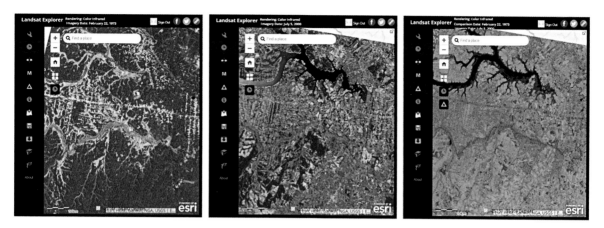

Figure 11.7. Esri's Landsat Explorer app showing land-use changes in Paraguay between 1973 and 2000 as forests are converted to farms and a reservoir on the Lago del Rio Yguazú is filled (esriurl.com/IG117)

Returning to our Las Vegas example, figure 11.8 displays the resulting panchromatic band from differencing the near-infrared bands of 1993 and 1998 Landsat imagery. Once the difference image is created, it must be separated into map classes correlated with no change in spectral response, increases in spectral response, and decreases in spectral response. Usually, this is done by examining the difference image and deciding how landscape change is related to spectral change. Examining the histogram of the difference-image pixels can be helpful in determining which pixel values represent change. Figure 11.9 shows the histogram of the eight-bit difference image of figure 11.6, which has been rescaled from its original values (−127 to +127) to have a range from 0 to 255. Most of the change-image pixels fall in the no-change range around the mean of 127, with the change pixels composing the tails of the histogram. Creating a map from the difference image requires "slicing" the histogram into classes of change and no change. The image analyst must determine which difference values represent change and which don't, as shown in figure 11.10.

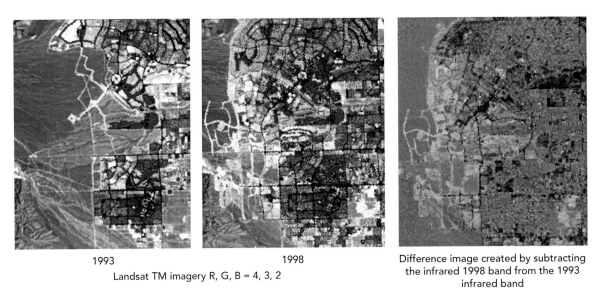

1993

1998

Landsat TM imagery R, G, B = 4, 3, 2

Difference image created by subtracting the infrared 1998 band from the 1993 infrared band

Figure 11.8. Example of a difference image over west Las Vegas (esriurl.com/IG118)

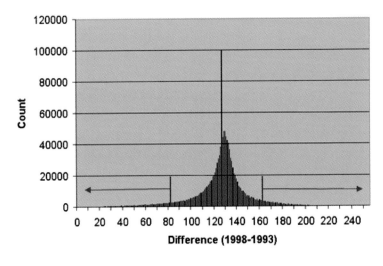

Figure 11.9. Histogram of the difference-image pixels

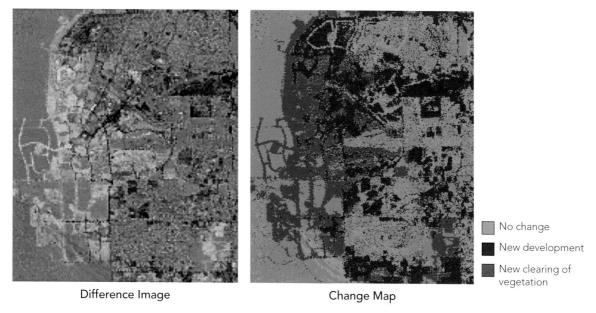

Difference Image   Change Map

■ No change
■ New development
■ New clearing of vegetation

Figure 11.10. A change map created by slicing the difference-image histogram. Pixels to the left of the blue slice in figure 11.9 are labeled "new development." Pixels to the right of the red slice in figure 11.9 are labeled "new clearing of vegetation." Pixels between the red and blue slices are labeled "no change." (esriurl.com/IG1110)

Image differencing is easy to do. However, it is somewhat of a "sledgehammer" approach to change detection, allowing for the identification of several classes of change (e.g., increase in vegetation, decrease in vegetation, etc.), but only one class of nonchange. Therefore, in the nonchanged areas, you know only that the areas have not changed, but you do not know what they are. You also know little about the change areas. They can be denoted as positive change and negative change, but little information about changes in specific map classes is available.

Multitemporal Unsupervised and Supervised Classification

As you learned in chapter 10, unsupervised classification algorithms statistically group or cluster the spectral variation in the imagery. Unlike image differencing, which is restricted to one band from each date, unsupervised classification can use more bands, which can support mapping the entire landscape and may allow for distinguishing more subtle classes of change. In figures 11.11 and 11.12, three bands from each Landsat scene have been combined into a six-band multitemporal image of our west Las Vegas area. The six bands were then classified using an ISODATA classifier and the clusters labeled based on the analyst's knowledge of the area. Unsupervised classification of multitemporal, multispectral imagery is more difficult to do than image subtraction, because the clusters must be analyzed and labeled across the entire landscape (not just the changed areas). However, because of the amount of information contained in a multitemporal, multispectral image, subtle changes can often be discerned, and sometimes to—from classes can be identified.

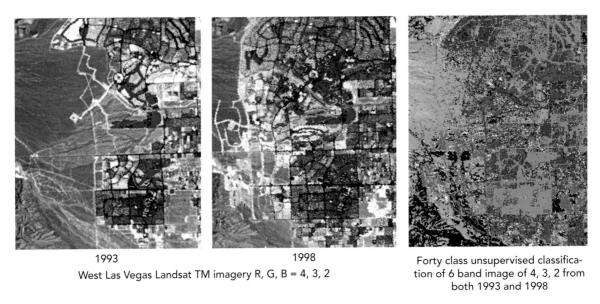

1993
1998
West Las Vegas Landsat TM imagery R, G, B = 4, 3, 2

Forty class unsupervised classification of 6 band image of 4, 3, 2 from both 1993 and 1998

Figure 11.11. Unsupervised classification of multispectral, multitemporal imagery of west Las Vegas

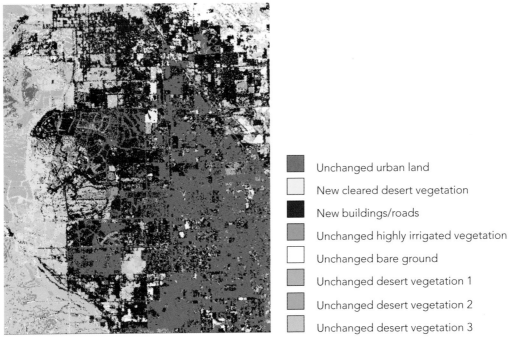

- Unchanged urban land
- New cleared desert vegetation
- New buildings/roads
- Unchanged highly irrigated vegetation
- Unchanged bare ground
- Unchanged desert vegetation 1
- Unchanged desert vegetation 2
- Unchanged desert vegetation 3

Figure 11.12. Unsupervised clusters labeled into a map of the entire landscape

Similarly, supervised classification of multitemporal imagery can also be used to produce a detailed map of both changed and unchanged areas. Any of the supervised methods discussed in chapter 10 can also be employed in change detection. However, training sites must be carefully chosen to capture all the change and nonchange variation in the multiple dates of imagery. Table 11.2 presents a comparison between image differencing, supervised classification, and unsupervised classification showing the pros and cons of each.

Table 11.2. Pros and cons of image differencing, unsupervised classification, and supervised classification for change detection analysis.

Types of Multitemporal Image Comparisons	Pros	Cons
Image Differencing	Simple to implement. Focuses only on changed areas.	Only provides change and nonchange classes. Very difficult to identify to–from classes. Because only one band or index is used, not powerful for identifying spectrally subtle or complex types of change.
Unsupervised Classification	Able to detect more subtle classes of change than image differencing. Results in a map of the entire landscape as well as identification of changed and nonchanged areas. Often possible to identify some of the change clusters as to–from classes.	Substantially more difficult to implement than image subtraction. Requires identifying clusters across the entire landscape, not just for the changed areas.
Supervised Classification	Able to detect more subtle classes of change than image differencing. Results in a map of the entire landscape as well as identification of changed and nonchanged areas.	Requires training samples of all landscape and change classes which means that the type of change occurring on the ground must be well understood.

Continuous Change Analysis

Given the historically large expense of high-to-moderate-spatial-resolution imagery, until recently, most change detection analysis was limited to a few selected dates in time. Some large-area (continental) change/trend analysis was performed using low-spatial-resolution imagery such as AVHRR or MODIS (Tucker et al. 1985). However, with the advent of free Landsat imagery in 2008 and the launches of Sentinel 2 A and B, it is now possible to conduct moderate resolution continuous change analysis (Zhu et al., 2016).

Similar to multitemporal pattern analysis using many dates of imagery (image cube) described in chapter 9, continuous change analysis also uses large volumes of imagery to look at trends and detect changes. Using many dates of imagery with as many images per year as possible allows the analyst to look at trends in the area of study. For example, measures of NDVI, other indices such as greenness, or even just simple ratios of NIR/red can show patterns of vegetation growth over many years. Other measures/indices may show patterns of water loss or gain, or shrinkage or expansion of deserts. Many of these analyses use machine learning to identify change.

An issue when looking at trends over long periods is what to do if the area experiences a large change that disrupts the trend. For example, looking at a coniferous forest over a 30-year period beginning when the trees were just planted until they have grown into a closed canopy forest would show a positive upward trend in NDVI. However, if some partial harvesting or thinning of the trees occurred during the 10th year, this abrupt change would cause a disruption in the trend.

Therefore, many studies using an image cube with many image dates to conduct trend analyses are done in areas that have not experienced abrupt change. However, because change is inevitable, it would be far better to perform a continuous change analysis that can incorporate abrupt change. Several scientists have explored ways to incorporate abrupt change. One method is the Continuous Change Detection and Classification method proposed by Zhu and Woodcock (2014). This method uses an image cube of all available cloud-free imagery to determine trends, but it can also detect abrupt changes. When a change occurs, the trend is interrupted and a new trend is begun starting from the abrupt change and using the imagery available by moving forward from that point. In this way, this method incorporates both continuous change and abrupt change into the same analysis.

As with all classification algorithms, many algorithms are available for continuous change analysis, and they often do not agree. For example, Healy et al. (in review) discovered significant differences between change detection algorithms for monitoring forest disturbance. They suggest using an ensemble approach that uses quantitative measures of success and failure to weigh and merge algorithm results.

Multitemporal Image-to-Map Comparisons

A final method of multitemporal change analysis compares an older map to new imagery and/or old imagery. An operational example of manual map-to-image change detection is the Humanitarian Open Street Map Team (**https://hotosm.org/about**), which relies on its vast network of volunteers worldwide to map the impacts of disasters online by comparing Open Street Map layers to new imagery captured after the disasters. In automated change detection, two commonly used methods for image-to-map comparisons are masking and cross correlation.

Masking or Postclassification Change Detection

Masking or postclassification change detection, also often referred to as the C-CAP (NOAA Coastal Change Analysis Program Protocol), relies on the use of multitemporal imagery to create a binary mask of change versus no change. The Time 1 imagery is the same date as an existing map, and, in the best case, is actually the imagery used to create the existing map. The Time 2 imagery is from the date to which the existing map is to be updated. Multitemporal image comparison (either image differencing or unsupervised classification) is used to identify the areas of change. Only the changed areas are mapped, and the newly mapped areas are superimposed on the existing map, updating that map in areas that changed. Tasks to complete masking change detection are as follows:

1. A map of changed and nonchanged areas is created using one of the several methods mentioned in the earlier concepts (e.g., image differencing, unsupervised).
2. The map of changed and unchanged areas is converted into a mask.
3. The mask of change/no change is combined with the existing map to separate the existing map into areas of change and no change.
4. The areas of no change are left as is in the existing map.
5. The areas of change are classified with the Time 2 imagery, most preferably using the same techniques as were used to create the Time 1 map.
6. The Time 2 map of the changed areas is then overlaid with the existing map to create an updated map.

Some of the advantages of this type of change analysis are that the new map created carries with it the to–from classes for the areas that have changed. Not only do you know what map class an area is today, you also know whether it has changed during the period of analysis, and, if it has changed, what it was before the change. Additionally, because only the changed areas are mapped in Time 2, the level of effort is less than that required to create complete landscape maps from Time 1 and Time 2 and then comparing them to detect change using multitemporal map analysis. Finally, because the changed areas are identified spectrally in the multitemporal imagery, the potential for map error (from either the Time 1 or Time 2 maps) being confused with change is reduced.

Cross-Correlation Analysis

Cross correlation is a standard method of estimating the degree to which two datasets are correlated. Cross correlation in change detection allows using only one date of imagery to update an existing map (Civco et al., 2002). First, new imagery is registered to the map to be updated, and then the expected average spectral value of each landscape class on the map is determined using cross-correlation analysis between the existing map classes and the new imagery. Using the results of the cross-correlation analysis, a Z statistic is derived for each pixel or segment in the new image, which indicates how close the pixel's spectral response is to the expected spectral response for its corresponding landscape class on the map. If the change is spectrally significant, the spectral response of changed pixels will be significantly different than the expected value for that land-cover/use class. A map of probability of change can then be created for each pixel and used to identify areas of probable change in the existing map. The new imagery can then be used to update the existing map in only the areas that have changed. Table 11.3 presents a comparison between masking and cross-correlation change analysis showing the pros and cons of each.

Table 11.3. Pros and cons of masking versus cross correlation for change detection analysis

Types of Multitemporal Image to Map Comparisons	Pros	Cons
Masking	Only requires new mapping for the changed areas. Results in not only updated map, but also to–from classes for changed areas. Minimizes the potential of map error being mistaken for change.	Requires complex manipulation of multiple data sets.
Cross Correlation	Eliminates nonchange variance caused by seasonal or atmospheric conditions. Uses only one image rather than two or more, minimizing costs.	Requires that change be a rare event for the statistics to work. Time 1 map must be very accurate. Inaccuracies in the Time 1 map result in false expected pixel values.

Managing the Nonchange Differences

Multitemporal analysis of GIS layers or digital airborne or satellite imagery is effective for change detection only because a high correlation exists between variation between multitemporal maps and/or images and land and land-use cover change. Hopefully, when the ground changes, the GIS coverages or imagery also change as specified by the classification scheme. Change detection necessitates understanding what caused the changes on the ground and understanding how the imagery or maps respond to those changes. As with image classification, a critical step in any type of change detection analysis is to remove or neutralize any nonchange image or map differences that may be misinterpreted as changes in land cover or vegetation (see chapter 6 for an in-depth discussion of techniques for removing unwanted variation in the imagery). For example, misregistration of multitemporal images or maps will result in a map of misregistration rather than a map of change. Seasonal differences in sun angle can create shadows in one image that are not present in the second and may be mistaken as land-cover differences (figure 11.13). Variations in atmospheric conditions can produce radiometric differences between two images that were acquired on separate dates. The requirement of change analysis is to isolate the change of interest from random, uninteresting, or spurious variation, and from areas of no change. Accurate change detection requires that all such variation be removed or controlled.

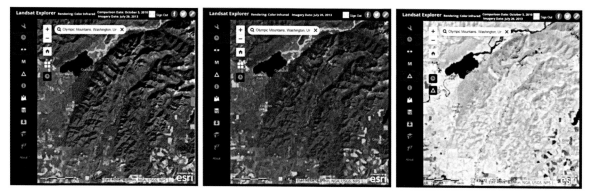

Figure 11.13. A change image of the Olympic Mountains in Washington showing how differences in sun angle can result in changes in spectral response that are unrelated to land-cover change. The figures show a comparison of imagery captured in 2010 (left) and 2013 (middle). The change image (right) shows a spectral difference (in green) on the north-facing aspects of the mountains that is caused by sun angle differences resulting from the 2010 image being captured in the fall, and the 2013 image being captured in the summer.

Four major types of nonchange variation need to be controlled:
1. Registration differences
2. Classification-scheme differences
3. Map errors
4. Image variance

Multitemporal map change analysis is potentially affected by several types of nonchange variation including misregistration, classification-scheme differences, map error, and image variation. Multitemporal image change analysis is affected only by misregistration and image variation. As a result, control of nonchange variation is often easier in image-to-image change analysis.

Registration Differences

Registration differences are caused by misregistration of the multitemporal maps or imagery to one another. The result is areas of misregistration between the two maps or two images being mapped as changes. Misregistration is a significant problem when using old maps in map-to-map change detection. The accuracy of map registration decreases with the age of maps created before the use of calibrated aerial photography, precise survey equipment, and especially GPS (i.e., the older the map, the more likely there are significant problems with registration). Registration differences can be controlled through careful registration of the datasets to one another, as discussed in chapter 6.

Classification-Scheme Differences

Classification-scheme differences occur only in multitemporal map comparisons when the maps being compared were developed using different classification schemes. The differences can be caused by using different labeling rules for the map classes, by using different minimum mapping units, or by both. The impact of classification-scheme differences is that areas can be mapped as change, when in reality the differences are actually caused by differences in the classification schemes used to create each map. For example, some classification schemes define forests as areas with more than 10 percent crown cover in trees, while other classification schemes describe them as areas with more than 30 percent crown cover in trees. If one attempted to conduct change detection using two maps with these two different classification schemes, one would end up with a map that showed differences in forest, but also a map that showed differences in the definition of forest. All of the unchanged areas with crown closure between 10 percent and 29 percent would show up as changed areas, and thus greatly overestimate change. Similarly, a map with a minimum mapping unit of 100 acres cannot be directly compared to a map with a minimum mapping unit of 1 acre.

The only way to control classification-scheme differences is to fully understand both the labels and the rules used to create both maps being compared. If the rules are well known, then in many cases a translation (often called a crosswalk) can be created between different classification schemes to minimize the differences.

Map Error Differences

Map error differences occur only in multitemporal map comparisons when one or both of the maps being compared contain mapping errors. The impact of map errors is that areas can be mapped as change that really are different only because one or both maps contain map errors. For example, if the earlier map correctly identifies an area as a wetland, and the later map incorrectly identifies the same area as an agricultural field (when it is really a wetland), then the change analysis will indicate that a change has occurred (wetland to agriculture), when in fact no change has occurred. Because few (if any) land-cover or vegetation maps are 100 percent correct, the probability of introducing this type of confusion into map-to-map change detection is relatively high. The only way to control map error differences is to know the accuracy of all the maps used in the change analysis. This requires that an error matrix exists for all the maps (see chapter 12), so that the analyst understands the class accuracies of each map, and, therefore can understand in what classes map error confusion might occur.

Image Variance

Image variance occurs in multitemporal change analyses when some of the changes between the images are caused by factors other than the land-cover or vegetation change of interest. Differing atmospheric conditions, sun angles, seasons, or tides between the two or more images being compared are common causes of image variance. The impact of image variance in change analysis is that areas can be mapped as change, when in reality the differences exist in the images and are not related to change. For example, clouds in one image can confuse the change analysis, often resulting in the clouded areas being mapped as change when the only reason a difference exists is because of the clouds. Differences in sun angle can also appear to be a change. Image variance is controlled through careful selection of the multitemporal imagery to be used in change analysis. Image variance is especially important in coastal studies, where tide levels can make a huge impact. Because satellites are restricted by their orbits, it is extremely difficult to acquire cloud-free multitemporal imagery of a coastal area of the same season and tide level. Image variance can be a huge source of confusion in per-pixel, high-spatial-resolution change detection.

Summary—Practical Considerations

Change detection is a type of image classification that focuses less on an inventory of map classes and more on identifying processes occurring over time within and between map classes. This chapter discusses the various methods for using imagery to detect, measure, and identify landscape change. You have learned that change detection requires special attention to controlling differences between different dates of imagery that are not related to map class change.

Section 4
Managing Imagery and GIS Data

Chapter 12
Accuracy Assessment

Introduction

Any new technology, in its infancy, experiences an explosion in use coupled with tremendous excitement about the potential of the technology. Little thought is initially given to the quality of the results produced, because all efforts are devoted to experimentation and exploration of the new technology. As the technology begins to mature, more thought and effort are dedicated to assessing the value and accuracy of the results. This process is a natural progression in any new field, and remote sensing and geospatial analysis are no exceptions.

Remotely sensed imagery (i.e., aerial photography) came into being as a result of World War I and especially World War II. Nonmilitary applications of aerial photography such as crop mapping, vegetation health, forest type mapping, and many others benefited from the large, trained, workforce after the wars. However, given that photo interpretation is such a time-honored skill, little effort was devoted to evaluating the thematic accuracy of maps derived from aerial photographs for these nonmilitary applications. All this changed with the 1972 launch of Landsat, the first civilian digital image satellite for earth observing. During the next 10 years or so, much excitement and effort surrounded this digital imagery's plethora of uses including land-use and land-cover maps that were impractical to produce at a landscape level before Landsat and before the maturing of the computer era that we now enjoy. However, beginning in the early 1980s, some researchers began to think about evaluating the accuracy of the maps produced from this digital imagery (e.g., Congalton et al., 1983). Techniques for assessing the accuracy of maps derived from remotely sensed imagery have steadily increased in complexity and usefulness since that time (Congalton, 1991; Congalton and Green, 2009).

The same process can be demonstrated for other geospatial data. While the concept of analyzing spatial data has been around for a very long time, the science and application of GIS has only been popular since the mid-1980s. Again, early in the development of GIS, little was done to consider the accuracy of the spatial data. Instead, the excitement and enthusiasm surrounding this amazing technology ruled the day. Once the technology matured a little, it was natural for some to begin to ask about the quality of the data and the accuracy of the decisions that could be made using GIS. Today, the quality of much geospatial data is recorded in the metadata and can readily be viewed by the data user. In addition, qualitative measures of completeness and logical consistency have been used to evaluate geospatial data layers.

This chapter presents an overview of the concepts, considerations, and techniques behind quantitatively assessing the accuracy of maps created from remotely sensed imagery. The goal is to provide an overview for the GIS analyst to be aware of the need for and general methods to assess the accuracy of imagery and maps generated from remotely sensed data. The chapter begins with a section on map accuracy followed by positional and then thematic assessment considerations and techniques. The chapter concludes with a discussion of the practical considerations necessary to conduct an assessment. Much more detail about everything presented in this chapter can be found in Congalton and Green (2009).

Assessing Map Accuracy

Before beginning to think about assessing map accuracy, it is important to understand the concepts of accuracy and precision. These terms are often incorrectly used interchangeably and it is important to understand the difference. Accuracy is how close one is to the correct value/answer while precision is a measure of repeatability. In most cases, high accuracy and high precision are desirable. Shooting at a target provides the best analogy to easily understand these concepts. Figure 12.1 demonstrates both accuracy and precision. Accuracy is shown on the target as those shots that hit near the bull's-eye, because that was where the shooter was aiming. Precision is demonstrated when shots are closely grouped. A shooter may be very precise, but not very accurate. This condition might indicate that the scope of their rifle is not properly aligned, forcing him or her to miss the bull's-eye. The shooter may also be accurate, but not very precise. In this case, the shots would all be near the bull's-eye, but not in a nice, tight grouping. Finally, if the shooter is both very accurate and very precise, the target would show a very close grouping of shots at the bull's-eye.

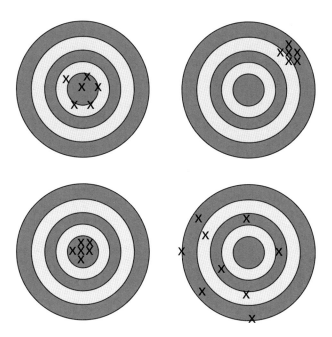

Figure 12.1. Demonstration of the concepts of accuracy and precision using targets and a bull's-eye. The upper-left target demonstrates accuracy, but not precision. The upper-right target demonstrates precision, but not accuracy. The lower-left target is both accurate and precise, while the lower-right target is neither.

The first step most analysts take when assessing map accuracy is to visually evaluate the map and conclude either "This map looks good" or "This map looks bad." Actually, four qualities or characteristics of a map can be evaluated. They are 1) logical consistency, 2) completeness, 3) positional accuracy, and 4) thematic accuracy (Bolstad, 2012). Both logical consistency and completeness are qualitative measures of map accuracy that are more useful than the simple idea that the map "looks good." Of course, it is important to note that if the map does not look good to the analyst, there is little reason to continue with the assessment until the map is revised enough to at least meet these minimum criteria.

Logical consistency evaluates whether the map makes sense. Are objects in the correct places? For example, are lamp posts or fire hydrants shown in the middle of the road? Is still water mapped on the slope of a mountain? Asking these questions to make sure the map is consistent is a valuable step in assessing its accuracy.

Completeness helps the map user assess whether everything is on the map that is supposed to be represented. Are components missing? Have areas been left out that obviously should have been included? It is possible that a map could be very accurate for what is included on the map. However, if much of what is supposed to be included is absent, the overall completeness of the map will be low. Again, careful study of the map to qualitatively assess it for completeness is a useful step before beginning the more rigorous quantitative measures of accuracy. If a map is inconsistent or incomplete, it will usually not look good.

Using these two qualitative assessment methods are important to make sure that the map is ready for more quantitative measures of accuracy.

In addition to these methods for qualitatively evaluating a map, there are two very important quantitative accuracy assessment techniques: positional map accuracy assessment and thematic map accuracy assessment. Positional map accuracy assessment evaluates the location of objects on a map. In other words, is everything in the correct place? Assessing positional accuracy to verify that the data all lines up and is in the correct place is key to using any spatial data in a GIS. A spatial data layer created using remotely sensed imagery must line up with the other data layers in the GIS to be useful.

Thematic map accuracy assessment evaluates whether the object has been given the correct label (i.e., theme). Thematic accuracy assessment is a critical component of maps created from remotely sensed imagery, because these maps tend to be thematic (e.g., land-cover types, crop maps, forest fire maps). Positional and thematic accuracy are inherently linked when assessing map accuracy. If an object is labeled correctly but is in the wrong place, an error will occur. If an object is in the right place but is labeled incorrectly, an error will occur. Therefore, when assessing map accuracy, it is important to consider both positional and thematic accuracy.

A number of components and considerations are common to assessing both positional accuracy and thematic accuracy. These include 1) some initial considerations related to sources of error and the classification scheme used, 2) procedures for collecting the reference data including sampling and other data considerations, and 3) computing descriptive statistics and other analysis techniques. Conducting a positional accuracy assessment is somewhat simpler than conducting a thematic accuracy assessment and has had more standards developed for it over the years. Thematic accuracy assessment is more complex and requires balancing statistical validity with what can be practically achieved. This chapter next describes the process of positional accuracy assessment, and then follows with thematic accuracy assessment.

Positional Map Accuracy Assessment

Assessing positional map accuracy means determining whether the location on the map agrees with the position on the ground. In other words, is the map location in the correct place? Figure 12.2 shows an example of a road map overlaid on an orthocorrected image. The green dots represent the location of ground survey points of road intersections. As you can see, some of the intersections align well with the survey locations and others do not. It is important to determine the amount of positional error using a positional map accuracy assessment and compare the results to some standard to see if the map is acceptable. In most cases, positional accuracy is measured using a series of points. However, it should be noted that linear features could also be used to determine accuracy.

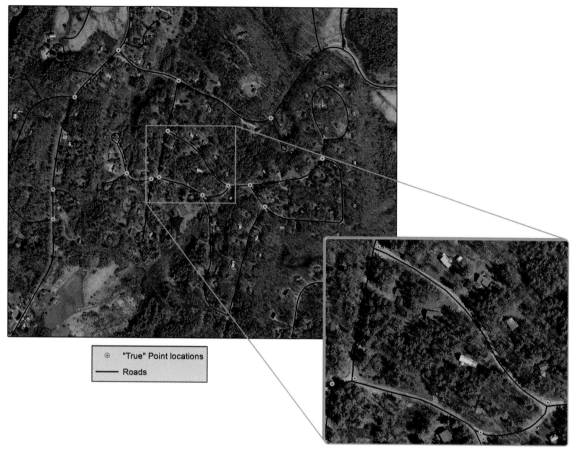

Figure 12.2. Example of positional accuracy assessment of a road network where the points on National Agriculture Imagery Program (NAIP) imagery show the surveyed ground location and the positional error in the geospatial road network layer

Positional map accuracy assessment has always been achieved through a series of standards that set limits or guidelines for determining whether the positional accuracy of a map was sufficient for that map to be used. Initially, those standards were for paper maps. Only recently have the standards expanded to address digital mapping. The very first standard was called the National Map Accuracy Standards (NMAS) and was developed by the United States Bureau of the Budget in 1947 (United States Bureau of the Budget 1947). NMAS is easy to apply, but is quite limited in its usefulness, as it does not specify any method for providing statistical bounds around the positional error. Instead, it simply uses a percentile calculation for accepting or rejecting the map.

The next standard was suggested by Greenwalt and Schultz (1962 and 1968) in a report titled "Principles of Error Theory and Cartographic Applications." This standard provided equations—still important today—to estimate the maximum positional error interval for a given statistical probability level. Guidance was provided for both a vertical or one-dimensional map accuracy standard and a horizontal (x and y) or 2D circular map accuracy standard.

The American Society for Photogrammetry and Remote Sensing (ASPRS) produced a standard in 1989 called the ASPRS Interim Accuracy Standards for Large Scale Maps (ASPRS 1989). That method computes a mean positional error from a set of samples, and then the standard provides a stipulated value that this cannot exceed if the map is to be accepted. The big improvement in this standard was that the errors are expressed in ground units instead of map units, which was an important first step toward considering digital data and not just paper maps.

In 1998, the National Standard for Spatial Data Accuracy (NSSDA) was developed by the Federal Geographic Data Committee (http://www.fgdc.gov/standards/projects/FGDC-standards-projects/accuracy/part3/chapter3). This is still used today as the recognized standard for assessing positional map accuracy. The NSSDA, unlike previous standards, does not use a maximum allowable error. Instead, it determines some allowable error threshold in ground distance at the 95 percent confidence level. The threshold is determined using the equations from Greenwalt and Schultz. However, some confusion between computing standard deviations and standard errors has caused some issues with the NSSDA. This issue has been minimized in recent times, because digital imagery has achieved significantly improved positional accuracies well above the standard set by the NSSDA.

Given recent improvements in digital imagery along with the development of lidar, new guidelines have been suggested that go beyond any of these existing standards. These guidelines address the need to assess positional accuracy (i.e., take samples) in various vegetation conditions (e.g., bare ground, nonwoody vegetation, and forests). There are guidelines from the Federal Emergency Management Agency (Federal Emergency Management Agency, 2003) called "Guidelines and Specifications for Flood Hazard Mapping Partners," which specifies sampling in major vegetation types. A similar document was produced by ASPRS (2004) called the "ASPRS Guidelines for Reporting Vertical Accuracy of Lidar Data." Finally, there is a document called "Guidelines for Digital Elevation Data" from the USGS National Digital Elevation Program (NDEP 2004). Each of these documents deals with paper maps rather than the digital maps prevalent today.

A new standard was developed by the ASPRS Accuracy Standards Subcommittee in 2014, called the "ASPRS Positional Accuracy Standards for Digital Geospatial Data" (ASPRS 2014). This standard fully incorporates the current state of mapping technology including digital imaging and nonimaging sensors, airborne GPS, inertial measurement units, and aerial triangulation. The new standard uses the root mean square error (RMSE) as previous standards have and allows cross-referencing to previous standards. Finally, the new standards are independent of scale and contour unit, which provides flexibility for incorporating future technologies as they are developed. The full standard is available at https://www.asprs.org/pad-division/asprs-positional-accuracy-standards-for-digital-geospatial-data.html.

Initial Considerations

In thinking about conducting a positional accuracy assessment, there are two initial considerations: sources of error and classification scheme. Actually, it is impossible to think about one of these without the other because they are closely linked. Historically, positional accuracy standards made no mention of where samples should be taken (i.e., reference data collected) to assess a map's accuracy. It was usually assumed that bare ground would be used. However, recent standards, especially in this century, have better considered new technologies including digital imagery, lidar, and GPS. Therefore, these newer standards have insisted on sampling to assess positional accuracy in these different land-cover types to in turn assess the magnitude of error associated with these varying conditions. While the various guidelines differ slightly on the exact land-cover types (classification scheme) to sample in, all are in agreement that a minimum number of samples (usually 20) should be taken in each to account for the magnitude of error variation by land-cover type.

Collecting Reference Data

Factors to consider when collecting reference data include independence, source, timing, distribution, number, and consistency. If the data is not properly collected, the positional accuracy assessment is likely to be invalid. Data used to evaluate the positional accuracy of the map must be independent of the data used to register the map to the ground. This is a well-known concept in any type of model building/statistics/map making. It is critical to have an independent set of sample reference data with which to test the results. Unfortunately, there are still far too many situations in which the data used to register the map to the ground is then used to report the positional accuracy of the map. This data is not independent and represents an optimistically biased and invalid evaluation of the map's actual positional accuracy.

The reference data also must be more accurate than the map being assessed. Some have suggested the data be from one to three times more accurate than the anticipated accuracy of the map being tested (e.g., Ager, 2004; NDEP, 2004; ASPRS, 2004). While different sources can be used, such as a map of larger scale or a GPS or ground survey, NSSDA recommends that the reference data be the most accurate that can possibly be collected (Federal Geographic Data Committee, 1998).

Although typically more important in thematic accuracy assessment, when and who performs the data collection is also important. If more than a single individual/team is collecting the reference data, procedures must be put in place to minimize human bias and use the most objective procedures possible. It is also important that changes have not occurred on the ground that may alter the positions of sample locations (e.g., from earthquakes or other natural processes). While a change in position is rare, it is possible that land-cover changes may occur that would alter the number of samples taken per land-cover type.

Assessing the positional accuracy of a map requires sampling. It is impossible to evaluate every location on the map, so a valid sample is collected and used in the assessment. A good sampling approach includes determining the sample size (i.e., the number of samples), the sample unit, and the sampling scheme (i.e., the distribution of the samples). A poor sampling approach results in an invalid assessment. The NSSDA (Federal Geographic Data Committee, 1998) states that a minimum of 20 samples must be used in a valid assessment. Unfortunately, many projects have not used this number, and their assessments are suspect. Recent standards that have included sampling in various land-cover types have required 20 samples per land-cover type. In a large number of statistical analyses, 30 samples is the minimum number recommended. While collection of the reference data is time consuming and costly, the effort must be made to collect enough samples to conduct a valid assessment.

The sample unit used in positional accuracy is a well-defined and easily identifiable point. There can be no confusion about the exact location of the sample on both the map and the source of the reference data. Any issues with locating these points will cause unacceptable errors in the estimation of the map accuracy. In urban areas, those points are easily selected as corner points or other clearly identified points. Sometimes, lines are painted on the street to mark points. In more rural and vegetated areas, fewer potential targets means that locating points is more problematic. In that case, targets are sometimes placed on the ground to augment any existing locations.

Finally, the samples must be distributed appropriately throughout the entire map. Samples cannot be selected only where they are easy to locate. They must include a full range of the variation in the entire map including not only the elevation changes but also the land-cover types. Figure 12.3 shows an approach suggested by ASPRS (1989) that ensures that the samples will be spread through the map. As can be seen in the figure, the map is divided into four quadrants, and then a minimum of at least 20 percent of the samples are randomly placed in each part. Spatial autocorrelation, which becomes an issue when the samples are placed too close together, is minimized by making sure that no two points are closer together that the length of the diagonal (D) divided by 10. This method is a very effective way of making sure that the reference data samples are well distributed throughout the map.

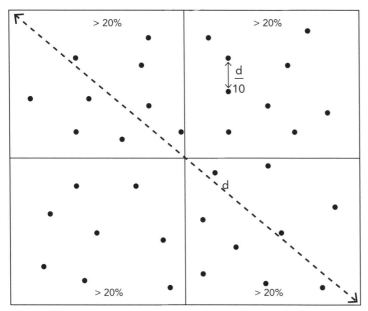

Figure 12.3. An effective method of ensuring that reference samples for positional accuracy assessment are well distributed throughout the map. Source: ASPRS

Computing Descriptive Statistics

Positional accuracy is assessed by comparing a sample of data (points assumed to be of higher accuracy than the map being assessed) to the same locations on the map and computing a number of statistics. These statistics are then compared to whichever standard the user has selected to see whether the map is acceptable. The most important statistic that is computed for positional accuracy assessment is the RMSE. The RMSE is defined as the square root of the mean of the squared differences between the samples (i.e., locations) on the map and those exact same samples on the reference data. The reason that this number is squared is to ensure that all values are positive, because the actual differences between the map and the reference samples can be either positive or negative. If the original values were summed, the positive and negative values would cross out, resulting in a number not representative of the true accuracy. While squaring the values works well, the absolute value can also be used as an effective alternative.

Because position on a map occurs horizontally (i.e., in the x and y directions) and also vertically (i.e., in the z direction), two equations are used depending on which type of positional accuracy is to be computed. The vertical equation is simpler than the horizontal equation because it measures in only a single direction. The equation for RMSE for vertical accuracy is:

$$RMSE_v = \sqrt{\sum_i^n (e_{vi})^2 / n} \qquad (1)$$

where

$$e_{vi} = v_{ri} - v_{mi} \quad \text{and} \qquad (2)$$

v_{ri} equals the reference elevation at the ith sample point,
v_{mi} equals the map elevation at the ith sample point, and
n is the number of samples.
The equation for RMSE for horizontal accuracy is

$$RMSE_h = \sqrt{\sum_i^n (e_{hi})^2 / n} \qquad (3)$$

where

$$e_{hi}^2 = (x_{ri} - x_{mi})^2 + (y_{ri} - y_{mi})^2, \text{ and} \qquad (4)$$

x_{ri} and y_{ri} are the reference coordinates, x_{mi} and y_{mi} are the map coordinates for the ith sample point, and n is the number of samples.

NSSDA is the current national standard (and is related to the new 2014 ASPRS standards) and requires that positional accuracy be reported at the 95-percent level, defined as "95% of the locations in the data set will have an error with respect to the reference position that is equal to or less than the computed statistic" (Federal Geographic Data Committee, 1998). The equation for computing NSSDA for vertical accuracy is given in the guidelines as

$$NSSDA\ Vertical\ Accuracy_v = 1.96\ (RMSE_v) \qquad (5)$$

The equation for computing the NSSDA for horizontal accuracy is, again, a little more complicated because of the two dimensions (x and y). It is possible that the errors in the x direction and the y direction have different distributions (i.e., are not circular) causing the errors around the sample locations to be oblong. While this condition can easily occur, most analysts use the simplified equation, ignore the distribution of errors, and compute NSSDA for horizontal accuracy as

$$NSSDA\ Horizontal\ Accuracy = 1.7308 * RMSE_h \qquad (6)$$

Table 12.1 shows an example of the computations of NSSDA and RMSE for a small horizontal dataset (sample size = 23). A quick look at this table demonstrates why the values are squared to eliminate the positive and negative values so that these errors do not cross out (i.e., sum to zero or close to zero). Computation of RMSE and NSSDA can quickly be computed using a spreadsheet or may be part of a mapping analysis software package. For more details on positional accuracy assessment, please see Congalton and Green (2009).

Table 12.1. An example of assessing the positional accuracy of a small dataset by computing RMSE and NSSDA

Sample no.	x (ref)	x (map)	x diff	(x diff)2	y (ref)	y (map)	y diff	(y diff)2	Σ xdiff2 + ydiff2
1	33.2	39.4	-6.2	38.44	843.5	843.9	-0.4	0.16	38.60
2	10.5	11.1	-0.6	0.36	900.0	897.0	3.0	9.00	9.36
3	65.8	66.7	-0.9	0.81	1010.1	1009.3	0.8	0.64	1.45
4	3.9	4.2	-0.3	0.09	786.3	782.4	3.9	15.21	15.30
5	9.2	8.5	0.7	0.49	656.8	658.2	-1.4	1.96	2.45
6	76.5	79.0	-2.5	6.25	676.0	672.4	3.6	12.96	19.21
7	113.1	113.1	0.0	0.00	656.1	666.2	-10.1	102.01	102.01
8	99.0	99.7	-0.7	0.49	688.1	689.2	-1.1	1.21	1.70
9	54.5	54.9	-0.4	0.16	744.1	737.1	7.0	49.00	49.16
10	102.1	110.5	-8.4	70.56	810.7	811.2	-0.5	0.25	70.81
11	6.9	6.8	0.1	0.01	845.6	847.2	-1.6	2.56	2.57
12	200.5	198.2	2.3	5.29	902.1	902.4	-0.3	0.09	5.38
13	252.1	250.9	1.2	1.44	937.4	939.2	-1.8	3.24	4.68
14	260.8	262.0	-1.2	1.44	945.5	940.1	5.4	29.16	30.60
15	300.9	299.5	1.4	1.96	960.8	962.0	-1.2	1.44	3.40
16	264.2	267.1	-2.9	8.41	978.3	978.2	0.1	0.01	8.42
17	142.1	135.8	6.3	39.69	1000.3	999.1	1.2	1.44	41.13
18	96.7	96.5	0.2	0.04	1006.8	1008.2	-1.4	1.96	2.00
19	42.0	36.9	5.1	26.01	1101.3	1110.0	-8.7	75.69	101.70
20	266.8	267.0	-0.2	0.04	971.2	971.1	0.1	0.01	0.05
21	86.7	86.5	0.2	0.04	1006.8	1008.1	-1.3	1.69	1.73
22	44.4	36.9	7.5	56.25	1101.3	1109.8	-8.5	72.25	128.50
23	12.2	15.1	-2.9	8.41	999.9	1002.0	-2.1	4.41	12.82
								Sum	653.03
								Ave.	28.39
								RMSE	5.33
								NSSDA	9.22

Thematic Map Accuracy Assessment

Assessing thematic map accuracy involves most of the same issues and considerations as positional accuracy assessment. That is, there are some initial considerations, then collecting reference data, and finally computing descriptive statistics. Thematic map accuracy assessment is more complicated than positional accuracy and includes some basic analysis techniques that can be computed as well.

As previously discussed, any new technology experiences an explosion in use where little time or effort is devoted to issues of quality or accuracy. While the use of remotely sensed data (analog imagery) has been prevalent since the end of World War II, the use of digital imagery has occurred only since the early 1970s with the launch of the first Landsat satellite. In the late 1970s and early 1980s, some researchers began to explore assessing the accuracy of maps derived from remotely sensed data. Initially, photo interpretation of analog imagery was used to compare with maps generated from digital imagery. The use of photo interpretation for reference data may not be appropriate in all situations. Since that time, great effort has been put into developing methods to effectively assess the accuracy of maps derived from remotely sensed imagery. The convergence of remote sensing as a source of geospatial data with the ever-growing field of geographic information systems has further fostered the need for assessing the accuracy of all data used as part of the spatial data decision-making process.

Map accuracy assessment has developed over three periods. Initially, no quantitative assessment was performed, and as long as the map looked good it was accepted as accurate. The second period incorporated some quantitative assessment using a non-site-specific approach. There, map totals were compared to ground totals to see how accurately the map predicted the results. For example, the number of acres of corn for a county in Iowa could be computed from a map generated from remotely sensed imagery and compared to the corn acreage as reported by the farmers in the county to the Farm Bureau. While quantitative, this method evaluated only total areas and indicated nothing about a specific location on the map.

The third and current method of thematic map accuracy assessment uses a site-specific approach or error matrix. As the name implies, site-specific assessment uses a quantitative method to evaluate the accuracy at specific sample locations on the map. The technique used to conduct this assessment is called an error matrix (see figure 12.4). An error matrix is called a contingency table in statistics and is a square table or cross-classification set out in rows and columns, in which the rows represent the map and the columns represent the reference data. The size of the matrix is determined by the number of map categories (e.g., land-cover classes, forest types) that are being mapped. A map with six land-cover classes would be evaluated with a 6 × 6 error matrix. Sample units are selected, and the label at that location is recorded for both the reference data and the map. This information is then used to tally in the appropriate place in the error matrix. For example, referring to figure 12.4, if the reference sample unit is labeled as forest and the map agrees that the location is a forest, then a tally would be indicated along the major diagonal of the error matrix where F (forest) for the reference data (down the column) and F for the map (across the row) intersect. However, if the reference sample unit is labeled as Forest, and the map disagrees and indicates that the location is Other, then the tally in the error matrix would occur where the F for reference data (down the column) and the O (other) for the map (across the row) intersect.

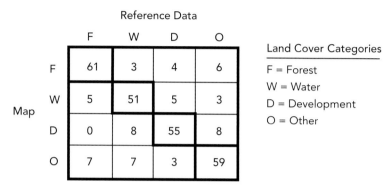

Figure 12.4. An example error matrix used for assessing thematic accuracy

It is important to note that the term for the data that is compared to the map to determine its accuracy is referred to in this chapter as "reference data." Reference data is often also referred to as "ground truth," although this term should be avoided because it is not always possible, even when on the ground, to know the answer with 100 percent certainty. Also, in many cases the data used for comparison is not collected on the ground, but rather using some other high-spatial-resolution imagery. Referring to this data as ground truth is inappropriate as there was no visit to the ground. Therefore, the authors suggest that terms such as "reference data," "ground-collected data," "field data," or the like be used instead.

Initial Considerations

As with positional accuracy assessment, it is important to consider both the classification scheme and the sources of potential error when conducting a thematic accuracy assessment. However, unlike positional accuracy, these two considerations are not closely linked. Each will be discussed separately in this section.

The characteristics and importance of a good classification scheme have already been discussed in chapter 7 of this book. Failure to make sure that the scheme is 1) composed of labels and rules, 2) mutually exclusive, 3) totally exhaustive, and 4) hierarchical can cause serious issues with the project and potentially invalidate the results. It is also absolutely critical that the same classification scheme used for making the map is used for collecting the reference data. While this statement may seem quite obvious, it is disturbing how often the same scheme is not applied when collecting the reference data. In that situation, further and unnecessary uncertainty is introduced into the assessment process. Finally, when selecting the appropriate classification scheme to be used for the mapping and accuracy assessment, the appropriate minimum mapping unit (MMU) must be determined. As discussed later in "Collecting Reference Data," the choice of a valid MMU is important. Selecting an MMU smaller that the sample unit used in the accuracy assessment process prevents allowing for positional error on the map, which makes it unclear when an error is the result of a true thematic mistake or simply the result of being in the wrong location.

There are many sources of error in a thematic map. Errors can occur during image acquisition, during data processing, during data analysis, and even in the accuracy assessment process itself (Lunetta et al., 1991). Some of these sources of error are easy to control, while others can be quite difficult. Every effort should be made to minimize the sources of thematic errors on the map. For example, as discussed above, an inappropriate classification scheme may be selected for a mapping project. Also, a different scheme may be used to collect the reference data than was used to make the map. In either case, errors will result. Another example involves conversion of the map from raster to vector format. While software exists to easily convert between these formats, the data is changed in the process and errors can result. Careful planning of not only the mapping process but also the thematic accuracy assessment can help significantly reduce these errors.

Collecting Reference Data

Collecting reference data for assessing the accuracy of thematic maps is typically more complex and time-consuming than collecting reference data for positional accuracy assessment. As a result, even more care must be taken to ensure that sufficient and valid data is acquired. Factors for collecting good reference data include independence, proper sources, timeliness of collection, objectivity/consistency, and sampling (including the sample unit, the sample size, the sampling strategy, and spatial autocorrelation). Each of these factors is discussed below.

Just like reference data collected to perform a positional accuracy assessment, reference data for conducting a thematic accuracy assessment must be independent. A valid assessment cannot be done with the same data that was used to create the map. The reference data can be collected at a different time from the training data used to make the map. It is important to use the same classification scheme. It is also possible to collect the reference data at the same time as the training data. In this case, the data is collected, and then randomly divided so that the data selected as reference data for the accuracy assessment is put aside and not viewed until after the map is complete. Collecting the reference data for the assessment simultaneously with the training data used for mapping is often a more effective approach, because going to the field is an expensive and time-consuming activity. Sometimes, insufficient samples are taken, and it may be necessary to augment the reference data collection at a later time.

Ideally, all reference data used for thematic accuracy assessment is collected in the field (i.e., ground-reference data). However, this is not always possible. Other sources of reference data are possible depending on the level of detail on the map (i.e., classification scheme) and the budget of the project. Sometimes, existing reference data is available for the mapping area. It is important that the same classification scheme is used for labeling the reference data. If not, it may be possible to crosswalk the reference data to the map scheme. Usually, the crosswalk is imperfect, at best, and a trade-off between confidence in the reference data versus costs of collecting new data may be necessary. Often, the existing reference data used a sampling unit smaller that the sampling unit (or possibly the MMU) of the map. In that situation, the existing reference data cannot be used to assess the accuracy of the map. In every case, the reference data is assumed to be more accurate than the map. Sometimes, interpretation of high-spatial-resolution imagery can be used as reference data. Sometimes, field visits are necessary. When in the field, it may be necessary to observe the physical conditions to determine the map classes. In other situations, it may be necessary to make exact measurements to ensure accurate reference data. Each project is different, and deciding on the appropriate source of reference data is critical to a valid thematic map accuracy assessment. Careful thought must go into the selection, and it is often necessary to find a balance between what is affordable and what is accurate.

Another factor important in collecting reference data is the timing of the collection. For some thematic maps, the reference data must be collected simultaneously with the image collection. For example, mapping agricultural crops can be very time sensitive and if the reference data is not collected at or very near the image collection, the crop might change. Other reference data may be collected within some reasonable time of the image collection, because the land-cover types in that map do not change as rapidly as crops. Finally, some reference data may be generated long after the imagery was collected. Again, timing depends on the classification scheme and other details required on the map. For example, if the map classes are simple (e.g., general land-cover types such as Forest, Water, Urban) then some time can exist between collecting the image and the reference data. If changes occur between the time periods (e.g., forest harvesting or urban expansion), those changes will be quite distinct and easy to identify to avoid collecting reference data from those areas. However, if the changes are more subtle, for example, the aging of a forest such that the size or species distribution changes, those changes would be more difficult to detect, and therefore the reference data should be collected as closely as possible to the collection date of the imagery used to make the map.

Consistency or objectiveness is the last nonsampling factor that must be considered when collecting reference data. Consistency can be ensured using a number of techniques. The use of a field form ensures that all data is recorded the same way. This field form can be anything from a simple sheet of paper to a digital data logger or collector application on an iPad. Every field form should include basic information such as date, collectors' names, location, and classification scheme. Customization should be done to record the additional information specific to that project. In addition, all anomalies, issues, or special findings that occur at a location must be recorded. All field procedures for collecting the reference data should be completely documented. Those procedures can then be easily followed regardless of whether a single person or a large group of collectors is collecting the field data. Carefully following established procedures reduces collector bias and further ensures objectivity.

Finally, proper collection of reference data requires understanding some statistical concepts to ensure that the data collected is valid. These concepts include 1) the selection of the proper sampling unit, 2) the calculation of the proper sample size, 3) the determination of the proper sampling scheme, and 4) the consideration of spatial autocorrelation. Thematic map accuracy assessment, like positional accuracy assessment, involves sampling to test the map. It is not possible to evaluate every place on the map. Instead, a valid, representative sample must be obtained. Understanding these four statistical concepts is key to obtaining the proper sample needed to assess thematic accuracy.

In selecting the appropriate sample unit to assess thematic accuracy, it is important to consider the effect of positional accuracy on the map as well. Selecting a sample unit that is so small that it is subject to high positional error ensures that the assessment cannot determine whether the error was an actual thematic error or was due to being in the wrong location. An example helps illustrate this point. Landsat imagery has a spatial resolution of 30 meters. It is well-recognized that Landsat imagery can be accurately registered to the ground (positional accuracy) to about half a pixel (i.e., 15 meters). It is also well known that a typical GPS unit can locate a place on the ground to within about 5 to 15 meters. Therefore, if a single pixel is selected as the sampling unit, it will be impossible to know whether the error on the map is thematic or positional (because of the errors in image registration and the GPS unit). However, if a 3 × 3 grouping of pixels that are homogeneous (i.e., the same thematic map class) is selected as the sampling unit, and the reference data is collected from the center of this grouping, then any positional error will be eliminated. Any error on the map can then be known to be a result of thematic error.

Therefore, as can be seen from this example, a single pixel should not be selected as the sampling unit for collecting reference data. Some grouping of pixels or a homogeneous polygon should be chosen to minimize positional error. The issue is even more important when assessing high-spatial-resolution image maps. If the image has a spatial resolution of 1 meter, the reported positional accuracy of the registration to the ground is 7 meters, and the same GPS unit is used, then it is easy to see that a 3 × 3 grouping of pixels will not work at all. A sampling unit closer to 15 × 15 pixels or greater is required. It is important to note here that the measurement is in pixels and not meters. A 15 × 15-pixel grouping would still only be 15 × 15 meters for 1-meter spatial-resolution imagery, while a 3 × 3 Landsat pixel grouping would be 90 × 90 meters. Unfortunately, there are far too many examples where a single pixel was used to assess the thematic accuracy of a map. Those assessments do nothing to compensate for positional error and result in an invalid assessment.

Once the sampling unit has been established, the next step is to determine the number of sample units that must be taken to generate the error matrix. Unlike positional accuracy, where the standard requires that the analyst collect 20 (30 would be better) samples to conduct the assessment (for the newer standards, 20 samples in each land-cover class), thematic accuracy assessment requires considerably more samples to be taken. There are simple statistical equations for computing the sample size to determine overall accuracy. The binomial equation (i.e., right or wrong) applies. However, to generate a valid error matrix requires the use of the multinomial sample size equation, because there is one right answer but $n-1$ (where n is the number of map classes) wrong answers. In addition to this equation, a guideline has been published (Congalton, 1988) that proposes 50 sample units per map class. A minimum number of sample units must be collected for each map class, and then proportionately more samples taken in the map classes that encompass the largest map areas.

Balancing what can be practically collected and statistical validity is a large part of the sample-size component of reference data collection. If insufficient sample units are collected, the assessment may become invalid. In some cases with smaller projects, it is not possible to collect 50 samples per class. In those cases, the analyst should collect as close to 50 as possible with a minimum of 30. The assessment must be carefully planned so that the effort results in useful information. If it is not possible to collect enough sample units for a valid assessment, then effort would be much better spent on improving the map with more training data and not conducting an assessment at all. Documentation of the process and the decisions made is imperative for the user to understand the project.

Like positional accuracy assessment, the sampling scheme provides the mechanism to distribute the sample units across the map. While many strategies are possible, the use of a stratified random sampling approach is most effective. This scheme ensures that samples are taken in each stratum (map class) and that they are distributed across the map. However, in many cases involving ground collection of the reference data, these strata are limited by access to the potential sample locations. Simple random sampling (not within strata) is not effective, because small but important map categories tend to be undersampled. Cluster sampling is sometimes employed to maximize the collection of the reference data over the smallest distances possible. However, care must be taken because sample units that are too close together will exhibit spatial autocorrelation, rendering them invalid. Finally, systematic sampling in which a sample is taken at some regular interval is often used, especially when the reference data is being collected using high-spatial-resolution imagery instead of the ground.

As mentioned above, the concept of spatial autocorrelation is an important factor when considering sampling schemes. Spatial autocorrelation occurs when the presence, absence, or degree of a certain characteristic affects the presence, absence, or degree of that same characteristic in neighboring units (Cliff and Ord, 1973). In other words, if sample units are selected that are too close to each other, the information from one sample is not independent of that of the nearby sample. This means that an error at a certain location will affect whether there is an error at nearby locations. It is important to have sufficient distance between the sampling units so that one sample has no impact on the other.

Computing Descriptive Statistics

Once all the initial considerations and factors for collecting valid reference data have been accounted for, the result should be a valid error matrix indicative of the thematic accuracy of the map. The matrix is then the beginning point for a series of descriptive statistics that can be computed to provide more insight into the thematic accuracy. These statistics include the calculations of the overall accuracy, the producer's accuracy, and the user's accuracy. Figure 12.5 shows the exact same error matrix that was used as an example in figure 12.4, but now shows the computation of these three descriptive statistics. This error matrix is an effective way to represent thematic accuracy, because careful study of the matrix reveals information about each map class including both errors of inclusion (commission errors) and errors of exclusion (omission errors) present on the map. Just as a coin has tails and heads, map errors also have two components: commission error and omission error. A commission error is defined as including an area into a thematic class when it doesn't belong to that class, while an omission error is excluding that area from the thematic class in which it does belong. As can be seen in the error matrix, every error is both an omission from the correct thematic map class and a commission to a wrong thematic map class.

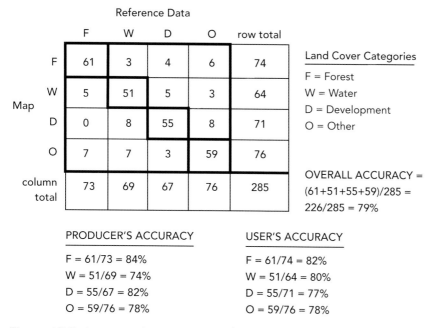

Figure 12.5. An example error matrix showing the calculations for overall, producer's, and user's, accuracies

Figure 12.5 shows that 61 sample units were labeled forest (F) on the map and were also labeled Forest (F) on the reference data (tallied along the major diagonal of the matrix). However, three times the map said it was Forest when the reference data said it was Water (i.e., commission error to Forest and omission error from Water). It is important to understand omission and commission errors before looking at the other descriptive statistics.

The first descriptive statistic is called overall accuracy and is the easiest to understand. Overall accuracy is just the sum of the major diagonal of the error matrix (i.e., the agreement between the map and reference data—the correct sample units) divided by the total number of sample units taken in the assessment. In figure 12.5, these values are 226/285 or 79 percent. Therefore, the overall accuracy of this map is 79 percent. This value is useful, as it represents the entire matrix. However, the next obvious question is, what about the accuracies of each map class alone? To know this answer, the analyst must compute the producer's and user's accuracies (Story and Congalton, 1986). The map producer would like to know how accurate the map is (the producer's accuracy). This value is computed by dividing the value from the major diagonal (the agreement) for that map class by the total number of sample units in that map class as indicated by the sum of the reference data for that map class. Figure 12.5 shows that the map producer labeled 61 areas as Forest, while the reference data indicates a total of 73 Forest areas. Twelve areas were omitted from the Forest (five were committed to Water and seven were committed to Other). Therefore, 61/73 sample units were correctly called Forest for a Forest producer's accuracy of 84 percent. However, this is just the producer's accuracy. It is important to consider the user's accuracy, as well. To compute the user's accuracy for the map, divide the 61 sample units that were called Forest on the map and were actually forest by the total number of sample units called Forest on the map. In other words, the user's accuracy is 61/74 or 82 percent. It should be noted that the map labeled three samples Forest that were actually water, four samples Forest that were actually developed, and six samples Forest that were actually other. The map, therefore, called 74 samples Forest, but only 61 were actually forest. There was commission error of 13 samples into the Forest that were not forest. In evaluating the accuracy of an individual map class, it is important to consider both the producer's and the user's accuracies.

Basic Analysis Techniques

In addition to these descriptive statistics that can easily be computed from the error matrix, some additional analyses of the matrix can be performed. These analysis techniques allow the analyst to learn even more about the error matrix. Two common techniques have been applied called Margfit and Kappa (Congalton et al., 1983; Congalton, 1991; Congalton and Green, 2009). The Margfit technique uses an iterative proportional fitting routine to normalize the error matrix so that it can be directly compared to another error matrix regardless of the number of samples used to create the matrix. Once the matrix has been normalized, a new accuracy measure akin to overall accuracy can be computed, called the normalized accuracy, by summing the major diagonal and dividing by the number of map classes. The second technique is Kappa and represents a third method of representing the accuracy of the map (i.e., overall accuracy, normalized accuracy, and Kappa). However, the real power of Kappa is that this statistic can be used to test whether two error matrices (and therefore maps) have a statistically significant difference from each other. This test is useful in evaluating whether one classification algorithm is better than another, whether one analyst is better than another, or almost any other comparison imaginable given two or more error matrices.

An example helps to show the usefulness of both of these analysis techniques. Figure 12.6 shows the results of normalizing the original error matrix that was presented in figure 12.5. The normalized matrix shows the results of the iterative proportional fitting algorithm, such that all the rows and columns now sum to 1. The value in each cell of the error matrix can then be compared to that of the same cell in another matrix regardless of the sample size used to create either matrix. Also, normalized accuracy can be computed by summing the major diagonal and dividing by four (the number of map classes). Having a second error matrix will further demonstrate the usefulness of Margfit.

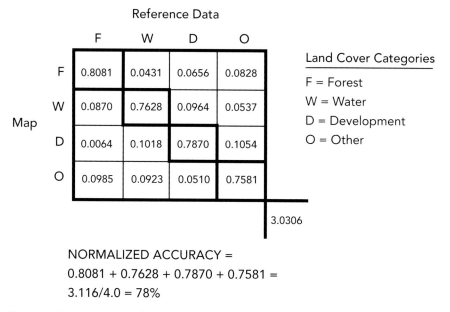

Figure 12.6. An example showing the results of normalization (Margfit) on the error matrix in figure 12.5

Figure 12.7 is an error matrix of a map generated from the same imagery as figure 12.5, but by a different analyst. Note that the number of sample units in the error matrix in figure 12.7 is 348, while the analyst that created the error matrix in figure 12.5 used only 285 sample units. As a result, it is not possible to directly compare these matrices. Yes, the descriptive statistics (overall, producer's, and user's accuracies) can be computed, but individual cell values cannot be compared, because the sample size used to create the matrices was different. However, if the error matrix in figure 12.7 is normalized as shown in figure 12.8, then the cell values are directly comparable. For example, the value in the major diagonal for Water in figure 12.6 is 0.7628, or about 76 percent, while the same value in figure 12.8 is 0.8772 or about 88 percent. Therefore, it is clear that analyst #2 did a better job of mapping water than did analyst #1. Just looking at the original error matrices (values of 51 versus 83) is not meaningful because different sample sizes were used to create the matrices.

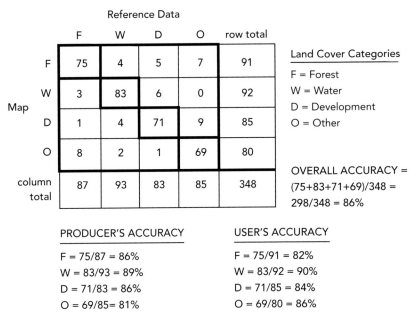

Figure 12.7. An example error matrix showing the calculations for overall, producer's and user's accuracies for a map made from the same imagery as in figure 12.5, but by a different analyst

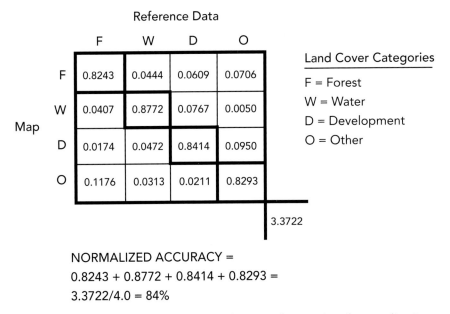

NORMALIZED ACCURACY =

0.8243 + 0.8772 + 0.8414 + 0.8293 =

3.3722/4.0 = 84%

Figure 12.8. An example error matrix showing the results of normalization on the error matrix in figure 12.7

As mentioned above, three accuracy measures are now possible once the Margfit and Kappa analyses are performed. These three accuracies are overall accuracy computed from the original error matrix, normalized accuracy computed from the normalized error matrix, and Kappa. Table 12.2 summarizes these results for the error matrices created to assess the maps generated by two separate analysts. The question of which of these measures of accuracy is most useful is open to discussion. Each measure incorporates different amounts of information about the matrix itself. Overall accuracy is the most common and simply sums the major diagonal of the matrix and divides by the total number of samples. Kappa incorporates a little more information about the matrix, because it is calculated using the sums of the rows and columns. Finally, normalized accuracy incorporates the entire matrix because of the iterative proportional fitting algorithm that directly uses each cell value in the matrix. If the error matrix is reported, then the analyst evaluating the accuracy of the thematic map has a choice of which descriptive statistics and measures of accuracy to consider in the analysis.

Table 12.2. Three measures of thematic map accuracy

Accuracy Measure	Error Matrix by Analyst #1	Error Matrix by Analyst #2
Overall Accuracy	79%	86%
Kappa	72%	81%
Normalized Accuracy	78%	84%

Finally, Kappa is not only a measure of accuracy, but more importantly can be used as a test to see whether one error matrix has a statistically significant difference from another. In other words, in the example in this chapter, did analyst #1 produce a better map as indicated by the error matrix than did analyst #2? Using the Kappa statistics computed for each matrix and the corresponding test of significance, the resulting test value for comparing error matrix #1 with #2 was computed to be 2.08. At the 95 percent confidence level, any resulting test value greater than 1.96 indicates a statistically significant result. Therefore, matrix #1 has a statistically significant difference from matrix #2. Because the accuracy (overall, Kappa, and normalized) of matrix #2 is higher, analyst #2 produced a statistically more accurate map than did analyst #1.

Summary—Practical Considerations

Assessing map accuracy is a key component of any project when generating a map from remotely sensed imagery. However, conducting the actual assessment does not follow a simple step-by-step process, but rather is a series of considerations used to balance statistical validity with practical application. It does not make sense to begin an assessment if insufficient resources are available to collect enough samples. However, while there is no single correct way to conduct an accuracy assessment, there are a great number of wrong ways to do it. Therefore, it is critical that the accuracy assessment process be thoroughly designed from the beginning of the project and well documented to show how each consideration was made and what analyses were conducted.

Positional accuracy and thematic accuracy have many considerations and techniques in common. Both require an understanding of the errors involved and careful consideration of the classification scheme to be used. Both require sampling, because it is not practical to assess every place on the map. Sampling for thematic accuracy is more complicated, because it requires more samples and consideration of positional accuracy when selecting the appropriate sampling unit. Careful consideration of the practical issues related to the assessment can lead to greater efficiencies and the more effective collection of reference data.

It is important to remember that the positional or thematic accuracy of a map must be measured against what the map is going to be used to accomplish. A certain positional accuracy might be appropriate for one objective, but not usable for a different objective. The same is true for thematic accuracy. Therefore, not only is it important when planning an accuracy assessment to consider the various factors to balance the statistical validity with what can practically be achieved, but it is also important to consider the use of the map.

Many software packages have some components to aid the analyst in accuracy assessment. The tools for assessing positional accuracy tend to be further developed because of the simpler process involved. The ArcGIS Data Reviewer is such a tool. These tools are helpful to the analyst who understands their strengths and limitations. It is great to be able to compute the RMSE quickly. However, simply using any tool without a strong understanding of how it works can be quite harmful. Plugging in incorrect values will still result in a computed RMSE, but it is up to the analyst to supply the correct values. Given the many considerations and options available to conduct an accuracy assessment, it is especially prudent to use these tools with care.

An even stronger warning is appropriate for the use of many software tools that attempt to help the analyst with thematic accuracy assessment. It is not that difficult to generate an error matrix. The question is, is this a valid error matrix? If it is not, then even the most sophisticated analysis of the matrix will be meaningless. Thematic accuracy assessment is complex and requires lots of good decisions that balance statistical validity with practicality to obtain a valid error matrix. No software currently exists that can do it all. It is very helpful to have software to compute RMSE or Kappa or to fill in an error matrix. However, it is up to the analyst to ensure that the assessment is done correctly and is well documented to prove it.

Chapter 13
Managing and Serving Imagery

Managing ever-growing, ever-important collections of imagery can be complex and challenging. There are many ways to manage imagery—this chapter focuses on Esri's solution for imagery management, the mosaic dataset, and on Esri's solution for providing access to imagery web services hosted in ArcGIS Enterprise.

The mosaic dataset provides a robust, scalable, and flexible solution for imagery management. Chapter 5 introduced the mosaic dataset and basic information on mosaic dataset structure, mosaic dataset properties and methods, and mosaic dataset functions. The first part of this chapter provides a more in-depth discussion of the mosaic dataset.

Imagery is of little value unless it can be easily shared to users within an organization and across the Internet. Mosaic datasets published as web services provide a mechanism to unlock an organization's investment in imagery, making it accessible to everyone across the Internet via desktops, browsers, web clients, and GIS software. This chapter provides a more detailed discussion of managing and serving imagery using mosaic datasets. The second part of the chapter discusses publishing and distributing mosaic datasets as image services online.

The chapter ends with a discussion of creating and publishing geoprocessing services. Geoprocessing services, which were introduced in chapter 5, allow you to expose the powerful analytic ability of ArcGIS to the World Wide Web.

Managing Image Collections with Mosaic Datasets

This chapter provides an in-depth discussion of the mosaic dataset, which was introduced in chapter 5. The next section elaborates on mosaic dataset structure and the ArcGIS tools for creating and managing mosaic datasets.

Mosaic Dataset Structure

Introduction

As discussed in chapter 5, mosaic datasets typically reside inside a file geodatabase but can reside in enterprise geodatabases. Mosaic datasets have the following components:
- A catalog of metadata about each image (each of the individual scenes or tiles of imagery). Each entry in the catalog is referred to as an item, and each item in the catalog contains a footprint of the image. This footprint defines the extents of each image and contains pointers to the pixel data. The item also stores attributes of the image and defines on-the-fly processing steps to be applied to each image by the mosaic dataset.
- A feature class that defines the boundary of the entire mosaic dataset.
- Mosaic dataset properties that include mosaic methods that define rules for the required order of overlapping images as well as a range of other properties.

Items, Footprints, and the Mosaic Dataset Attribute Table

Footprints

Each component image of a mosaic dataset (each item) is represented by a polygon in the mosaic dataset's footprint vector feature class. Typically, the extent of each footprint represents the valid area of pixels for the raster the footprint represents. The green polygon in figure 13.1 represents a footprint for a mosaic dataset component image (in this case, the mosaic dataset has one Landsat scene).

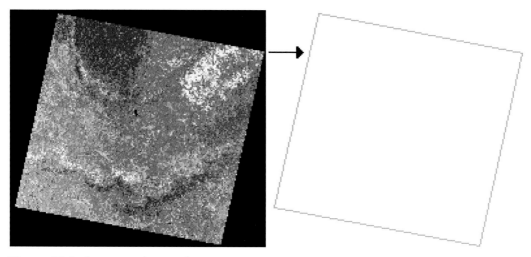

Figure 13.1. A mosaic dataset footprint

When the mosaic dataset is displayed, component images can be clipped to the extent of their footprints, or shown entirely (including NoData). This option is configured in the mosaic dataset's properties (change the Always Clip the Raster to its Footprint property). NoData values can also be removed from display in a mosaic dataset using the Define Mosaic Dataset NoData tool. This tool will allow you to specify more than one NoData value.

Footprints are useful for excluding NoData when a mosaic dataset is displayed, but they can also be used to remove other unwanted areas such as the collars of scanned maps, fiducial marks around scanned aerial photographs, or even areas of clouds. Like polygons from any vector polygon layer, footprint polygons can be edited or reshaped using the editing tools in ArcGIS.

Footprints can be recalculated at any time using the Build Footprint tool. This tool can use the radiometry of the imagery to build footprints, which is very useful if, for example, the footprint includes "collar" or border areas of known pixel values, and you want these areas to be excluded from the mosaic dataset.

Mosaic Dataset Attribute Table

The attribute table of the items is called the mosaic dataset attribute table. The mosaic dataset attribute table includes a row for each item. Each item has a footprint polygon (one polygon for each component image in the mosaic dataset). The mosaic dataset attribute table contains many fields that characterize the mosaic dataset's component images. The fields include a Raster field, which contains a link to the component image and properties, functions, and metadata for the component image. Other fields include low pixel size (LoPS) and high pixel size (HiPS), which define the range of display scales at which a component image is displayed in the mosaic dataset. The LoPS and HiPS fields include the range of pixel sizes for a component raster. For example, a raster dataset that contains a pyramid (or internal overview) will have a range of pixel sizes—the low value represents the base pixel value, and the high value represents the top pyramid that is being used. For raster datasets with no pyramids, the low and high pixel sizes may be the same value. In addition to the standard fields mentioned above, more fields can be added to the mosaic dataset attribute table. Added fields are often used to store metadata about the images, such as the component image's acquisition date or the percentage of the component image covered by clouds.

Mosaic Dataset Overviews

Mosaic datasets point to large collections of imagery, yet users still expect to be able to zoom out and see the mosaic render quickly at a small scale. For displaying images at small scales, mosaic datasets make use of image pyramids and service overviews (the concept of image pyramids is introduced in the Image Storage and Formats section of chapter 5). Each component image in the mosaic dataset can include image pyramids; the pyramids are shown when the image is viewed at lower resolutions. If pyramids exist for the component images, the mosaic dataset takes advantage of them for quick display. When the mosaic dataset is viewed at very small scales, a large number of component image pyramids need to be displayed, slowing down the display of the mosaic dataset. For this reason, overviews are created at these very small scales. Overviews act like pyramids for the complete mosaic dataset, instead of just for the component rasters. Figure 13.2 helps illustrate the way that image pyramids and mosaic dataset overviews work together to speed up dynamic display of the mosaic dataset.

The pyramids and overviews, which are much smaller files than the full-resolution source imagery, are compiled from source images and displayed when a user zooms to a small scale. Using pyramids and overviews, the mosaic dataset provides faster access to imagery. Figure 13.2 shows the combined use of image pyramids and overviews.

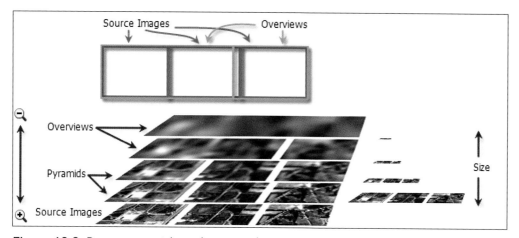

Figure 13.2. Raster pyramids and mosaic dataset overviews

Image pyramids are associated with individual rasters and typically have the same file name as the image file, but with a suffix of .ovr or .rrd. File-geodatabase-based mosaic dataset overviews are stored in a folder; the folder name is the name of the mosaic dataset, but with an .Overviews suffix. Mosaic datasets stored in an enterprise geodatabase can have their overviews stored internally in the geodatabase or in separate directories in the file system.

Pyramids can be created for the individual rasters that compose a mosaic dataset at any time. An image's pyramids are typically created before adding the image to a mosaic dataset. For example, pyramids are often created automatically when a raster is created by a geoprocessing tool. However, if pyramids don't already exist for a raster, they can be created automatically for each raster when the rasters are added to a mosaic dataset.

To create overviews in a mosaic dataset, the overviews are defined and then generated. When they have been defined, the application analyzes the mosaic dataset and, using the parameters set for the overviews, it defines how many are needed, at what levels, and where. Then, pointers to the soon-to-be-created overviews are added as items in the mosaic dataset attribute table, appearing as new rows. Next, the overview rasters themselves are generated. Both defining overviews and generating them can be done with one tool, Build Overviews. However, if you need to modify any properties, such as defining a new output location or tile size for the overviews, then you must run Define Overviews first (to define the properties and add the items to the attribute table), and then run Build Overviews to generate the overview files.

The mosaic dataset keeps track of any changes made to it, such as changes or additions of component images or alterations to footprints. When the analyst runs the Build Overviews tool or the Synchronize Mosaic Dataset tool with the appropriate options, the overviews are updated to reflect any changes to the mosaic dataset's component images or footprints.

Mosaic Methods—Dynamic Mosaicking

As discussed in chapter 5, dynamic mosaicking is the ability to define or refine the order in which images should be merged or blended when displayed in a mosaic dataset. The mosaic methods (such as Closest to Center) control the drawing order of the mosaic dataset's component images, determining in areas of overlap which image is displayed on top. Working side by side with mosaic methods, mosaic operators control how overlapping pixels are displayed. Common mosaic methods (discussed in chapter 5) include First, Mean, and Blend. First is the most commonly used mosaic operator and results in an output pixel value equal to the topmost overlapping image's pixel value.

Mosaic Dataset Properties

The mosaic dataset includes a set of properties that apply to all the rasters referenced by the mosaic dataset. These properties control the mosaicking. Figure 13.3 shows the mosaic dataset properties window, which is accessible through ArcGIS Desktop. Mosaic dataset properties fall broadly into three groups: image properties, catalog properties, and download properties. The properties are listed below by group.

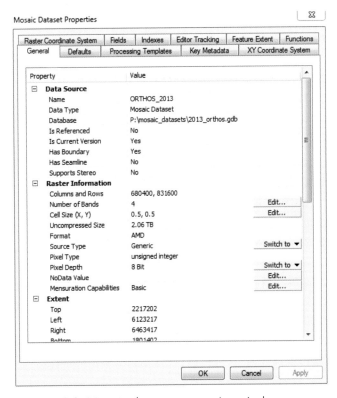

Figure 13.3. Mosaic dataset properties window

Image Properties

- **Maximum Size of Requests**—This property applies only when the mosaic dataset is published and accessed as an image service. It defines the maximum number of rows and columns allowed for a single request.
- **Allowed Compression Methods**—Defines the method of compression used to transmit the mosaicked image from the server to the client.
- **Default Resampling Method**—Defines the default sampling method of the pixels, which are sampled to match the resolution of the user's display (or client request if published).
- **Maximum Number of Rasters Per Mosaic**—Prevents the server from mosaicking an unreasonably large number of rasters if, for example, the client zooms to an overview scale in a nonoptimized image service dataset that has no overview tiles generated. The default is 20.
- **Allowed Mosaic Methods**—Defines the order of the rasters that are mosaicked together to create the image. You can choose one or more mosaic methods and select one as the default. The end user can choose from the methods you select.
- **Default Sorting Order**—Controls the expected ordering of the images defined by the mosaic methods.
- **Default Mosaic Operator**—Lets you define how to resolve the overlapping cells, such as choosing a blending operation.
- **Blend Width**—Defines the distance in pixels (at the display scale) used by the Blend mosaic operator.
- **Always Clip the Raster to its Footprint**—You can choose whether to limit the extent of each raster to its footprint.
- **Footprints May Contain NoData**—Controls how the mosaic dataset deals with NoData when there are overlapping images. If Yes, then if the mosaic method uses rasters that contain NoData, the mosaicked image will contain NoData values (and the application will not search for an overlapping raster that contains different pixel values). If No, then the application will try to find values to fill in the NoData using overlapping rasters.
- **Always Clip the Mosaic Dataset to its Boundary**—You can choose to limit the image extent to the geometry of the boundary or to the extent of the boundary.
- **Apply Color Correction**—If a color correction has been computed for the mosaic dataset, you can choose to apply it.

Catalog Properties

- **Raster Metadata Level**—Defines how much metadata will be transmitted from the server to the client.
- **Maximum Number of Records Returned Per Request**—Limits the requested number of records that will be returned by the server when viewing the mosaic dataset as a published image service.
- **Allowed Fields**—Defines which fields from the attribute table will be visible to the client when the mosaic dataset is served.
- **Time**—If the mosaic dataset contains attribute fields that define time, you can create a mosaic dataset that will automatically be time aware, meaning the time properties in the layer will be defined by default.
- **Geographic Coordinate System Transformation**—If the spatial reference system of the mosaic dataset is based on a different spheroid than the spatial reference system of the source raster data, you may need to specify a specific geographic transformation.

Download Properties

- **Maximum Number of Items Downloadable per Request**—Limits the number of rasters that a client can download from an image service.

Creating and Maintaining Mosaic Datasets

Creating Mosaic Datasets

Mosaic datasets can be authored directly in ArcGIS Desktop and ArcGIS Pro. They can be used in all ArcGIS Desktop applications, with processing being applied as the data is accessed. For providing optimized access to a larger number of users, mosaic datasets can be served as image services through ArcGIS Image Server.

Adding Rasters

Rasters are added to a mosaic dataset using the Add Rasters to Mosaic Dataset tool. Many rasters have minimal metadata, such as georeferencing, the number of bands, and bit depth. These rasters can simply be added to a mosaic dataset by referring to the directories where they reside. Mosaic datasets also support the ingestion of a wide range of raster products from imagery providers that contain sensor-specific metadata. These rasters are added to mosaic datasets using "raster types" that crawl through directories to extract all the relevant metadata from vendor products, as well as setting up the appropriate functions such as orthorectification and pan sharpening used to transform the raw imagery into higher-value products. The use of raster types simplifies the ingesting process. Many organizations also have existing databases or tables that provide metadata and other parameters about the rasters to be added to a mosaic dataset. Such tables can also be quickly added to mosaic datasets using the "table raster type." The metadata ingested is used to aid in defining the correct processing as well as to enable queries and filters.

Raster Functions

Esri raster functions allow for the on-the-fly rendering of derivative image products from a mosaic dataset, transforming the pixels from data values stored on disk to values required by the end user or application. For example, when the analyst applies a slope raster function to a mosaic dataset of digital elevation model (DEM) tiles, the mosaic dataset appears as a slope image instead of a DEM. The concept of raster functions was introduced in chapter 5.

Raster functions are applied in the "function chain," which exists for each individual raster in a mosaic dataset, as well as for the mosaic dataset as a whole. As a result, raster functions can be applied to the individual rasters that compose a mosaic dataset, or to the mosaic dataset as a whole. Figure 13.4 shows a function chain for a Landsat 7 scene. In this example, the functions are applied to an individual mosaic dataset component raster, not to the entire mosaic dataset.

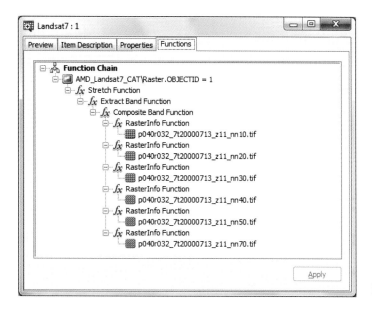

Figure 13.4. A function chain

On-the-Fly Processing with the Mosaic Dataset Function Chain

As mentioned above, function chains apply on-the-fly processing to component rasters of a mosaic dataset or to the mosaic dataset as a whole. A wide range of functions is available. The geometric functions transform the location of the pixels and are used to perform processing such as reprojection of imagery between different coordinate systems or orthorectification of satellite or aerial imagery. The radiometric functions transform the pixel values and are used for processes such as image enhancement, the pan sharpening of satellite imagery, or the computation of vegetation indices or band arithmetic. These functions are able to work with the full bit depth of the imagery, enabling the full dynamic range of the imagery to be exploited. This is especially important when working with newer satellite and aerial imagery or with elevation and scientific datasets. ArcGIS provides a large library of functions that can be chained together. The functions can also be extended using the Python Adapter function, which enables new function chains to be written using Python, allowing the end user to exploit a wide range of NumPy- and SciPy-based image processing and analysis functions.

The function chain can be edited manually in ArcGIS Desktop or modified with the Edit Raster Function tool, which can be used to add a new function, such as image enhancement, to a selection of images in a mosaic dataset. The Edit Raster Function tool can also be used to change the parameters of existing functions in the function chain. For example, the accurate orthorectification of imagery requires appropriate sensor model parameters and elevation models. Those parameters can be set during the creation of the mosaic dataset and later refined using the Edit Raster Function tool.

Advanced Tools

Processing Templates

Processing templates are powerful because they allow a single mosaic dataset to house multiple function chains. The end user or client application chooses the function chain that they want the mosaic dataset to display. For example, a mosaic dataset of DEM rasters could include three processing templates: one for slope, one for aspect, and one for elevation. The end user of the mosaic dataset simply selects the processing template they want to view—slope, aspect, or elevation—and the mosaic dataset appears in that format.

Processing templates are defined and configured in the mosaic dataset properties window in the Processing Templates.

Mosaic Dataset Block Adjustment

Block adjustment is a set of tools designed to improve the georeferencing accuracy of overlapping but misaligned imagery. Block adjustment is a technique used to reduce the shifts between images. It is used in mosaicking collections of satellite images or aerial photographs for an area or a project, which is called a block. Block adjustment is the process of computing adjustment (or transformation) based on the internal relationship between overlapping images (referred to as tie points) and ground control points, which define the spatial relationships between points in the image and points on the ground. The block adjustment process determines refined transformation parameters that reduce the shifts between images and applies the adjustment to the images within a block. Block adjustment is one of the most important steps in creating a seamless orthomosaic from a collection of images.

In ArcGIS, the block adjustment solution contains a set of geoprocessing tools to apply a block adjustment. Also, there is a Block Adjustment window for the quality control of tie points and control points, which includes tools to edit points that have high errors. The ArcGIS 10.4 block adjustment tools provide a solution based on applying polynomial transforms to each of the images. The tools in ArcGIS 10.5 and ArcGIS Pro 1.4 provide rigorous adjustment of frame camera and rational polynomial coefficients that take into consideration the digital terrain model to refine the orientation of the sensors.

Mosaic Dataset Color Correction

Sometimes, a mosaic dataset's images are collected at different times or on different dates, so tonal differences between the component images result in visible faults in the mosaicked results. Color balancing removes those tonal differences, creating a seamless mosaic or reducing the severity of the seams. The mosaic dataset color correction tools offer robust functionality for color balancing.

The color correction tools apply adaptive enhancements to each image in the mosaic dataset, resulting in a more seamless-looking image mosaic. There are three options for color balancing in the color correction tools: Dodging, Histogram, and Standard Deviation. Dodging typically provides the best results. There is also an Exclude Area option in the color correction tools, which allows areas to be excluded from the color correction algorithm. These exclude areas ensure that outlier areas, such as houses with bright red roofs or strong blue water bodies, do not influence the correction process.

Mosaic Dataset Seamline Generation

Seamlines can be used instead of footprints when mosaicking the raster data in a mosaic dataset. They define the line along which the rasters in the mosaic dataset will be mosaicked when using the seamline mosaic method.

When using the seamline mosaic method, the mosaicked image will not change as you move around the image, as can happen when using mosaic methods like Closest to Center. Also, if you create seamlines, you can ensure the mosaicked image is the best possible mosaic by editing the seamlines so that features, such as buildings, are not along the line and that the seam between images is not obvious. The seamline generation tools can be used to automatically determine the best locations for seamlines that define where to blend images together.

The Build Seamlines tool creates seamlines for a mosaic dataset. Seamlines are like footprints, in that one polygon represents each image. The shape of the polygon represents the part of the image that will be used to generate the mosaicked image when viewing the mosaic dataset. Once the seamlines are built, a seamline layer will be present in the table of contents each time you add the mosaic dataset to ArcMap.

There are many options for how the Build Seamlines tool creates seamlines, including an option for automatically creating seamline polygons from the radiometry and spectral patterns of the areas of overlap.

In areas of image overlap along a seamline, blending can be applied, which results in a smooth transition between one image and the other. By default, the Blend Width is defined in the mosaic dataset's default properties.

Use Cases for Mosaic Datasets

Scaling to Massive Collections

Mosaic datasets are scalable. They can be used to work with a few images or many millions of images. The images can be added, edited, or removed at any time, enabling the use of mosaic datasets in dynamic environments where new imagery is frequently received.

Mosaic datasets work with a wide range of imagery and can be scaled to house massive image collections. An example is a US state housing its collection of statewide orthophotography using mosaic datasets. In this case, there is orthophotography from multiple years of collection in various spatial resolutions and projections. Some of the older imagery is panchromatic; the new orthophotos are either true color (red, green, and blue) or 4-band (red, green, blue, and near infrared). A single mosaic dataset or multiple mosaic datasets (one per statewide date of collection) could be used to manage the entire statewide set of imagery, providing seamless access to the imagery for desktop and web-based users without duplicating or reprocessing any of the raw data.

Integrating Satellite Data with Different Levels of Processing

The source and resolution of the imagery in a mosaic dataset may be varied. For example, a single mosaic dataset can include collections of different high-resolution satellite imagery, such as a mix of DigitalGlobe's WorldView-3 and Airbus Pleiades imagery. In this example, the mosaic dataset's component rasters are a mix of level 2 products that have already been orthorectified and level 1 products that require orthorectification and pan sharpening. In this case, the level 2 products would be added directly to the mosaic dataset and the level 1 products would be added and include on-the-fly orthorectification and pan-sharpening functions. Atmospheric correction, color balancing, or seamline-based blending can be applied between the different images.

Imagery from Unmanned Aerial Systems and Aircraft

Massive collections of aerial and unmanned aerial system imagery can be added to a mosaic dataset. Imagery from such aerial cameras could be preorthorectified unprocessed imagery directly from the frame cameras. As with satellite imagery, the parameters for orthorectification and pan sharpening can be defined in the function chains of the component rasters and applied as required. The ability to orthorectify and pan sharpen the imagery on the fly enables significant savings in storage and processing and still provides access to the original nonrectified pixel data. This is especially useful for oblique imagery stored in a mosaic dataset, which can be accessed by the user of the mosaic dataset either as orthorectified imagery in map space or as the raw, undistorted, oblique imagery.

Elevation Data

Mosaic datasets work with all forms of elevation data. Mosaic datasets can house elevation data in the form of gridded DEMs, lidar (LAS or zLAS) point clouds, or terrain datasets. By applying raster functions in the mosaic dataset's function chain, many derivatives can be produced on the fly from elevation data. These include slope, aspect, and hillshades. Using a mosaic dataset, global datasets such as those from the Shuttle Radar Topography Mission (SRTM) can be combined with higher-resolution local datasets collected by photogrammetric or lidar methods. The user of the mosaic dataset can get access to the data as a single, virtually blended dataset at any resolution with the SRTM data appearing at smaller scales and the local data showing up at a larger scale. Alternatively, the mosaic dataset user can query the mosaic dataset based on attributes and filter the result, so as to view a single dataset or some custom selection of the data.

Scientific Multidimensional Rasters

The management of scientific multidimensional rasters is another compelling use case for mosaic datasets. Such rasters are often stored in NetCDF, hierarchical data format, and general regularly distributed information in binary form formats. These formats can be added to a mosaic dataset with each slice of the dataset becoming a separate, uniquely identified record. It is then easy to query the mosaic dataset's attribute table to get any individual slice and obtain temporal profiles or data cubes for powerful visualization and analysis.

Full-Motion Video

Mosaic datasets can be created from full-motion video using the ArcGIS Full Motion Video (FMV) add-In. Instead of each frame existing as a separate component in the mosaic dataset, the FMV extension enables the creation of a mosaic dataset from key frames that are based on time or movement. In this way, a virtual mosaic of the imagery can provide a quick overview of different missions as well as the ability to quickly filter and find suitable sections of footage.

Publishing Mosaic Datasets as Image Services

Introduction

Image services were introduced in chapter 5. This section elaborates on the details of configuring and managing image services. Here we discuss optimizing image services for performance, client-defined processing of mosaic datasets, open standards, considerations when publishing image services in the cloud, and big raster processing.

Image services can be consumed by web maps, applications such as ArcGIS Desktop and ArcGIS Pro, and by mobile and web apps that have access to the image services through Esri's JavaScript API.

Mosaic datasets are excellent data models for cataloging and managing data. They are also a great tool for disseminating imagery across the web as image services. A server administrator can modify many properties of a mosaic dataset, such as the maximum image size, the level of metadata, the compression method, or the maximum number of downloads according to user needs. When web clients connect to a mosaic dataset published as an image service to see a mosaicked image, their application can control the same mosaic methods and other properties that a directly connected user will have access to, along with the ability to select raster datasets and download them to their local disk.

As discussed in chapter 5, there are two types of image services: dynamic and tile cache services. Tile cache services are useful for displaying static images that won't be used for analysis and don't require enhancing for visual display. Tile cache services do not change once they are published and don't afford the end user access to the raw pixel data. Tile cache services are fast, versatile, and are well suited to, for example, serving up orthophoto base images for visual reference. They do not provide access to the source pixel values and require preprocessing to create. Dynamic image services, on the other hand, provide full access to the image pixels and can be used for analysis, stretched, or even downloaded locally. Dynamic services are in many cases not as fast to get access to as tiled services as the server needs to process the pixels being accessed, but dynamic services provide complete access to the raw imagery.

Image services published without a cache are dynamic by default. Image services published with a cache can be accessed either as dynamic services or as cached services.

Tile Cache Services

About Tile Cache Services

Image service caching improves the performance of image services when they are displayed in client applications. When you cache an image service, the server pregenerates tiles at different levels. These static tiles can be delivered from the server to the client faster and use less server processing power than delivering the actual pixels from the mosaic dataset. An image service cache does not serve imagery that is processed on the fly; it preprocesses the imagery to create the cached tiles and then serves the cached tiles.

Some applications or users require imagery as a simple background for visual context. In cases where such access needs to be scaled to large numbers of users, it is often advantageous to use tile cache services.

The tile cache can then be generated in ArcGIS Desktop or, for large projects, ArcGIS Enterprise, which can run across multiple servers. The processing generates a very large number of tiles that are stored in an optimized format, enabling the server to quickly return requested tiles with no additional processing. Applications that need tile cached imagery obtain from the server the schema of the tile layout and can request only the tiles required to cover the screen at the appropriate scale.

Tiles can be preprocessed for all areas covered by a mosaic dataset, but ArcGIS also provides the option for on-demand tile cache generation, so that cache tiles are created only when first used.

Image service caching results in a dual-purpose image service that is accessed depending on its purpose. One purpose is to provide the fastest access to the image as a tiled service. The other purpose is to provide access to the data for queries, downloading, access to individual rasters, and for use in processing and analysis.

When you display an image service that has been cached, it will have limited image service capabilities by default. However, if you need to work with the image service dynamically and get access to the full pixel data or change the mosaic method, then you can change the layer's mode. To change the mode, you right-click the image service layer and uncheck Enable Cache View Mode. To use the image service as a cache service, you just check this option back on.

Cache Configuration

The cache tiles are processed and stored in a selected format (see Cache Formats for Image Services below) so that the server can distribute these images whenever there is a request.

For all its performance benefits, caching comes with some overhead. You need time and server power to create the cache tiles and the hardware to store them. You may also need to perform cache updates if, for example, your source mosaic dataset is edited or changed. If your application offers imagery for a vast area at a large scale, you may decide that the time and storage required to build and maintain the cache outweighs the performance benefit.

Tiling Scheme

The scales that you pick and the properties you set for the cache constitute the tiling scheme. The tiling scheme should be consistent with the other layers you may be integrating. For example, you can choose to use the well-known tiling scheme of ArcGIS Online/ Google Maps/Bing Maps so that you can easily overlay your caches with these online mapping services, or you can create your own tiling scheme to be consistent within your own web application. Each cache has a tiling scheme file that you can import when you create new caches so that all your caches use the same tile sizes and scales.

If your tiling scheme doesn't match the one used by the other layers in your application, then you may not see your cached layers. This is because web clients often cannot resample your data to display it at a different level.

When caching imagery in ArcGIS the system generates the highest-resolution data first. Subsequent levels can be generated by using the imagery from the mosaic dataset or by resampling the higher-resolution data. If your source data has a mix of spatial resolutions, then you will want to cache from the mosaic dataset; if the source has a single resolution, or overviews have not been built, then you can let the system resample the higher-resolution data. This is controlled by entering a value for the Maximum Source Cell Size on the Advanced Settings page for caching.

Cache Formats for Image Services

Common formats for cached tiles include PNG, JPEG, MIXED, and Limited Error Raster Compression (LERC); these raster formats are discussed below:

- **JPEG**—Use this format for base image services that have large color variation and do not need to have a transparent background. The JPEG format works well for caching continuous raster data, such as orthophotos. JPEG is a lossy image format. It attempts to selectively remove data without affecting the appearance of the image (see chapter 5).
- **PNG (PNG, PNG8, PNG24, and PNG32)**—PNG is a lossless compression that works well for data that is not continuous. PNG is typically used to cache vector data, but it can also be used to cache thematic raster data such as the results of a classification. It also has the advantage of transparency, because NoData areas can be displayed as blank or transparent using the PNG format. Due to its lossless compression, PNG tiles can be significantly larger than JPEG when used for continuous imagery. PNG is not recommended for caching imagery such as orthophotos.
- **MIXED**—A mixed cache uses both JPEG and PNG tiles. Tiles without transparent pixels are cached using the JPEG format; remaining pixels are coded using PNG. In many projects, the result is that nearly all tiles are JPEG except for the ones at the edge. Use of Mixed mode is typically recommended for continuous imagery.
- **LERC**—LERC is an efficient, controlled, lossy compression method. LERC compression is used for creating cached tiles of elevation data that is typically 32-bit float and would not compress well using either JPEG or PNG.

On-Demand Cache

On-demand caching lets you set up the tiling scheme and publish the image service, but generate the cache only when and where you get access to the image service. The first user to navigate to an uncached area must wait while the corresponding cache tiles are created by the server. The tiles are then added to the service's cache folder and remain on the server until updated or deleted by the server administrator. This means that subsequent visitors to the area will not have to wait for the tile to be created.

When used wisely, on-demand caching can save you processing, time, and disk space. Many image services contain areas that are barren or uninteresting to the end user, especially at large scales (zoomed in). Caching on demand relieves you of the burden of creating and storing these unneeded tiles but leaves the possibility that a user could still view the area if needed.

Cache Best Practices

Test before Deploying

Before committing to an image format for a large cache, build a small cache of a representative area of your map, and examine the tile quality and performance in a test application. If you'll be working with multiple caches, build a small test cache for each, and add them to a test application to make sure they overlay as expected. This will allow you to make adjustments before you create the entire cache.

To make a small test cache, use the editing tools in ArcGIS Desktop or ArcGIS Pro to create a new feature class consisting of a small rectangle around the area you want to test. Then use the option at the bottom of the Manage Map Server Cache Tiles tool dialog box that allows you to create tiles based on the boundary of a feature class. Browse to the feature class containing your test area and create the tiles.

Optimizing Services for Performance

Mosaic datasets published as image services may have little or no processing applied by their function chains or may have lengthy processing chains applied. Complex on-the-fly processing chains can place heavy processing loads on a server and contribute to sluggish performance. Most processes such as reprojection, orthorectification, or pan sharpening have relatively little effect on performance. The biggest contributor to slow performance is typically not the processing, but the speed of access to the source data. Huge performance improvements can be achieved in some cases by preprocessing the data into a more efficient format for resampling and on-the-fly processing.

For applications that use tiled base image layers, some performance improvements can be achieved by aligning the pixel sizes of the overviews with the scales of the base image tiling scheme. Figure 13.5 shows the recommended pixel sizes to use as the base pixel size when creating overviews using the ArcGIS Online/Bing Maps/Google Maps tiling scheme.

AGOL Tiling Scheme		AGOL Tiling Scheme Pixel Sizes			Suggested Pixel Sizes for Data		
Tile Level	Map Scale	Meters	Decimal Degrees	Feet	Meters	Decimal Degrees	Feet
0	1 : 591,657,527.591555	156542.7208	1.408603	513591.6038	153600	1.31072	512000
1	1 : 295,828,763.795777	78271.36042	0.704301	256795.8019	77800	0.65536	256000
2	1 : 147,914,381.897889	39135.68021	0.352151	128397.9010	38400	0.32768	128000
3	1 : 73,957,190.948944	19567.84010	0.176075	64198.95048	19200	0.16384	64000
4	1 : 36,978,595.474472	9783.920052	0.088038	32099.47524	9600	0.08192	32000
5	1 : 18,489,297.737236	4891.960026	0.044019	16049.73762	4800	0.04096	16000
6	1 : 9,244,648.868618	2445.980013	0.022009	8024.868810	2400	0.02048	8000
7	1 : 4,622,324.434309	1222.990007	0.011004	4012.434405	1200	0.01024	4000
8	1 : 2,311,162.217155	611.4950033	0.005502	2006.217202	600	0.00512	2000
9	1 : 1,155,581.108577	305.7475016	0.002751	1003.108601	300	0.00256	1000
10	1 : 577,790.554289	152.8737508	0.001376	501.5543006	150	0.00128	500
11	1 : 288,895.277144	76.43687541	6.878e-4	250.7771503	75	6.4e-4	250
12	1 : 144,447.638572	38.21843770	3.439e-4	125.3885751	38	3.2e-4	125
13	1 : 72,223.819286	19.10921885	1.719e-5	62.69428757	19	1.6e-4	62.5
14	1 : 36,111.909643	9.554609426	8.597e-5	31.34714379	9	8.0e-5	31.25
15	1 : 18,055.954822	4.777304713	4.299e-5	15.67357189	4.5	4.0e-5	15.5
16	1 : 9,027.977411	2.388652357	2.149e-5	7.836785947	2.0	2.0e-5	7.80
17	1 : 4,513.988705	1.194326168	1.075e-5	3.918392973	1.0	1.0e-5	3.90
18	1 : 2,256.994353	0.5971630892	5.373e-6	1.959196487	0.5	5.0e-6	1.95
19	1 : 1,128.497176	0.2985814448	2.687e-6	0.9795982435	0.25	2.5e-6	0.97

Figure 13.5. Recommended pixel sizes to use as the base pixel size when creating overviews

The file format and compression type of source data can also have a huge impact on performance. GeoTIFF with internal tiles is recommended for situations where it's necessary to reformat the imagery from a slower format. Pixel compression has the potential to either increase or decrease performance depending on the how the data is stored and accessed by the server. Some formats, such as JPEG, are fast to decompress and provide reduced file size, resulting in increased performance.

Client applications can also play a role in dynamic image service performance. There is a direct relationship between service performance and the size of the map window (extent of the imagery shown) in an application. Applications running full screen on very-high-resolution monitors will make much larger requests to the server, resulting in higher data access and server processing.

Client-Defined Processing

Image services enable the creation of advanced, lightweight web applications that fully leverage the power and configurability of the mosaic dataset that underpins the image service. The image service's representational state transfer (REST) API enables client applications to get access to and control all aspects of how the server gets access to and processes imagery. Function chains configured on the image service to accomplish different processing tasks can be exposed along with the image service via the REST API. An application consuming the service can apply one of the preconfigured processing function chains. For example, an elevation image service could have preconfigured function chains for elevation, slope, and aspect; the client application would specify in its call to the service which processing chain should be applied. The image service's function chains can contain collections of functions to be processed by the server, enabling applications to specify and control a wide range of dynamic, on-the-fly processing to be applied to the service that the client consumes. Client applications can also transmit individual functions or chains of functions to the server to be applied on the imagery before it is transmitted back to the client. This enables the analyst to create a wide range of applications while using the server to get access to and process the imagery. For users requiring very advanced raster analysis, the Python raster adapter can be used. This enables Python-based functions to be developed that use the extensive libraries of both NumPy and SciPy. Such Python functions can be associated with services and invoked by client applications. Python raster functions are useful for the development of, for example, more advanced change detection algorithms or the processing of multidimensional scientific data.

The image service REST API is rich and enables the configuration of a wide range of processing parameters in a client application. For example, a client application using the API can specify the resampling method applied to the image service that it consumes (the choices are nearest neighbor, bilinear interpolation, or cubic convolution). Via the API, a client is able to request information other than the imagery itself. For example, if the source image service includes sensor orientation information, the client can perform mensuration to determine the heights of objects. In this case, for example, the application could request that the user define the displayed locations of the top and base of a building (with two mouse clicks), and then use the service to compute and return the building height. Similarly, image services can be used by client applications to return image statistics and other information about polygons digitized in the client application. These client-defined processing functions enable the development of applications that invoke advanced image-processing capabilities beyond just visualizing the data.

Open Standards

In addition to the ArcGIS REST API, ArcGIS Enterprise also supports Open Geospatial Consortium (OGC), Web Map Service, Web Map Tile Service, web coverage service, Wi-Fi Protected Setup, and Keyhole Markup Language standards. These OGC standards can be used to quickly get access to imagery from all OGC conforming applications to aid in interoperability. Although the OGC standards support some features such as temporal control, they do not provide the rich API provided by ArcGIS REST.

Cloud Considerations

Traditionally, organizations have stored imagery on direct access storage, enterprise network attached storage, or storage area networks. Such storage systems are typically optimized for fast, low-latency access and provide block level access through file systems such as network file systems. These storage systems are relatively expensive to scale, resulting in significant costs for large volumes of imagery. The advent of cloud computing has promoted the abundance of object storage such as Amazon S3 or Azure Blob storage, which can be accessed simultaneously by large numbers of computers. Such object storage is available from most cloud infrastructure providers, but can also be implemented in on-premise and hybrid cloud environments. Object storage offers simple storage for massive data collections with significantly lower storage costs while providing very high durability. However, object storage does not provide the same levels of access performance or latency as traditional storage and cannot be effectively accessed through a file system, so it is typically not suitable for image-processing applications. ArcGIS includes technology that overcomes these limitations. To optimize performance, the cloud-hosted imagery should be converted to a cloud-optimized format such as metaraster format as it is transferred to the cloud. Fast access is achieved by minimizing the number and size of requests as well as using tile-based caching. ArcGIS can also get access to tiled formats such as TIFF or JP2000 stored in object storage, although performance is not as good. These techniques enable organizations to manage their imagery holdings on inexpensive cloud storage and then provide fast access to the imagery as image services. Making use of the elasticity in cloud infrastructure also enables the systems to efficiently handle the often varying loads.

An example of such a cloud-based image service is the Landsat archive service hosted by Esri (http://www.esri.com/landsatonaws). These publicly accessible image services provide global access to hundreds of thousands of full-resolution, multispectral Landsat 8 and global land survey scenes. The services are updated daily with new scenes. Access to all the multispectral bands with various band combinations and indices are provided with all processing being applied on the fly. The source multispectral Landsat scenes are stored on Amazon S3 object storage as tiled GeoTIFF files as part of the Amazon Public Datasets. Each band of each scene is a separate file. A mosaic dataset defines the processing to be applied as the data is accessed, but client applications can also refine the processing and analysis. The mosaic dataset is served as an image service using elastic Amazon Machine Images that can scale up or down depending on the server load. These services can be used in a range of web and desktop applications to quickly visualize and analyze this phenomenal dataset made available by the US Geological Survey.

Spatial Analysis in the Cloud—Geoprocessing Services

Overview of Geoprocessing Services

Geoprocessing services contain geoprocessing tasks; a geoprocessing task takes simple spatial data captured in a web application, processes it, and returns meaningful and useful output in the form of features, maps, reports, and files. A task could calculate the probable evacuation area for a hazardous chemical spill, the predicted track and strength of a gathering hurricane, a report of land cover and soils within a user-defined watershed, a parcel map with historical details of ownership, or a permitting application for a septic system.

ArcGIS geoprocessing tasks can be served through ArcGIS Enterprise. Geoprocessing services have access to the extensive range of geoprocessing tools within ArcGIS as well as tools developed by Esri partners. Unlike image services, geoprocessing tasks are asynchronous and can return all forms of geospatial data, tables, or maps. A task may return a result within a second, but for larger processing tasks the response may take longer (hence the asynchronicity). Geoprocessing tasks can get access to raster data directly from a mosaic dataset, an image service, or some other raster data type. Geoprocessing can be used to perform simple tasks such as the computation of a viewshed, but also be the interface to Big Data analytics where a geoprocessing task can initiate a set of processing steps on many machines, and then compile and return the results.

Geoprocessing Service Examples

An example of a geoprocessing service is the Trace Downstream analysis tool available on ArcGIS Online web maps. This tool enables a user to define a starting point on their web map and submit the starting point to a geoprocessing service. The geoprocessing service uses the user-submitted point to run a trace against a geometric network of flowlines, returning the resulting geometry—a line feature class representing areas downstream of the submitted point—to the web map user. Another geoprocessing service example is the Clip and Email Viewer shown in figure 13.6. This viewer enables access to high-resolution imagery for Sonoma County, California, providing users with clipped orthophotography and lidar-derived DEMs clipped to their area of interest. To use it, the viewer logs onto a JavaScript application (http://sonomavegmap.org/imagery) in a browser, digitizes an area of interest (polygon), and enters a desired contour interval and an email address. The shape of the polygon and the user's information is passed to the geoprocessing service via Esri's JavaScript API. The geoprocessing script, which exists as a Python script on the ArcGIS Server, takes the geometry of the user-defined polygon and clips the countywide mosaic datasets of the requested high-resolution rasters to the user's area of interest. The resulting clipped images are zipped and uploaded by the geoprocessing service to an FTP server. As a final step, the geoprocessing service sends an email to the user of the application, providing a link for downloading the clipped and zipped raster data.

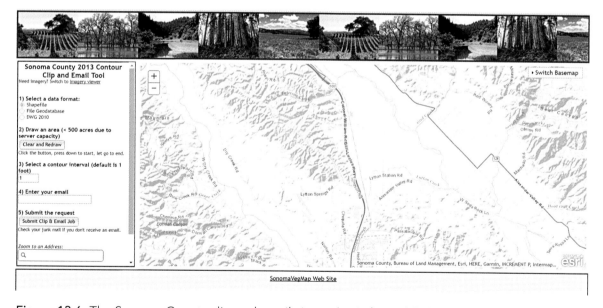

Figure 13.6. The Sonoma County clip and email viewer (esriurl.com/IG136). Source: Sonoma County Agriculture Preservation and Open Space District

Summary—Practical Considerations

The mosaic dataset offers a revolutionary approach for managing and serving imagery and can scale to massive collections of raster data. It is important to remember that the mosaic dataset doesn't store pixels of imagery; instead, it provides a framework for cataloging and managing imagery. The raw imagery itself isn't part of the mosaic dataset but instead is referenced by the mosaic dataset.

This chapter reviewed the structure of the mosaic dataset and provided information on how to implement mosaic datasets. Several use cases for mosaic datasets are included in this chapter; they include information on best practices for implementing mosaic datasets for various types of imagery (such as videos and elevation data).

The final part of the chapter focuses on the details of serving mosaic datasets as web services. The section begins with a discussion of tile cache services and instructions and best practices about how to implement them. Cache services can greatly improve the performance of image services used for visual display. The discussion of cache services is followed by more information on the capabilities of and best practices for serving mosaic datasets as web services, including sections on optimizing services for performance, client-defined processing, and cloud considerations. The chapter concludes with a discussion of creating and publishing geoprocessing services, which provide client access to powerful geospatial processing through web services.

Chapter 14
Concluding Thoughts

This book provides you with the foundational knowledge needed to use imagery within a GIS to support informed decisions. You now know how imagery can bring value to your GIS projects and how GIS can enhance imagery classification. The level of involvement and importance of imagery varies project to project depending on the goals and objectives of each project. Often, imagery is used only as a base layer providing context for other geospatial layers used in a project. In that case, only basic understanding about where the imagery came from, how it was processed/corrected, and how it can be displayed is needed to make effective use of it. In other situations, imagery can be used to add additional information to the project. The elements of imagery discussed in this book can be combined to manually identify and map objects on the ground. They also can be processed using any of the classification algorithms discussed in the book to create new geospatial layers. Using imagery to identify objects or make a map requires a solid understanding of the imagery itself, exploring the imagery to determine the relationship between the image and the ground, developing an effective classification scheme, choosing the appropriate classification methodology, and assessing the accuracy of the results. This book has provided you with the knowledge needed to integrate imagery into GIS projects, and for extracting information from imagery. This chapter concludes the book by offering some nuggets of wisdom that the authors have learned through their decades of remote-sensing research and operational mapping projects. These nuggets are presented below in bullet form with a brief explanation of each. It is our hope that this book has provided you with the knowledge you require to obtain the most value from imagery in your geospatial analysis projects.

- *Imagery technology will continue to rapidly improve, and the supply and accessibility of imagery products will continue to expand.* Over the last 20 years, the supply and demand for imagery have changed remarkably. These changes will continue and the pace of change will increase. Unmanned aerial vehicles, small satellites, lighter electronics, higher Internet speeds, global Internet access, and new sensor technologies will combine to increase the supply of imagery. As the barriers to entry into imagery markets dissolve, the markets will become more competitive and transparent, prices will continue to fall, access will increase, and use will soar.

- *Building a robust classification scheme is vitally important to project success.* The classification scheme sets the objectives regarding the exact information that the analyst wishes to acquire from the imagery. In some cases, specific items need to be identified. In others, very general map classes such as Water, Vegetation, and Developed are sufficient. In all situations, it is vital that the map classes are well defined to ensure there is no confusion about what composes each class. It is easy for analysts and map users with different backgrounds and training to make assumptions about what is meant by a particular map label. Defining each class clearly by establishing rules for class membership avoids these biases and eliminates confusion. Once defined, the classification scheme becomes the starting point for data exploration where the relationship between what is recorded on the image and what is actually on the ground is investigated. Detailed, specific map classes require significantly more data exploration than general classes. Therefore, determining the scheme at the beginning of the project is vital to all the steps that follow. The classification scheme also dictates the classification algorithms that are most appropriate for creating an accurate map of those particular map classes. Spending time developing or selecting the appropriate classification scheme at the beginning of the project is well worth the effort and ultimately produces a more cost-effective and accurate map.

- *Use your classification scheme to drive the choice of imagery for your project.* The abundance of imagery available can make the choice of the most appropriate imagery an overwhelming task. Focus your decision making by matching your classification scheme's requirements to the technical and organizational characteristics of available imagery products, thereby using your classification scheme to guide your selection of imagery. Be open to relaxing the requirements of your classification scheme to meet your budget constraints, if needed.

- *Imagery sources and products will continue to rapidly expand.* An abundance of excellent imagery is available and served online such as Landsat, Sentinel, and the National Agricultural Imagery Program. Every day, new sources of imagery are added to our geospatial world. More and more countries and public agencies are flying their own sensors. and there is a trend toward making most civilian government acquired imagery available for free on the web. Imagery is being collected by sensors in orbit, but also on airplanes, balloon, helicopters, and most recently unmanned aerial systems. This growing body of imagery is becoming more and more easily accessible as Internet and search technologies improve. Abundant imagery is available to aid your project. Great synergies can be created by using more than a single source of imagery. Spend the time needed to become familiar with all the imagery that is available or could be acquired for your study area.

- *Lidar imagery will significantly improve many projects.* Lidar imagery brings the image element of height into semiautomated image classification. When a lidar instrument is flown over an area, the data collected provides a comprehensive and detailed picture of the elevation of the earth's surface as well as the height of vegetation, buildings, and other features. The data provides a detailed view of the vertical structure of our environment. High-resolution elevation data and forest structure metrics, such as tree height and canopy density, significantly enhance our ability to assess and monitor carbon stocks, document sea-level rise, map groundwater, and assess vegetation and habitat. Lidar data improves almost any mapping effort and has myriad applications across many disciplines.

- *Landsat and Sentinel imagery can add valuable information to all mapping projects including high-spatial-resolution image classification.* Even if you are creating a map from high- or very-high-spatial-resolution imagery, moderate-resolution Landsat or Sentinel imagery can add value to your project because of its high temporal resolution and the increased information provided by its short-wave infrared and thermal (in the case of Landsat) bands, which are not available from most high-spatial-resolution systems. In remote sensing systems, there is usually a trade-off between spatial resolution and temporal and/or spectral resolution. As a result, much high- and very-high-resolution imagery is collected infrequently and captures only one to four bands in the optical and infrared portions of the electromagnetic spectrum. Landsat and Sentinel complement high-resolution imagery by providing a valuable spectral representation of the near-infrared, mid-infrared, and thermal, which are extremely useful for vegetation mapping and analysis. These sensors also produce remarkably consistent imagery and aren't prone to the spectral inconsistency often present in high-resolution aerial imagery. Because Landsat and Sentinel images are free and accessible via the web, it is easy to incorporate them into any project, and they often add value, even in high-spatial-resolution projects.

- *Spectral accuracy and calibration are critically important.* Not all sensors are the same. While many sensors may collect red, green, blue, and infrared bands, the spectral accuracy of those measurements may vary greatly depending on the quality of the sensor. As a result, data collected will vary from sensor to sensor, with some sensors offering limited separability of similar features. Additionally, few sensors are rigorously calibrated on a regular basis. If your goal is to use imagery to identify and label variation on the ground or to monitor change, you must be confident that the sensor's measurements are capable of distinguishing those features from one another. Make sure you learn about and understand the spectral accuracy and calibration of the sensors capturing your imagery.

- *Spatial accuracy is essential, especially in high-spatial-resolution mapping.* A key reason that imagery has become a valuable component of any geospatial analysis is that the imagery can now be very accurately registered to the ground and the other geospatial data layers. As imagery increases in spatial resolution and more and more detail can be seen, it is increasingly important that the imagery is registered with high spatial accuracy. If not, there will be a noticeable shift in the position of the image relative to the other geospatial layers and the ground. Spatial accuracy must be tested with a set of checkpoints that are independent of the ground control points used in the imagery registration process. The results of analyzing those checkpoints allow the analyst to know how well the imagery will register to the other geospatial layers and the ground.

- *Hyperspectral imagery will become more commonplace.* There are significant gains in spectral information available from hyperspectral imagery. Wavelengths of electromagnetic energy that may be represented by a single band in multispectral imagery are divided into much smaller portions, resulting in more refined spectral patterns. These patterns can reveal subtler relationships in the imagery than previously possible from multispectral imagery. While hyperspectral imagery is available now, it is far from prevalent. Soon, more hyperspectral imagery will be available, and there will be great advantages to the analyst who knows how to use it.

- *Licensing, pricing, and other organizational characteristics of imagery products will become more fluid.* The market for imagery is rapidly expanding as microelectronics are incorporated into sensors, resulting in decreasing costs for sensor launch and operation. As the commercial market evolves from a few to multiple providers, imagery producers will start to compete on organizational characteristics. The current business model for commercial high- and very-high-resolution imagery, which restricts the sharing of imagery through licensing agreements, will probably erode as more and more competitive imagery providers enter the market. If you acquire imagery, especially large amounts of imagery, push for licensing restrictions to be relaxed and prices to be lowered.

- *Data exploration is the key to understanding the variation in the imagery and how it relates to variation on the ground.* Remote sensing is an effective tool for identifying objects and creating maps because what is recorded in an image is highly correlated with what is occurring on the ground. Therefore, the image is often a good, but not perfect, representation of the ground. Anything that detracts from the correlation between the image and the ground should be known and investigated. Many powerful tools can be used to understand the variance in the imagery and how it relates to the ground. Exploring this relationship allows the analyst to select the appropriate classification scheme for the project, the most suitable imagery, and the classification algorithms that are likely to produce the most accurate map. Failure to explore this relationship between the imagery and the ground results in a less accurate and more expensive map.

- *Nonimagery geospatial data can add significantly to the effective use of imagery.* The power of a GIS is the ability to combine many layers of geospatial data and consider all of them simultaneously for decision making. Imagery is just another source of geospatial information. When using imagery, especially when creating maps from the imagery, adding other geospatial data layers can significantly improve the effectiveness of the imagery. As described in this book, the confluence of evidence of the elements of imagery result in the analyst developing a label for the map. However, not all the elements can always be derived from imagery alone. Slope, elevation, and aspect are very useful geospatial layers that can improve the accuracy of a map. Other layers such as fire history, weather patterns, previously created land-cover maps, tidal maps, and crop calendars can also be very helpful. Think about your classification scheme and spend some time determining and locating other geospatial data layers that can help produce the best possible map from the imagery.

- *Classification of high- and very-high-spatial-resolution imagery requires segmentation and object-oriented classification.* Per-pixel classification of high-resolution imagery is not recommended at the pixel level because there is so much information and noise. For example, making a per-pixel map of vegetation types from high-resolution (six-inch) pixels, would be very difficult because a stand of oaks trees, for example, might include hundreds of pixels that represent the very detailed components of the stand including bare ground, leaf litter, illuminated oak canopy, dark shadows, tree branches, etc. None of these pixels alone represents the oak class being mapped, but collectively the pixels do. Hence, it is essential when using high-resolution imagery to map vegetation type and land cover to group the pixels into objects (this process is called segmentation) and to classify the objects, not the individual pixels that compose them.

- *Semi-automated image classification makes sense only if economies of scale can be captured.* Collecting training data, performing data processing, and choosing and tuning algorithms for automated classification is time consuming. Most automated classifications require manual editing even after they are run to achieve acceptable accuracies. Consider the time it will take to perform a semi-automated workflow, and be sure that your project is large enough that significant economies of scale can be captured. If you will spend the same amount of time to perform an automated classification as you would to perform a manual classification of your entire project area, then choose manual interpretation over a semi-automated approach. In general, don't use semi-automated methods on a project smaller than 10,000 acres.

- *Combined classification algorithms can increase classification accuracy.* Just as there is no longer any reason to use only one imagery source in your project, there is no reason to use only one classification algorithm. All classification algorithms have strengths and weaknesses. Ensemble classification, which involves the use of multiple classification algorithms, can exploit the strengths and identify the weaknesses of each algorithm, thereby increasing the robustness of the classification.

- *Automating image-processing workflows with Python or other scripting languages opens doors in terms of workflow repeatability, scalability, and efficiency.* Image analysis and image processing is often complex and requires many steps. A typical workflow uses many functions applied in a sequence that produces intermediate results and eventually a final result. Performing a sequence of functions manually is laborious and prone to error. Often a sequence of functions needs to be rerun because—after reviewing the results—an additional function is added to the workflow or a change is made to one of the functions. Automating image-processing workflows—even very small ones—is easily done. Python is an easy-to-learn language that is deeply integrated into Esri products and is ideal for imagery workflow automation. Automation is desirable because it makes the workflow repeatable, provides documentation of the workflow, and brings much greater efficiency to workflows. It also makes it easy to quickly rerun existing workflows or easily adapt already written code to new situations. In addition to Python, other great additions to the image analyst's toolset include R, MATLAB, and interactive data language (IDL).

- *Web services have significant advantages.* Over the past decade, web services for providing Internet-based access to GIS data and imagery have come of age. With the ability to easily serve massive volumes of high-resolution imagery and data, imagery is no longer cloistered within the GIS department. Esri's mosaic dataset provides a framework for managing imagery collections, and ArcGIS Enterprise provides a mechanism for these collections as image services, unlocking these critical imagery investments for use throughout organizations.

- *Change analysis offers significant new information about a project area.* When an analyst works to create a map from imagery for a single date in time, she or he learns a great deal about that area through field visits, developing the classification scheme, performing data exploration, performing the classification, and conducting the accuracy assessment. When an analyst gets to work on a change analysis, the effort taken and information gained are not simply doubled, but increase significantly. Not only are the two or more different dates of imagery investigated, but also how and often why the imagery (and the ground) have changed over that period. Determining what has changed can be extremely valuable. The change map may show the results of poor zoning on urban development, the impact of a hurricane on a coastal town, the effect of sea-level rise, or the loss of wetlands due to poor management. Using the imagery from multiple dates forces the analyst to thoroughly understand the dynamics of the project areas.

- *Geospatial analysis and imagery are vibrant.* One of the most exciting things about the field of geospatial analysis is that it is constantly changing—improving all the time. New hardware, software, and methods are always emerging and changing the way you do your work. It is vital to stay abreast of these developments by reading about new work, attending conferences and user-group meetings, participating in webinars, and having a cohort of colleagues that you can share ideas and challenges with. It a challenge to keep up, but it is more than worth it in productivity and satisfaction in doing a great job.

- *Technology has now made communicating your results easier.* With the rapid development of web technology over the past decade and the maturation of web mapping, it is easier than ever to communicate spatial information with your organization, your stakeholders, or the public. Esri's ArcGIS Online provides a suite of tools that makes it simple to publish spatial data and imagery in the form of web maps, apps, and story maps. Geoprocessing services expose the full power of Esri's ArcGIS suite of GIS analytics to end users via web maps and apps.

Acronyms

µm	micrometer
3DEP	3D Elevation Program
6S	second simulation of the satellite signal in the solar spectrum
ACIC	Aeronautical Chart and Information Center
ACORN	atmospheric correction now
AFPO	Aerial Photography Field Office
ALI	Advanced Land Imager
AMI	Amazon machine image
ANN	artificial neural network
API	application programming interface
ASPRS	American Society of Photogrammetry Remote Sensing
ASTER	Advanced Spaceborne Thermal Emission and Reflection Radiometer
AT	aerial triangulation
ATCOR	atmospheric correction
AVHRR	Advanced Very High Resolution Radiometer
AVIRIS	Airborne Visible/Infrared Imaging Spectrometer
BigML	big machine learning
BIL	band interleaved by line
BIP	band interleaved by pixel
BLM	Bureau of Land Management
BRDF	bidirectional reflectance distribution function
BSP	bispectral plots
BSQ	band sequential
CART	classification and regression tree
CBP	Chesapeake Bay Program
C-CAP	Coastal Change Analysis Program
CCD	charged coupled device
CCDC	continuous change detection and classification
CE90	circular error 90 percent
CEOS	Committee on Earth Observation Satellites
CHM	canopy height model
CIR	color infrared
CLASS	Comprehensive Large Array-Data Stewardship System
CMAS	circular map accuracy standard
CPU	central processing unit

CRF	cloud readiness format
DA	divergence analysis
DAAC	Distributed Active Archive Center
DAS	direct access storage
DEC	deciduous
DEM	digital elevation model
DHM	digital height model
DN	digital number
DOS	dark object subtraction
DOT	Department of Transportation
DOY	day of year
DRA	dynamic range adjustment
DSM	digital surface model
DTM	digital terrain model
ECW	enhanced compression wavelet
EG	evergreen
EM	electromagnetic
EO	earth observation
EOL	end of life
EROS	Earth Resources Observation and Science
ESA	European Space Agency
ETM	Enhanced Thematic Mapper
EVI	Enhanced Vegetation Index
FAA	Federal Aviation Authority
FAO	Food and Agricultural Organization
FEMA	Federal Emergency Management Agency
FGDC	Federal Geographic Data Committee
FLAASH	fast line-of-sight atmospheric analysis of spectral hypercubes
FMV	Full Motion Video
FOV	field of view
FRAP	Fire Resource Assessment Program
FSA	feature space analysis
FTP	file transfer protocol
GAP	Gap Analysis Program
GCP	ground control point
GDAL	Geospatial Data Abstraction Library
GEDI	Global Ecosystem Dynamics Investigation
GIBS	Global Imagery Browse Services

GIS	geographic information system
GloVis	Global Visualization Viewer
GLS	global land survey
GPS	global positioning system
GRIB	general regularly-distributed information in binary form
GSD	ground sample distance
HAC	height above channel
HAR	height above river
HDF	hierarchical data format
HH	horizontal transmit and horizontal receive
HiPS	high pixel size
HOT	Humanitarian Open Street Map Team
HV	horizontal transmit and vertical receive
HVH	high and very high
ICS	image coordinate system
IDL	interactive data language
IFOV	instantaneous field of view
IFSAR	interferometric synthetic aperture radar
IMD	image metadata
IMU	inertial measurement unit
IR	infrared
ISODATA	iterative self-organizing data analysis
JPEG	Joint Photographic Experts Group
KML	Keyhole Markup Language
LCCS	Land Cover Classification System
LERC	limited error raster compression
LoPS	low pixel size
LUT	lookup table
LZW	Lempel–Ziv–Welch
MAPPS	Management Association of Private Photogrammetric Surveyors
MAS	map accuracy standard
MIR	middle infrared
MMU	minimum mapping unit
MODIS	Moderate Resolution Imaging Spectroradiometer
MODTRAN	moderate resolution atmosphere transmission
MRF	meta raster format
MrSID	multiresolution seamless image database
MSS	multispectral scanner

NAARA	National Archives and Records Administration
NAIP	National Agricultural Imagery Program
NAPP	National Aerial Photography Program
NAS	network attached storage
NASA	National Aeronautics and Space Administration
NBR	normalized burn ratio
NCAR	National Center for Atmospheric Research
NDEP	National Digital Elevation Program
NDMI	normalized difference moisture index
NDS	native display scale
NDVI	normalized difference vegetation index
NED	National Elevation Dataset
NetCDF	network common data form
NFS	network file system
NGA	National GeoSpatial-Intelligence Agency
NGO	nongovernmental organization
NHAP	National High Altitude Program
NHD	National Hydrography Dataset
NIR	near infrared
NLCD	national land cover data
nm	nanometer(s)
NMAS	National Map Accuracy Standards
NOAA	National Oceanic and Atmospheric Administration
NRCS	Natural Resources Conservation Service
NSSDA	National Standard for Spatial Data Accuracy
NVCS	National Vegetation Classification Standard
OGC	Open Geospatial Consortium
OLI	Operational Land Imager
OOB	out of bag
OpenCV	open source computer vision
PCA	principal components analysis
PNG	Portable Network Graphics
RBF	radial basis function
REST	representational state transfer
RGBN	red, green, blue, and near-infrared
RMSE	root mean square error
RPC	rational polynomial coefficient
SAN	storage area network

SAR	synthetic aperture radar
SAVI	soil-adjusted vegetation index
SID	security identifier
SIR	shuttle imaging radar
SLC	Scan Line Corrector
SPA	spectral pattern analysis
SPOT	Satellite Pour l'Observation de la Terre (satellite for observation of Earth)
SQL	structured query language
SRTM	Shuttle Radar Topography Mission
sUAS	small unmanned aerial system
SVM	support vector machine
SWaP	Size, weight, and power
SWIR	short-wave infrared
TCA	tasseled-cap analysis
TCT	tasseled-cap transformation
TIFF	tagged image file format
TIN	triangular irregular network
TIR	thermal infrared
TIRS	thermal infrared sensor
TM	thematic mapper
TWI	topographic wetness index
UAS	unmanned aerial system
UAV	unmanned aerial vehicle
URL	universal resource locator
USDA	United States Department of Agriculture
USGS	United States Geological Survey
UTM	Universal Transverse Mercator
UVM	University of Vermont
VIIRS	Visible Infrared Imaging Radiometer Suite
VNIR	visible and near-infrared
VV	vertical transmit and vertical receive
WCS	web coverage service
WELD	web-enabled Landsat data
WMS	Web Map Service
WMTS	Web Map Tile Service
WPS	Wi-Fi Protected Setup

Glossary

A

accuracy assessment: Quantitative sampling and analysis that identifies and measures map error.

affine transform: Method of warping an image by applying scale, rotation and shear such that parallel lines remain parallel.

agility: Ability of a remote sensing platform to change position, including positioning itself over a target, remaining in the target area, or slewing across the target area.

airborne imagery: Data collected by sensors on airborne platforms; airborne platforms include airplanes, helicopters, balloons, and unmanned aerial vehicles.

Anderson classification scheme: Early land cover/land use classification scheme developed by the US Geological Survey for use with remotely sensed imagery.

annulus: The area bounded by two concentric circles, often used to determine a neighborhood for statistical calculations.

anthropogenic: Caused by or resulting from human activity.

arcsecond (as "arc second"): The distance traversed on the Earth's surface while traveling 1/3600th of a degree of latitude or longitude.

aspect: The downslope direction of the maximum rate of change in value from each cell to its neighbors.

B

band: The information stored in one raster, often recording reflectance or radiation in a specific range of the electromagnetic spectrum. An image may be composed of one or more bands.

basemap: a collection of orthorectified imagery or vector data used as the background for a map.

bathymetry: The science of measuring and charting the depths of water bodies to determine the topography of a lake bed or seafloor.

bispectral plot: Graph where samples from two bands are plotted simultaneously, with one band on the x axis and the other on the y axis, used to evaluate the linkage between the variation in the image and the ground.

block adjustment: A technique used in photogrammetry to align and accurately georeference satellite images or aerial photographs of an area or a project (i.e. a block). The process produces the best statistical fit between images, for the whole contiguous block, minimizing errors with the ground control and between images .

breakline: Linear feature that defines the location of a sudden change in the slope of the terrain.

C

classification: The manual or semi-automated process of assigning the pixels or objects of an image to a set of categories or classes as determined by the classification scheme.

classification scheme: Hierarchical rules that determine the class categories to which image pixels or objects will be assigned when an image is classified. Classification schemes are driven by user requirements.

collection characteristics: Attributes of a remote sensing system that determine how imagery is collected, including spectral, radiometric, spatial, and temporal resolutions, viewing angle, and extent.

color (image element): The intensity of spectral response of an object across more than one band; used to help identify the object.

color balancing: A technique used to adjust the color rendition between images to make transitions from one image to an adjoining image appear seamless.

compression: The process of reducing the size of a digital file or database. Compression can improve data handling, storage, and database performance. See also lossy compression, lossless compression.

confluence of evidence: The convergence of image elements that indicates the most likely label for an image pixel or object as defined by the classification scheme. The concept of the confluence of evidence is used in both manual and semi-automated image classification.

context (image element): The neighbors of an object of interest, used to help identify the object

convolution filter: A function applied to the pixel values in an image used for sharpening, blurring, detecting edges, or other kernel-based image enhancements.

crosswalk: Table illustrating equivalent categories in two or more classification schemes.

D

date (image element): The date and time an image was acquired, used to help identify an object of interest.

datum: Defines the origin and orientation of a coordinate system, providing a frame of fixed reference for measuring location on the earth's surface.

dendrogram: Diagram with a tree-like structure representing hierarchical clustering.

derivative bands: The result of processing imagery to create transformed bands representing information or characteristics different from the original bands.

digital elevation model (DEM): A raster whose pixel values represent elevations of a surface, most commonly the elevations of the ground.

digital ortho quarter-quad (DOQQ): 1-meter resolution aerial image that has been orthorectified to remove image displacements, resulting in map-level accuracy; covers an area of 3.75 minutes longitude by 3.75 minutes latitude.

discrete cosine transform: Method of lossy compression; used by JPEG format, for example.

divergence analysis: Statistical technique used to determine which bands to use for creating the best thematic map for a given mapping project and classification scheme.

drone imagery: Still images and video gathered from sensors mounted on remotely piloted vehicles. Drones are also known as unmanned aerial vehicles (UAV) or unmanned aerial systems (UAS).

dynamic image service: Provides web access to a collection of images such that the calling application can define the processing to be applied to the source raster as well as how the resulting image is composed from overlapping rasters. Processing can affect both the radiometry and geometry of the pixels and the ordering can preference one raster over the other or have the pixels blended. See also image service.

E

electromagnetic energy: Energy (such as that emitted from the sun) that moves through space at the speed of light at different wavelengths. Types of electromagnetic wavelengths include gamma, x, ultraviolet, visible, infrared, microwave, and radio.

ellipsoid height: Distance above the reference ellipsoid used to approximate the earth's surface. (See also orthometric height).

envelope: The minimum bounding rectangle that defines the extent of a selected map or raster.

error matrix: Table used to determine thematic map accuracy by comparing the map classification to reference data, also called a contingency table in statistics.

extent: The minimum bounding area of a geospatial data layer (raster or vector).

F

feature space analysis: 2 dimensional plot of all of the values of two bands of an image dataset used to determine the degree of between-band correlation; similar to a bispectral plot.

float: A numeric data type capable of storing large numbers with an accuracy of seven digits after the decimal point.

flow accumulation: The number of cells (weighted or not) flowing into each cell in a raster. Represents a cell's upstream catchment area. Can be used to identify stream channels and local topographic height.

footprint: The extent of each image or raster dataset. A mosaic dataset contains footprints of all the rasters comprising the mosaic dataset.

function chain: An ordered list of image processing functions applied to a raster or mosaic dataset that are performed as the data is accessed.

G

Gamma: The degree of contrast between the midlevel gray values of a raster dataset (it does not affect the black or white values). By applying a gamma correction, you can control the overall brightness of a raster dataset, as well as the ratios of red to green to blue.

geoid: An approximation of the earth's mean sea level surface with respect to a defined ellipsoid. Orthometric heights are referenced to a geoid.

geoprocessing service: Geoprocessing services run on servers. They take requests and perform processing or analysis of data on defined inputs and then return the results in the form of features, maps, reports, and files.

georeferencing: Aligning geographic data to a known coordinate system so it can be viewed, queried, and analyzed with other geographic data. Georeferencing may involve shifting, rotating, scaling, skewing, and in some cases warping, rubber sheeting, or ortho-rectifying the data.

GeoTIFF: A Tagged Image File Format (TIFF) with spatial reference information.

Global Positioning System (GPS): A system of radio-emitting satellites used for determining positions on the earth. The orbiting satellites transmit signals that allow a GPS receiver anywhere on earth to calculate its own location through trilateration. Developed and operated by the U.S. Department of Defense, the system is used in navigation, mapping, surveying, and other applications in which precise positioning is necessary. GPS positions are referenced to the WGS84 ellipsoid and datum.

H

height (image element): Distance between the highest and lowest points of an object of interest, used to help identify the object.

histogram: A graph showing the distribution of values in a set of data. Individual values are displayed along a horizontal axis, and the frequency of their occurrence is displayed along a vertical axis.

I

image elements: All the characteristics of an image, including its tone/color, shape, size, pattern, shadow, texture, location, context, height, and date; also known as the elements of image interpretation.

image filter: On a raster, an analysis boundary or processing window which changes the pixel values at the center of the window.. Filters are used mainly in cell-based analysis where the value of a center cell is changed to the mean, the sum, or some other function of all cell values inside the filter. A filter moves systematically across the entire raster until each cell has been processed. Filters can be of various shapes and sizes, but three-cell by three-cell squares are common.

image service: A web service that provides access to raster data. Image services can be consumed in web maps, apps, and in ArcGIS Desktop. See also dynamic image service.

image statistics: Statistics that are calculated from the cell values of each band in a raster, including the minimum, maximum, mean, and standard deviation. Typically used to enable a suitable stretch to be applied to the raster for visualization. For multispectral imagery, covariance matrix values may be included. For thematic datasets, the number of classes are included. Statistics are required for many rendering and geoprocessing operations.

image stretch: A display technique applied to the histogram of raster datasets, most often used to increase the visual contrast between cells.

imaging surface: A device that measures the electromagnetic energy captured by a remote sensing device.

index: A ratio of original image bands, sometimes with other factors or coefficients included.

inertial measurement unit: An electronic device that measures the linear acceleration and angular velocity of a body, used to determine the precise location of the imaging surface when an image is captured.

infrared: The portion of the electromagnetic spectrum between visible and microwave wavelengths with wavelengths from 700nm to 1mm.

insolation: The amount of solar radiation received by an area over a given period of time.

J

JPEG: (1) file format for digital imagery, (2) method of lossy compression for digital imagery.

K

Kappa technique: Statistic representing map accuracy that can be used to test whether two error matrices (and therefore maps) have a statistically significant difference from one another.

key: An attribute or set of attributes in a database that uniquely identifies each record.

L

Landsat: Multispectral, earth-orbiting satellites developed by NASA and operated by USGS that gather imagery for numerous applications including land use and land cover mapping, forest management, agricultural monitoring, wetlands management, and change detection.

LERC: Limited Error Raster Compression. An image or raster compression method which supports rapid encoding and decoding. Users set the maximum tolerance per pixel while encoding, so the precision of the original input image is preserved within user defined error bounds.

light fall off: The reduction of brightness at the edges of an image compared to its center.

lidar: (light detection and ranging) An active optical remote-sensing technology that uses laser pulses to densely sample the surface of the earth, producing highly accurate x,y,z measurements.

location (image element): The x, y, and z coordinates of an object of interest, used to help identify the object.

lossless compression: Data compression that has the ability to store data without changing pixel values, but is only able to compress the data at a low compression ratios (typically 2:1 or 3:1). In GIS, lossless compression is often used to compress raster data when the pixel values of the raster will be used for analysis or deriving other data products. See also lossy compression, LERC.

lossy compression: Data compression that provides high compression ratios (for example 10:1 to 20:1), but does not retain all the information in the original pixel data. In GIS, lossy compression is used to compress raster datasets that will be used as background images, but is typically not suitable for raster datasets used for analysis or deriving other data products. See also lossless compression, LERC.

M

machine learning: A branch of artificial intelligence that is often applied to imagery and GIS datasets to perform semi-automated mapping. Machine learning is used to map all sorts of features and land use types.

map accuracy: The degree to which a value or label on the map matches values or labels measured on the ground or from other reference data. Accuracy is a measure of correctness. It is distinguished from precision, which measures exactness.

map precision: The closeness of a repeated set of observations of the same quantity to one another.

mapping key: See classification scheme.

Margfit technique: Method of analyzing an error matrix's accuracy that uses an iterative proportional fitting routine to normalize the error matrix so that it can be directly compared to another error matrix regardless of the number of samples used to create the matrix.

mensuration: Applying geometric rules to find the length of a line, area of a surface, or volume of an object using the information obtained from lines and angles.

microwave: The portion of the electromagnetic spectrum between infrared and radio wavelengths, with wavelengths from 1mm to 1 meter.

minimum mapping unit: The size of the smallest object to be identified on the ground; determines the required spatial resolution of imagery for a mapping project.

MIXED: Common file format for cached tiles that contains both JPEG and PNG files.

mosaic dataset: In ArcGIS, a data model within a geodatabase used to manage collections of raster datasets stored as a catalog and viewed as a single mosaicked image or individual images.

multitemporal: Characteristic of a collection of rasters or other features that have multiple time or date stamps.

MXD: File extension for an Esri ArcGIS map document.

N

nadir: In aerial photography or satellite imagery, the point on the ground vertically beneath the perspective center of the camera lens or scanner's detectors.

near-infrared: The portion of the electromagnetic spectrum from about 700 nm to 2500 nm.

NoData: pixels without data, often represented by a value that is not valid elsewhere in the dataset.

nonparametric: Describing statistics that do not make assumptions about variables' probability distributions.

O

occlusion: The phenomenon in which, due to viewing angle, one object blocks another object from view in an image.

off-nadir angle: The angle between nadir and a ray of light from the sensor to the object being observed.

off-nadir view: The view of any object not directly beneath a scanner's detectors or camera lens, but rather off at an angle; results in relief displacement.

on-demand caching: An Esri image service feature that allows you to set up a tiling scheme and publish an image service, but only generate the cache when and where a user accesses the service. The tiles are drawn by the server when an initial user navigates to an un-cached area, then added to the service's cache folder (where they remain until updated or deleted by the server administrator).

ordination: A type of multivariate statistical analysis used in community ecology and remote sensing to order observations along axes for each variable measured.

ortho: See orthophotography.

orthocorrected: See orthorectification.

orthogonal: Uncorrelated or independent.

orthoimage: See orthophotography.

orthometric height: Vertical distance above an approximation of the earth's mean sea level (called a geoid). Water should always run downhill on a terrain model with orthometric heights.

ortho-mosaic: Georeferenced image product mosaicked from an image collection, where the geometric distortion of the individual images has been corrected by orthorectification.

orthophotography: Photographs or images from which displacements due to camera orientation and ground relief have been removed. An orthophoto has the same scale throughout and can be used as a map.

orthorectification: Process by which imagery is geometrically corrected so that coordinates in the imagery accurately represent coordinates on the ground.

overall accuracy: The most common measure of a map's thematic accuracy; sums the samples on the major diagonal of the error matrix which is then divided by the total number of samples.

P

pan sharpening (as "pan-sharpening"): A radiometric transformation in which a higher-resolution panchromatic image is fused with a lower-resolution multiband raster dataset. It is used to increase spatial resolution and better visualize a multiband image.

parallax: The apparent displacement of the position of an object relative to a reference point due to a change in the point of observation; used to create stereo imagery.

parallelepiped: An algorithm used to label unknown pixels in an image based on training statistics that uses minimum and maximum values as a surrogate for variance.

pattern (image element): The spatial arrangement or configuration of objects, used to identify an object of interest.

phenology: The study of the seasonal cycles of plants, animals, and climate.

pixel value: Digital number (DN) representing information stored in one pixel (or cell) of a raster. The area summarized by one pixel value is determined by the spatial resolution of the imagery.

pixel: In remote sensing, the fundamental unit of data collection. A pixel is represented in a remotely sensed image as a cell in an array of data values.

platform: The vehicle that supports and transports sensors that gather remote sensing data.

PNG: portable network graphic. A lossless format for compressing images that is supported by all web browsers.

positional accuracy: The comparison of sample locations of an image or a map to the same locations surveyed on the ground.

principal component analysis: A data transformation method that rotates the axes of the input bands to a new multivariate attribute space in which the axes are uncorrelated/orthogonal. The main reason to transform the data in a principal component analysis is to reduce the number of necessary bands by eliminating redundancy.

processing template: A function chain in which the user uses a raster variable in place of a specific dataset so that the function chain can be applied to other raster datasets. Used to generate on-the-fly information layers.

pyramid: Reduced-resolution datasets stored for an individual image that are used to read and display imagery quickly at lower resolutions.

Q

quad: See digital ortho quarter-quad.

R

radar: radio detection and ranging. A device or system that detects surface features on the earth by bouncing radio waves off them and measuring the energy reflected back.

radiometric resolution: Describes the sensitivity of a sensor to differences in electromagnetic energy It is often represented by the bit depth of the sensor. Typical sensors have 8bit, 11bit 12bit or 16bit depth per band; the higher the bit depth the higher the sensitivity and radiometric resolution of the sensor.

raster function: Defines processing operations that can be applied to one or more rasters on the fly as the data is accessed and viewed, speeding up processing time. See also function chains.

raster: Matrix of cells (or pixels) organized into rows and columns where each cell contains a value representing data or information.

ratio bands: A derivative band generated by dividing one original band by another.

reference data: A baseline dataset used to compare processed or classified data to assess accuracy of results. It is primarily used to assess geo-positional or feature classification accuracy. Reference data is often collected with ground sampling.

reprojection: The mathematical conversion of a map or raster from one projected coordinate system to another, generally used to integrate maps from two or more projected coordinate systems into a GIS.

resolving power: The amount of detail the sensor can capture in each image, determined by the combination of the sensor's lens and the resolution of the imaging surface.

REST API: representational state transfer. An interface that simplifies interactions between GIS applications and web servers. The REST API allows applications to make a collection of standardized requests to a server and get well defined responses.

S

satellite imagery: Data (often images of the earth) collected by sensors on satellite platforms.

scale: The ratio or relationship between a distance or area on a map and the corresponding distance or area on the ground, commonly expressed as a fraction or ratio. A map scale of 1/100,000 or 1:100,000 means that one unit of measure on the map equals 100,000 of the same unit on the earth.

seamline: A polygon or polyline that defines the boundary between adjoining rasters in a mosaic. Overlapping rasters can be blended along the seamline by a specified width so as to reduce the sudden change from one raster to the next.

sensor: An electronic device for detecting electromagnetic energy and converting it into a signal that can be recorded as numbers and displayed as an image.

serpentine: A type of soil originating from ultramafic rocks, often associated with plants that tolerate extreme soil conditions.

shadow (image element): The consequence when the sensor's ability to capture reflectance or radiance of a feature on the ground is hindered by another feature; used to help identify objects of interest.

shape (image element): The form of the outline of an object of interest,; used to help identify objects of interest.

size (image element): The area of an object of interest,; used to help identify objects of interest.

slope: The incline, or steepness, of a surface, measured in degrees from horizontal (0–90), or percent slope (the rise divided by the run, multiplied by 100). The slope of a TIN face is the steepest downhill slope of a plane defined by the face; the slope for a cell in a raster is the steepest slope of a plane defined by the cell and its eight surrounding neighbors.

solar insolation: The amount of solar radiation received by an area over some user-defined period.

spatial resolution: (1) The smallest spatial element on the ground that is discernible on an image, (2) the smallest spatial unit on the ground that a sensor is able to image.

Spectral pattern analysis: A x, y plot of the mean value for each map class for each band in the imagery (original and derivative bands). The bands are on the x axis and the reflectance values (DNs) are on the y axis. SPA is used to determine the bands that show the most separability between the map classes.

spectral resolution: The wavelengths of the electromagnetic spectrum that an imaging system can detect; determined by the location of the bands, the number of bands and the wavelength range detected by each band.

stability: The ability of a sensor to resist changes in position and angle.

T

Tasseled Cap Transformation: Also known as the Kauth-Thomas transformation; uses an empirically derived transformation of image bands into new bands that measure brightness, greenness, and wetness.

temporal resolution: The frequency at which images are captured over the same location on the earth's surface.

texture (image element): The arrangement or repetition of tone/color across an image, often defined as the feel or appearance of the surface of an object of interest, used to help identify the object. Mathematically, it is usually calculated as the standard deviation of a defined window of pixels.

thematic accuracy: Measures whether map feature labels are different from the actual feature label as determined from the ground or other reference data.

tile cache service: Highly compressed, preprocessed imagery delivered from servers to applications and end users, used primarily for providing background imagery and unsuitable for analysis.

tiling scheme: Describes how clients should reference the tiles in a cache. The tiling scheme maps between the spatial reference of the source map document and the tiling grid. It also defines the scale levels at which the cache has tiles, the size of the tiles in pixels, and the screen resolution for which the tiles are intended to be displayed.

tiling: The internal subsetting of a spatial dataset, especially a raster, typically used to process or analyze a large dataset without consuming vast quantities of computer memory.

TIN: triangulated irregular network; a data structure used in a GIS to represent a surface.

tone (image element): Characteristic of an object of interest derived from the intensity of spectral response in each band of an image, used to help identify the object.

topography: The study and mapping of land surfaces, including relief (relative positions and elevations) and the position of natural and constructed features.

transform: A function that takes an image as input and generates an image as output by applying changes to the radiometry and or geometry of the pixels,

transformed band: The result of applying a function to a band.

U

UAV: unmanned aerial vehicles; Platforms used to collect remote sensing data which are piloted by users on the ground rather than in the aircraft; also referred to as drones or unmanned aerial systems (UAS).

V

vegetation alliance: A lower-level unit in the National Vegetation Classification Standard that identifies ranges with similar species composition and abundance.

visualization imagery: Imagery used to help the user understand the context of a location.

visible: The portion of the electromagnetic spectrum that human eyes can sense, typically wavelengths from 390 to 700nm.

W

wavelet transform: Method of lossy compression; used by JPEG 2000, for example.

web map (as "webmap"): Interactive, shareable display of geographic information accessed online.

web service: A software component accessible over the internet for use in other applications. Web services are built using industry standards such as XML and SOAP, and thus are not dependent on any particular operating system or programming language, allowing access to them through a wide range of applications.

X

xeric: Needing very little moisture, as in drought-tolerant plants.

References

Abdullah, Qassim. 2004. "Photogrammetric Platforms." In *Manual of Photogrammetry*, 5th ed. Edited by J. Chris McGlone, 677–730. Bethesda, MD: American Society of Photogrammetry and Remote Sensing.

Ager, Thomas. 2004. "An Analysis of Metric Accuracy Definitions and Methods of Computation." Unpublished memo prepared for the National Geospatial-Intelligence Agency. *InnoVision*. March.

Andelin, E., and S. Andelin. 2015. "Closing in on 500 Authorizations, Commercial UAS Operations Are Starting to Take Off." *LIDAR Magazine* 5 (5): 14–19.

Anderson, J., E. Hardy, J. Roach, and R. Witmer. 1976. "A Land Use and Land Cover Classification System for Use with Remote Sensor Data." *US Geological Survey Professional* Paper 964. Washington, DC: US Government Printing Office.

ASCE (American Society of Civil Engineers). 1994. *Glossary of the Mapping Sciences*. New York: ASCE.

ASPRS (American Society of Photogrammetry Remote Sensing). 1989. "ASPRS Interim Accuracy Standards for Large Scale Maps." *Photogrammetric Engineering & Remote Sensing* 54 (7): 1038–41.

ASPRS (American Society of Photogrammetry Remote Sensing). 2004. "ASPRS Guidelines, Vertical Accuracy Reporting for Lidar Data." *American Society for Photogrammetry and Remote Sensing*. May 24.

ASPRS (American Society of Photogrammetry Remote Sensing). 2014. "ASPRS Positional Accuracy Standards for Digital Geospatial Data." *Photogrammetric Engineering & Remote Sensing* 81 (3): 173–76.

Baldocchi, D., and E. Waller. 2014. "Winter Fog is Decreasing in the Fruit Growing Region of the Central Valley of California." *Geophysical Research Letters* 41 (9): 3251–56. doi: 10.1002/2014GL060018.

Ballanti, L., L. Bleskius, E. Hines, and B. Kruse. "Tree Species Classification Using Hyperspectral Imagery: A Comparison of Two Classifiers." *Remote Sensing* 8 (6): 445. doi: 10.3390/rs8060445.

Bolstad, Paul. 2012. *GIS Fundamentals*, 4th ed. White Bear Lake, MN: Eider Press.

Breiman, L. 2001. "Random Forests." *Machine Learning* 45 (1): 5–32. doi: 10.1023/A:1010933404324.

Campbell, James, and Randolph Wynne. 2011. *Introduction to Remote Sensing*, 5th ed. New York: Guilford Press.

Chuvieco, E., and R. Congalton. 1988. "Using Cluster Analysis to Improve the Selection of Training Statistics in Classifying Remotely Sensed Data." *Photogrammetric Engineering & Remote Sensing* 54 (9): 1275–81.

Civco, D. L., J. D. Hurd, E. H. Wilson, M. Song, Z. Zhang. 2002. "A Comparison of Land Use and Land Cover Change Detection Methods." 2002 ASPRS-ACSM Annual Conference and Technology Exhibition 2002 and XXII FIG International Congress. April 19–26, 2002.

Cliff, Andrew D., and J. K. Ord. 1973. *Spatial Autocorrelation*. London: Pion.

Congalton, Russell G. 1988. "A Comparison of Sampling Schemes Used in Generating Error Matrices for Assessing the Accuracy of Maps Generated from Remotely Sensed Data." *Photogrammetric Engineering & Remote Sensing* 54 (5): 593–600.

Congalton, Russell G. 1991. "A Review of Assessing the Accuracy of Classifications of Remotely Sensed Data." *Remote Sensing of Environment* 37 (1): 35–46. doi: 10.1016/0034-4257(91)90048-B.

Congalton, Russell G. 2010. "Remote Sensing: An Overview." *GIS Science & Remote Sensing* 47 (4): 443–59. doi: 10.2747/1548-1603.47.4.443.

Congalton, Russell G., and Kass Green. 2009. *Assessing the Accuracy of Remotely Sensed Data: Principles and Practices*, 2nd ed. Boca Raton, FL: CRC/Taylor & Francis.

Congalton, Russell G., Jianyu Gu, Kamini Yadav, Prasad Thenkabail, and Mutlu Ozdogan. 2014. "Global Land Cover Mapping: A Review and Uncertainty Analysis." *Remote Sensing* 6 (12): 12070–93. doi:10.3390/rs61212070.

Congalton, Russell G., R. G. Oderwald, and R. A. Mead. 1983. "Assessing Landsat Classification Accuracy using Discrete Multivariate Statistical Techniques." *Photogrammetric Engineering & Remote Sensing* 49 (12): 1671–78.

Cowardin, Lewis M., Virginia Carter, Francis C. Golet, and Edward T. Laroe. 1979. "Classification of Wetlands and Deepwater Habitats of the United States." Washington, DC: US Department of the Interior, US Fish and Wildlife Service.

Crist, E. P. and R. J. Kauth. 1986. "The Tassled-Cap Demystified." *Photogrammetric Engineering & Remote Sensing* 52 (1): 81–86.

Cutler, D. R., T. C. Edwards Jr, K. H. Beard, A. Cutler, K. T. Hess, J. Gibson, and J. J. Lawler. 2007. Random Forests for Classification in Ecology. *Ecology* 88: 2783–92. doi: 10.1890/07-0539.1.

Di Gregorio, Antonio, and Louisa J. M. Jansen. 2000. *Land Cover Classification System (LCCS): Classification Concepts and User Manual*. Rome: Food and Agriculture Org.

Dilts, Thomas E., Jian Yang, and Peter J. Weisberg. 2010. "Mapping Riparian Vegetation with Lidar Data: Predicting Plant Community Distribution Using Height above River and Flood Height." *ArcUser*. Winter 2010. http://www.esri.com/news/arcuser/0110/files/mapping-with-lidar.pdf.

Draeger, W. C., T. M. Holm, D. T. Lauer, and R. J. Thompson. 1997. "The Availability of Landsat Data: Past, Present, and Future." *Photogrammetric Engineering & Remote Sensing* 63 (7): 869–75.

FEMA (Federal Emergency Management Agency). 2003. "Guidelines and Specifications for Flood Hazard Mapping Partners." Washington, DC: FEMA.

FGDC (Federal Geographic Data Committee). 1998. "Geospatial Positioning Accuracy Standards." Washington, DC: FGDC.

FGDC (Federal Geographic Data Committee). 2008. "National Vegetation Classification Standard, Version 2." Washington, DC: FGDC.

Green, Kass. "Use and Value of Sonoma County's Vegetation Mapping and Lidar Program Products". A report prepared for the Sonoma County Agricultural Preservation and Open Space District in partnership with the Sonoma County Water Agency. 22pp.

Green, Kass, Keith Schulz, Chad Lopez, Alison Ainsworth, Meagan Selvig, Kathryn Akamine, Colin Meston, J. Woody Mallinson, Elizabeth Urbanski, Steve Fugate, Mark Hall, and Greg Kudray. 2015. "Vegetation Mapping Inventory Report: Haleakalā National Park." *National Parks Service Natural Resource Report NPS/PACN/NRR—2015/986*.

Greenwalt, Clyde R., and Melvin E. Schultz. (1962) 1968. "Principles of Error Theory and Cartographic Applications. United States Air Force. Aeronautical Chart and Information Center." *ACIC Technical Report Number 96*. St. Louis, MO. This report is cited in the ASPRS standards as ACIC, 1962.

He, Haibo, and Yunqian Ma. 2012. *Imbalanced Learning: Foundations, Algorithms, and Applications*. San Francisco: John Wiley & Sons.

Healey, S.P., W.B. Cohen, Z. Yang, C.K. Brewer, E.B. Brooks, N. Gorelick, A.J. Hernandez, C. Huang, M.J. Hughes, R.E. Kennedy, T.R. Loveland, G.G. Moisen, T.A. Schroeder, S.V. Stehman, J.E. Vogelmann, C.E. Woodcock, L. Yang, Z. Zhu. "Mapping Forest Change Using Stacked Generalization: An Ensemble Approach." In preparation.

Jensen, John R. 2000. *Remote Sensing of the Environment: An Earth Resource Perspective*, 4th ed. Upper Saddle River, NJ: Prentice Hall.

Jensen, John R. 2016. *Introductory Digital Image Processing: A Remote Sensing Perspective*. 4th Ed. Glenview, IL: Pearson.

Kakaes, Konstantin, Faine Greenwood, Mathew Lippincott, Patrick Meier, and Serge Wich. 2015. "Drones and Aerial Observation: New Technologies for Property Rights, Human Rights, and Global Development. A Primer." New America Weekly. July 22. http://www.newamerica.org/international-security/events/drones-and-aerial-observation/.

Kauth, R. J., and G. S. Thomas. 1976. "The Tassled-Cap—A Graphic Description of the Spectral-Temporal Development of Agricultural Crops as Seen by Landsat." *Proceedings of the Symposium on Machine Processing of Remotely Sensed Data*. West Lafayette, IN: Purdue University.

Lawrence, Rick L., and Christopher Jacob Moran. 2015. "The AmericaView Classification Methods Accuracy Comparison Project: a Rigorous Approach to Model Selection." *Remote Sensing of Environment* 170: 115–120. doi: 10.1016/j.rse.2015.09.008.

Lillesand, Thomas, Ralph W. Kiefer, and Jonathan Chipman. 2015. *Remote Sensing and Image Interpretation*, 7th ed. Hoboken, NJ: John Wiley & Sons.

Lunetta, Ross S., Russell G. Congalton, Lynn Fenstermaker, John R. Jensen, Kenneth C. McGwire, and Larry Tinney. 1991. "Remote Sensing and Geographic Information System Data Integration: Error Sources and Research Issues." *Photogrammetric Engineering & Remote Sensing* 57 (6): 677–87.

McGlone, J. Chris. 2013. *Manual of Photogrammetry*, 6th ed. Baltimore, MD: American Society for Photogrammetry and Remote Sensing.

Mountrakis, G., J. Im, and C. Ogole. 2011. Support vector machines in remote sensing: A review. *ISPRS Journal of Photogrammetry and Remote Sensing*, 66(3), 247-259.

National Gap Analysis Program. 2015. http://gapanalysis.usgs.gov/gaplandcover/data/.

National Map (2017). http://viewer.nationalmap.gov/basic.

NDEP (National Digital Elevation Program). 2004. "Guidelines for Digital Elevation Data." Version 1.0. Washington, DC: NDEP.

Paine, David P., and James D. Kiser. 2012. *Aerial Photography and Image Interpretation*. Hoboken, NJ: John Wiley and Sons.

Pajares, Gonzalo. 2015. "Overview and Current Status of Remote Sensing Applications Based on Unmanned Aerial Vehicles (UAVs)." *Photogrammetric Engineering & Remote Sensing* 81 (4): 281–329.

Rabben, E. L. (ed). 1960. "Fundamentals of Photo Interpretation." In *Manual of Photographic Interpretation*. 99–168. Bethesda, MD: American Society for Photogrammetry and Remote Sensing.

Renslow, M. (ed). 2012. *Manual of Airborne Topographic LIDAR*. Bethesda, MD: American Society for Photogrammetry and Remote Sensing.

Spurr, Stephen. 1948. *Aerial Photographs in Forestry*. New York. The Ronald Press Company.

Spurr, S. H. 1948. *Aerial Photographs in Forestry*. The Ronald Press Company. New York. 340pg.

Spurr, Stephen. 1960. *Photogrammetry and Photo-Interpretation*. New York: The Ronald Press Company.

Story, M., and Congalton, Russell G. 1986. "Accuracy Assessment: A User's Perspective." *Photogrammetric Engineering & Remote Sensing* 52 (3): 397–99.

Strobl, Carolin, Anne-Laure Boulesteix, Achim Zeileis, and Torsten Hothorn. 2007. "Bias in Random Forest Variable Importance Measures: Illustrations, Sources and a Solution." *BMC Bioinformatics* 8 (25). doi: 10.1186/1471-2105-8-25.

Tobler, Waldo. (1987). "Measuring Spatial Resolution." *Proceedings, Land Resources Information Systems Conference, Beijing*.

Tucker, Compton J., John R. G. Townshend, and Thomas E. Goff. 1985. "African Land-Cover Classification Using Satellite Data." Science 227 (2685): 369–75. doi: 10.1126/science.227.4685.369.

US Bureau of the Budget. 1947. "National Map Accuracy Standards." Washington, DC: US Bureau of the Budget.

Waske, Böjrn, Jon Atli Benediktsson, Kolbeinn Árnason, and Johannes R Sveinsson. "Mapping of Hyperspectral AVIRIS Data Using Machine Learning Algorithms." *Canadian Journal of Remote Sensing* 35: S106–S116. doi: 10.5589/m09-018.

Wu, Zhuoting, Prasad S. Thenkabail, and James P. Verdin. 2014. "Automated Cropland Classification Algorithm (ACCA) for California Using Multi-Sensor Remote Sensing." *Photogrammetric Engineering & Remote Sensing* 80 (1): 81–90.

Zhang, Caiyun, Donna Selch, and Hannah Cooper. "A Framework to Combine Three Remotely Sensed Data Sources for Vegetation Mapping in the Central Florida Everglades." *Wetlands* 36: 201–13. doi: 10.1007/s13157-015-0730-7.

Zhu, Zhe, and Curtis E. Woodcock. 2014. "Continuous Change Detection and Classification of Land Cover Using All Available Landsat Data." *Remote Sensing of Environment* 144: 152–71.

Zhu, Zhe, Yingchun Fu, Curtis E. Woodcock, Pontus Olofsson, James E. Vogelmann, Christopher Holden, Min Wang, Shu Dai, and Yang Yu. 2016. "Including Land Cover Change Analysis of Greenness Trends Using All Available Landsat 5, 7, and 8 Images: A Case Study from Guangshou, China (2000-2014)." *Remote Sensing of Environment* 185: 243–57.

Image Credits

Cover

Courtesy of NASA's Earth Observatory. On Sunday, October 26 [2003], the Advanced Spaceborne Thermal Emission and Reflection Radiometer (ASTER) captured this image of the Old Fire/Grand Prix fire burning on either side of Interstate 15 near the Cajon Pass in the San Bernardino Mountains, roughly 80 km (50 mi) east of Los Angeles, CA. When this image was acquired, the fire had burned more than 80,000 acres, consumed 450 structures, and caused 2 fatalities. Most of the local communities were evacuated as the fire continued to spread rapidly, fanned by the intense Santa Ana winds. The image combines ASTER bands 4, 3, and 1 to produce a thermal infrared look at the scene. The bright red-orange ribbon snaking along the northern side of the burn is the actively burning fire front, while the darker crimson patch shows the smoldering burn scar. ASTER is one of five instruments aboard NASA's Terra satellite.

Figure 1.1	Forest Areas in Europe © EEA, Copenhagen, 2012; EEA, Copenhagen 2014; Esri; DigitalGlobe; GeoEye; Earthstar Geographics; CNES/Airbus DS; USDA; USGS; AEX; Getmapping; Aerogrid; IGN; IGP; swisstopo; HERE; DELorme; MapmyIndia; © OpenStreetMap contributors, and the GIS user community. The Uprooted: Esri, UNHCR, Airbus Defense and Space, PesentiStory. El Nino Impacts on Florida: US Department of Commerce/NOAA, National Weather Service Tampa Bay/Ruskin, FL. Courtesy of Kentucky Geological Survey at the University of Kentucky. Under Construction: Development Projects Actively Under Construction in the City of Pflugerville, TX; Esri. Urban Observatory: © 2014 Esri, Richard Saul Wurman & RadicalMedia. Hurrican Sandy: The AfterMap: Esri, Microsoft, NOAA, HDDS, USGS, HERE, DeLorme, MapmyIndia; © OpenStreetMap contributors, GeoEye, Earthstar Geographics, CNES/Airbus DS, USDA, AEX, Getmapping, Aerogrid, IGN, IGP, swisstopo, and the GIS user community.
Figure 2.2	Teledyne Optech, Canada.
Figure 2.9	NASA.
Figure 2.11	Sonoma County Agricultural Preservation and Open Space District.
Figure 2.13	Sonoma County Agricultural Preservation and Open Space District.
Figure 3.9	NASA.
Figure 3.10	Dr. Maggi Kelly.
Figure 3.11	Image and analysis provided by Quantum Geospatial, Inc.
Figure 3.21	Image © DigitalGlobe, Inc. 2017.

Figure 3.22	Hexegon.
Figure 4.2	Euroconsult, Satellite-Based Earth Observation: Market Prospects to 2025, 2016 Edition.
Figure 4.4	USGS.
Figure 4.A	University of Vermont Spatial Analysis Lab.
Figure 5.20	Sonoma County Agricultural Preservation and Open Space District.
Figure 6.1	Reproduced with permission from the American Society for Photogrammetry & Remote Sensing, Bethesda, Maryland. www.asprs.org.
Figure 6.8	Image © DigitalGlobe, Inc. 2017.
Figure 7.4	Sonoma County Agricultural Preservation and Open Space District.
Figure 7.5	United States National Vegetation Classification.
Figure 9.5	Reproduced with permission from the American Society for Photogrammetry & Remote Sensing, Bethesda, Maryland. www.asprs.org.
Figure 9-19	Sonoma County Agricultural Preservation and Open Space District.
Figure 10.2	Sonoma County Agricultural Preservation and Open Space District.
Figure 10.6	Sonoma County Agricultural Preservation and Open Space District.
Figure 10B	Cassandra Pallai, Geopsatial Program Manager, Chesapeake Conservancy.
Figure 11.2	NOAA Office for Coastal Management, National Oceanic and Atmospheric Administration.
Figure 11.3	Image © DigitalGlobe, Inc. 2017.
Figure 12.3	Reproduced with permission from the American Society for Photogrammetry & Remote Sensing, Bethesda, Maryland. www.asprs.org.
Figure 13.6	Sonoma County Agricultural Preservation and Open Space District.

All other images Esri and/or the authors.

Index

Note: Page numbers in italic refer to photographs; page numbers followed by *f* or *f* refer to figures and tables, respectively.

Absolute atmospheric correction, 154
Absorption of electromagnetic energy, 147
Accessing imagery, 65–66, 73–75. *See also* Web services
    price of (See Pricing)
    as web services, 132, 133*f*
Accuracy, precision vs., 338, 339*f*
Accuracy assessment, 337–360
    map accuracy assessment, 338–340, 339*f*
        positional map accuracy assessment, 340–347, 341*f*, 345*f*, 347*f*
        thematic map accuracy assessment, 348–359, 349*f*, 355*f*, 357*f*–358*f*, 359*f*
Across-track scanners, 35
Active sensors
    passive sensors vs., 33–35, 34*f*
wavelengths sensed by, 38–42, 39*f*–41*f*
Active systems, 62
Additive primary colors, 231
Advanced Land Imager, 100
Advanced Very High Resolution Radiometer (AVHRR)
    date of maps, 240
    moderate- and low-spatial-resolution imagery, 95
    thermal imagery, 99
Aerial photography, origins of, 337
Affine transform, 169–170, 170*f*
Airborne Visible/Infrared Imaging Spectrometer (AVIRIS), 70, 100
Airbus, 92–93
Aircraft
    adding imagery collections to mosaic dataset, 373
    sensing platforms, 44–46
Along-track (push-broom) scanners, 35, 38*f*
Altitude
    and geometric correction, 158–159
    of platforms, 47
Amazon Machine Images, 382
Amazon Public Datasets, 382
Amazon S3, 381, 382
Amazon Web Services, 66
American Society of Photogrammetry Remote Sensing (ASPRS), 78, 85, 342–344
Ames Research Center, 20*f*
Anderson classification scheme, 186
Android, app development on, 141
Antennas, 30
APFO (Aerial Photography Field Office), 79–80
Apple, 71
Apps, using web services in, 140–141, 142*f*
ArcGIS
    Aspect tool, 256
    block adjustment, 371
    caching imagery in, 377
    cloud-hosted imagery, 381
    Collector for ArcGIS app, 291*f*
    DEM-derived products, 225–226
    DEMs from lidar point cloud, 224

flow accumulation raster creation, 259
geoprocessing tasks, 382
Hexagon HVH-resolution imagery, 92
and hybrid classification, 309
image coordinate system, 162, 163
mapping woody debris in Great Brook (Vermont), 82
mosaic datasets, 128
NDVI and, 248
NoData areas, 116
ratio bands, 243
Reclass tool, 256
Solar Radiation toolbox, 258
Spatial Analyst (See Spatial Analyst)
stretch/gamma optimization, 123
SVM and, 304
swipe tool, 315, 315f
and TCT, 246
unsupervised object classifier, 280
web services for imagery, 133, 133f
ArcGIS Desktop
    for creating mosaic datasets, 368
    downloading raster data from image services, 137–138, 138f
    image service access through, 135, 135f, 136, 137f
    machine learning algorithms, 299
    mosaic dataset query filters, 130
    and NDVI, 248
    raster products, 114
    tile cache generation, 376
ArcGIS Enterprise, 376
ArcGIS Full Motion Video (FMV), 374
ArcGIS Online, 7, 77, 98
    data for DEMs, 227
    free imagery on, 71
    HVH-resolution imagery, 85
    Living Atlas, 134
    moderate- and low-spatial-resolution imagery, 98
    and NDVI, 248
    publishing tools, 391
    web maps, 139, 139f
ArcGIS Professional, 299, 368
ArcGIS Runtime, 141
ArcMap, 299
Artificial neural networks (ANNs), 305
ASPRS. See American Society of Photogrammetry Remote Sensing
"ASPRS Guidelines for Reporting Vertical Accuracy of Lidar Data," 342
"ASPRS Positional Accuracy Standards for Digital Geospatial Data," 342–343
ASTER, 96, 97
ATCOR (Atmospheric CORrection), 156, 176
Atmospheric correction, 153–156, 154f, 155f
Atmospheric CORrection Now (ACORN), 156
Atmospheric satellite platforms, 44
Attribute table, 363–364
Automated feature extraction, 24
AVHRR. See Advanced Very High Resolution Radiometer

Banda Aceh, Indonesia, 320f
Band interleaved by line (BIL), 110, 112
Band interleaved by pixel (BIP), 110, 112
Bands, raster, 18–20, 19f, 20f
Band sequential (BSQ), 110, 112
Bare-earth DEMs, 220, 221. *See also* Digital terrain models (DTMs)
Base image, 21, 22f

Basemap Service, 92
Bathymetric lidar, 40
Bayer filter, 36, 37f
Bidirectional reflectance distribution function (BRDF), 150–151
Big Data, 7–8
Bilinear interpolation, 165, 166f
Bing, 7, 71
Bispectral plots (BSPs), 251–252, 252f, 296, 296f
Block adjustment
    bundle block adjustment, 49, 174
    mosaic dataset, 371
Bodies, 43
Body, of remote sensor, 30, 30f, 43
Breiman, Leo, 301, 302
BSPs. See Bispectral plots
Build Footprint tool, 363
Build Overviews tool, 365
Build Seamlines tool, 372
Bundle block adjustment, 49, 174
Bureau of Land Management Aerial Photo Archive, 80, 94

Calibration, Landsat, 73–75
California Department of Fish and Wildlife, 187, 189
Cameras, 30, 30f, 159
Canada, radar platforms, 102
Canopy height models (CHMs), 220. *See also* Digital height models (DHMs)
CART analysis, 300–301, 301f
Cartesian coordinate system, 161
Categorical information, 23t
C-CAP (NOAA Coastal Change Analysis Program), 245, 316, 316f, 329
CCDs. See Charged coupled devices
CE90, 74
Cells, 15
Cell size, 20f, 21
Change analysis, 313–334
    characterization of change, 314–316, 315f, 316f
    classification scheme differences in, 332
    cross-correlation analysis, 329–330, 330t
    image variance in, 333
    managing nonchange differences, 330–333
    map error differences in, 333
    multitemporal image analysis, 318–328, 320f–326f, 327t
    multitemporal image-to-map comparisons, 328–330, 330t
    multitemporal map-to-map comparisons, 317, 318t
    nonchange differences in, 331f
    registration differences in, 332
    significant information produced by, 391
Change detection, 73
Change images, 320–322, 321f, 331f
Charged coupled devices (CCDs), 31–33, 36, 42
Chesapeake Bay, watershed mapping, 292–293, 294f
CHMs (canopy height models), 220. *See also* Digital height models (DHMs)
Classification and Regression Tree (CART). See CART analysis
Classification schemes, 179–190
    combined classification algorithms, 390
    creating new, 187–189, 188f
    defined, 180
    hierarchical, 184, 184f, 185, 185f
    importance of, 180, 386
    and labels, 181, 182f, 183
    managing differences in, 332
    and minimum mapping units, 180

  mutually exhaustive, 183–184
  and rules, 181, 182*f*, 183
  Sonoma County vegetation map classification key, 189, 191–218
  and thematic map accuracy assessment, 349–350
  totally exhaustive, 183
  using existing, 186, 187*t*
Client-defined processing, 380
Climate and weather study, 6
Clip and email viewer, 383, 383*f*
Cloud (Internet)
  geoprocessing services, 382–383
  storage, 7, 381–382
Clouds (meteorology), 156–157, 157*f*, 158*f*
Cloud shadows, 156–157, 157*f*, 158*f*
Clustering algorithms, 282
Clusters (unsupervised classification), 280, 281*f*, 282, 283*f*, 284–285, 284*f*
Coastal Change Analysis Program. *See* C-CAP
Coast live oaks (*Quercus agrifolia*), 235*f*
Cold War, 4
Collection characteristics, 50–63
  extent, 62–63
  radiometric resolution, 54–55
  spatial resolution, 55–57, 56*f*–57*f*
  spectral resolution, 51–54, 52*f*–54*f*, 53*t*
  temporal resolution, 62
  viewing angle, 57–59, 60*f*–61*f*
Collector for ArcGIS app, 291*f*
Color (image element), 231
Color correction, 371
Color infrared composites, 231
Committee on Earth Observation Satellites Earth Observation (EO) Handbook and Database, 77
Complementary metal-oxide-semiconductor (CMOS) arrays, 31–33
Completeness of map, 339–340
Comprehensive Large Array-Data Stewardship System, 99
Compression, 115–116. *See also* specific compression formats
Compression, image. See Image compression
Confluence of evidence, 242
Context (image element), 238
Continuous change analysis, 327–328
Continuous Change Detection and Classification, 328
Continuous data, 13, 17, 17*f*, 23*t*, 295
Continuous information, 23*t*
Continuous raster data, 119–127
Contrast stretching, 120–122, 121*f*–122*f*
Control points, 168–169
Coordinate systems, 160–162
Copernicus program, 98
Corine Land Cover 2006 inventory, 5
C (cost) parameter, 304
Crops, 5, 241
Cross-correlation analysis, 329–330, 330*t*
Cubic convolution interpolation, 166–167, 167*f*
Cutler, Adele, 301

DA (divergence analysis), 253
Dark object subtraction (DOS), 154
Data
  continuous, 23*t*
  information vs., 23
Data exploration, 248–265, 389
  bispectral plots, 251–252, 252*f*
  divergence analysis, 253

elevation and landscape data, 254–258, 255f–258f
feature space analysis, 252, 253f
forest canopy metrics, 263
hydrology data, 258–261, 260f, 261f
precipitation and temperature data, 261–262
soil data, 262–263
spectral pattern analysis, 249–251, 250f, 251f
wildfire history, 263, 264f
Datasets. *See* Mosaic datasets
Date of image, 240, 240f, 241, 241f
Datum (reference surface), 162
Definition queries, 136, 137f
DEMs. *See* Digital elevation models
Dendrogram, 307, 308f
Derivative bands, 243–248
    indices, 246–248, 247f
    principle components analysis, 244, 245f
    ratios as, 243
    tasseled-cap transformation, 245–246, 246f
    transformations as, 243–244
DHMs. *See* Digital height models
Difference image. See Image differencing
Digital arrays, 7, 33
Digital cameras, 33
Digital Coast, 80, 227
Digital elevation models (DEMs), 24, 219–228, 254, 255, 256f
    creation of, 222–226
    defined, 219–220
    derivative products, 225–226
    flow accumulation, 259
    height models (*See* Digital height models)
    from lidar, 224
    mosaic dataset raster functions, 131
    from photogrammetry, 223–224
    quality/accuracy of, 227–228, 228f
    sources of data for, 226–227
    surface models (*See* Digital surface models)
    terrain models (*See* Digital terrain models)
    types of, 220, 220f
Digital Globe, 92–93
Digital height models (DHMs), 40, 221f, 222, 255, 256f
Digital number (DN), 51, 52f, 53f
    histograms and, 199, 200
    and PCA, 244
    radiance values and, 148
Digital surface models (DSMs), 40, 221f, 222, 255
Digital terrain models (DTMs), 40, 174–175, 220f, 221, 221f, 255
Dilts, Tom, 260
Discrete data, 17, 18f
Discrete return lidar, 41
Divergence analysis (DA), 253
DN. See Digital number
DOS (dark object subtraction), 154
Drone2Map, 80–81
Drones. See Unmanned aerial systems
DSMs. See Digital surface models
DTMs. See Digital terrain models
Dynamic image services
    about, 133, 375
    exporting and downloading imagery from, 137–138, 138f
Dynamic mosaicking, 138, 366
Dynamic range, 55

Earth Observing 1 (EO-1) Extended Mission, 100
Earth Resources and Science Center (EROS), 79, 85, 93–98
Edge (high-pass) filters, 126–127, 127f
Editing, 309–311, 310f
Einstein, Albert, 31
Electromagnetic energy
    bidirectional reflectance distribution function, 150–151
    and interactions involving, 147, 147f, 148
    reflectance vs. radiance, 148–150, 149f
    wavelength, 146
Electromagnetic spectrum, 31, 31f, 32, 32f
Electro-optical imaging, 30
Elevation contours, 225
Elevation data, 254–255, 256f, 373–374
Elevation profiles, 225
El Niño, 6
Engineering and construction, 6
Enhanced vegetation index (EVI), 247
EO-1 (Earth Observing 1) Extended Mission, 100
EOSAT, 67
EROS. *See* Earth Resources and Science Center
Error matrix, 348–349, 349f, 353–357, 355f, 357f, 358f, 359, 359t
Errors, thematic map accuracy assessment and, 350
ESA. *See* European Space Agency
Esri. *See also* ArcGIS
    imagery sources, 77
    and mosaic dataset, 129
    web services (*See* Web services)
Esri World Imagery, 71
European Space Agency (ESA)
    Copernicus program, 98
    EO Portal, 77
    free access to imaging from, 64
    moderate- and low-spatial-resolution imagery, 98
    radar platforms, 103
European Union, 7
EVI (enhanced vegetation index), 247
Extent, 62–63
Eye, 30, 30f

FAA (Federal Aviation Agency), 46, 81
FAO (Food and Agricultural Organization), 186
Fast Line-of-sight Atmospheric Analysis of Spectral Hypercubes (FLAASH), 156
Feature maps, 24
Features, imagery as attribute of, 21, 22f
Feature space analysis, 252, 253f
Federal Aviation Agency (FAA), 46, 81
Federal Emergency Management Agency (FEMA), 342
Federal Geographic Data Committee, 342
Film, 7, 33
Filtering, 123–127, 124f, 125f, 127f
Fire Resource Assessment Program (FRAP), 263
First-return elevation models, 222
FLAASH (Fast Line-of-sight Atmospheric Analysis of Spectral Hypercubes), 156
Flow accumulation, 259
FMV (Full Motion Video), 374
Fog, 262
Food and Agricultural Organization (FAO), 186
Footprint, mosaic dataset, 362–363, 363f
Forest canopy metrics, 226, 263
Forest fires, thermal imagery for detecting, 99

Forest maps, 183, 239f, 240f
Forestry, 5
Four-band airborne image sensors, 36
Frame camera model (orthorectification), 173–174, 173f
Framing cameras, 34–35
France. See SPOT (Satellite Pour l'Observation de la Terre)
FRAP (Fire Resource Assessment Program), 263
Full Motion Video (FMV), 374
Full waveform lidar, 41
Function chains, 369, 369f, 370

Gamma (of image), 123
Gamma (SVM parameter), 304
Gap Analysis Program, 300
GCPs (ground control points), 168–169
GDAL (Geospatial Data Abstraction Library), 113
GEDI (Global Ecosystem Dynamics Investigation) mission, 39
General aviation aircraft sensing platform, 45
Geometric correction, 158–175, 160f
    coordinate systems and map projections, 160–162
    georeferencing, 168–171, 169f–171f
    image registration, 162–163
    orthorectification, 172–175, 172f, 173f
    resampling, 163–167, 164f–167f
Geoprocessing services, 134, 382–383
Georeferencing, 113, 168–171, 169f–171f
Geospatial Data Abstraction Library (GDAL), 113
Geostationary systems, 62
Geosynchronous platforms, 44
GeoTiff, 379
geoTIFF files, 113
GIS layers, 14, 14f, 15
Global Ecosystem Dynamics Investigation (GEDI) mission, 39
Global Imagery Browse Services, 80
Global Land Surveys datasets, 97
Global position systems (GPSs)
    lidar and, 224
    positional accuracy, 352–353
    and semiautomated image classification, 277
Google, 66, 71
Google Earth, 7
GPS, 49
GPSs. See Global position systems
Great Brook (Vermont), 81–83, 83f
Ground-based sensing platforms, 45
Ground control points (GCPs), 168–169
Ground sample difference (GSD), 55–56
Ground truth, reference data vs., 349
GSD (ground sample difference), 55–56
"Guidelines for Digital Elevation Data" (USGS NDEP document), 342
Gunter's Space Page, 77

Height (image element), 239, 239f
Height above river (HAR), 260, 261f
Helios satellite, 44
Hexagon Geospatial, 92
Hexagon HVH-resolution imagery, 92
Hierarchical classification schemes, 184, 184f, 185, 185f, 189, 191–218
High- and very-high (HVH) spatial-resolution imagery, 84–94, 86t–92t
High-pass filters, 126–127, 127f
Hillshade rasters, 225, 228, 228f
Hillshades, 257, 258f

Histograms, 119–123, 120*f*–122*f*
Historical maps, 317, 318*t*
Humanitarian aid, 5
Humanitarian Open Street Map, 328
Hurricane Sandy, 6
HVH (high- and very-high) spatial-resolution imagery, 84–94, 86*t*–92*t*
Hybrid classification, 307–309, 308*f*
Hydro-enforced DTM, 221
Hydro-flattened DTM, 221
Hydrologic derivatives, 226
Hydrology data, 259–261, 260*f*, 261*f*
Hyperion sensor, 100
Hyperspectral data, 20*f*
Hyperspectral imagery, 100, 388
Hyperspectral sensors, 51

ICS (image coordinate system), 162
IFOV (instantaneous field of view), 57, 59, 60*f*
IFSAR (interferometric synthetic aperture radar), 223, 224
IKONOS, 159, 246
Image(s)
    georeferencing of, 113
    properties of, 113–114
    scale of, 107–109, 108*f*
    statistics and rendering of, 113–114, 114*f*
    storage formats for, 110, 111t, 112
Image classification, 23, 23t, 267–311
    manual image interpretation, 268–275, 269*f*, 271*f*–273*f*
    semiautomated image classification, 275–311
    supervised image classification, 307*f*
    unsupervised image classification, 307*f*
Image collection, 28–63
    and collection characteristics, 50–63
    via platforms, 43–50
    via sensors, 29–43
Image compression, 115–116. *See also* specific compression formats
Image coordinate system (ICS), 162
Image cube, 328
Image differencing, 322–323, 322*f*–325*f*, 325
Image display
    for continuous raster data, 119–127
    and enhancement/filtering, 123–127, 124*f*, 125*f*, 127*f*
    histograms, 119–123, 120*f*–122*f*
    for mosaic data sets, 128–132, 128*f*, 131*f*
    statistics and, 113–114, 114*f*
Image elements, 230–241
    and confluence of evidence, 242
    context, 238
    date, 240, 240*f*, 241, 241*f*
    height, 239, 239f
    location, 238
    pattern, 233, 234*f*
    shadow, 234, 235, 235*f*
    shape, 232, 233*f*
    size, 233
    summary of, 230t
    texture, 235–236, 236*f*–237*f*
    tone and color, 231, 232*f*
Image interpretation, 268
Image mosaics, 128, 128*f*. *See also* Mosaic datasets
Image pyramids, 117–119, 118*f*, 364, 365, 365*f*

Image registration, 162–163, 319
Image registration differences, managing, 332
Imagery, 3. *See also specific headings*, e.g.: selection of imagery
    as attribute of a feature, 21, 22*f*
    as base image, 21, 22*f*
    as data source, 21
    data structure for, 13, 13*f*, 14*f*
    defined, 11, 12
    future expansion of sources for, 386–387
    nonimagery geospatial data and, 389
    uses of, in GIS, 21, 22f, 23, 24
    using classification scheme to drive choice of, 386
    variables affecting, 27
    workflows, 25–26
Imagery processing, 145–176
    and bidirectional reflectance distribution function, 150–151
    and clouds/cloud shadows, 156–157, 157*f*, 158*f*
    and geometric correction, 158–175, 160*f*, 164*f*–167*f*, 169*f*–173*f*
    and interactions involving electromagnetic energy, 147, 147*f*, 148
    and mosaicking, 175
    and radiometric correction, 151–156, 152*f*, 154*f*, 155*f*
    and reflectance vs. radiance, 148–150, 149*f*
    and wavelength, 146
Image segmentation, 278–279, 278*f*
Image services
    client-defined processing, 380
    cloud considerations, 381–382
    dynamic, 133
    open standards, 381
    optimizing mosaic datasets for performance, 378–379
    publishing mosaic datasets as, 374–382
    tile cache services, 375–378
    using, in ArcGIS desktop, 135, 135*f*, 136, 137*f*
Image stretch, 120–122, 121*f*, 122*f*
Image-to-map comparisons, multitemporal, 328–330, 330t
Image-to-map registration, 162–163. *See also* Geometric correction
Image variance, managing, 333
Imaging surfaces, 30
Indices
    for change images, 322
    as derivative bands, 246–248, 247*f*
Inertial measurement units (IMUs), 49
Information
    continuous vs. discrete, 17
    data vs., 23
    turning data into map, 23
    types of, 23t
Informed decisions, making, 4
Instantaneous field of view (IFOV), 57, 59, 60*f*
Interferometric synthetic aperture radar (IFSAR), 223, 224
International Society for Photogrammetry and Remote Sensing, 78
International Space Station, 46
Interpretation key, 272, 272*f*, 273*f*

iOS, app development for, 141

Iterative self-organizing data analysis technique (ISODATA) clustering algorithm, 282

Japan, radar platforms, 103
JavaScript API, 141
Jet aircraft sensing platform, 44
JPEG file format, 113, 115, 116, 377

Kalaupapa National Historical Park (Hawaii), 158*f*
Kappa analysis technique, 356, 359, 359*f*
Kauth–Thomas transform, 245
Kentucky Geological Survey, 6
Kernel (filtering), 124*f*
Kernel (SVM), 304
K-means clustering algorithm, 282
Kodak, 84

Lambert conformal projection, 161
Land cover, slope/aspect correlation with, 256, 257*f*
Land Cover Atlas, 316, 316*f*
Land Remote Sensing Commercialization Act (1984), 66–67
Land Remote Sensing Policy Act (1992), 65, 67
Landsat (satellite), first launch, 337
Landsat 7
    Scan Line Corrector, 150–151, 151*f*
    spectral resolution, 95*f*
    WELD data, 98
Landsat 8
    image service, 136, 137*f*
    and NDVI, 248
    spectral resolution, 95*f*
    thematic map created from, 18*f*
Landsat 8 Operational Land Imagery sensor, 156
Landsat archive service, 381–382
Landsat Explorer, 322, 323, 323*f*
Landsat Global Land Survey, 98
Landsat imagery
    access to, 66
    calibration quality, 73
    and continuous change analysis, 327
    effects of price and licensing on use of, 66–67
    and Esri World Imagery, 71
    extent, 63
    first civilian digital satellite scanners, 84
    importance of, 387
    moderate- and low-spatial-resolution, 94–95
    multispectral, 19*f*
    multitemporal image analysis, 240
    positional accuracy, 352
    in public domain, 7
    public domain imagery, 75
    and Sonoma Vegetation Mapping project, 9, 70
    spatial resolution, 56, 56*f*
    spectral data accuracy, 74–75
    thermal imaging sensors, 99
    viewing angle, 58–59
Landsat Multispectral Scanner (MSS), 245
Landsat TM/ETM, 246
Landsat TM imagery
    and BSPs, 251–252, 252*f*
    and PCA, 244
    and SPA, 249, 250*f*
Landsat TM sensor, 99
Large-scale imagery, 108
LASer (LAS) data format, 101
Las Vegas, Nevada, vegetation change in, 320–321, 321*f*, 323, 324*f*–326*f*, 325–326
Latitude, 161
Leica Geosystems, 92
Lenses, 30, 30*f*, 42

LERC (Limited Error Raster Compression), 116, 377
LibSVM, 304
Licensing, 64–67, 388
Lidar imagery, 39–42, 40*f*, 41*f*, 100–102
    data formats, 101
    and DEMs, 223, 224, 255, 256*f*
    and DSMs, 222
    and elevation data, 254–255, 255*f*
    and forest canopy metrics, 226
    for forest canopy metrics, 263
    future of data collection by, 387
    and hillshades, 257, 258*f*
    point cloud (See Point cloud)
    point density, 40–41
    stream centerline imaging, 259
Limited Error Raster Compression (LERC), 116, 377
Lines, 15, 16*f*
Living Atlas, 134
Location (image element), 238
Logical consistency of map, 339
Longitude, 161
Lossless compression, 115
Lossy compression, 115–116
Louisiana, 2016 floods, 315, 315*f*
Low-pass filters, 126, 127*f*

Machine learning techniques, 298–306
    artificial neural networks, 305
    best practices, 305–306
    CART analysis, 300–301, 301*f*
    Random Forests, 301–304, 303*f*
    support vector machines, 304
Management Association of Private Photogrammetric Surveyors, 78
Managing image collections. See Mosaic dataset
Manual image interpretation, 268–275, 269*f*, 271*f*–273*f*, 319–322, 320*f*, 321*f*
Map(s)
    accuracy assessment (See Accuracy assessment)
    confluence of evidence and creation of, 242
    feature, 24
    image classification for, 23
    thematic, 24
    types of, created from imagery, 24
    validation and editing, 309–311, 310f
Map classes, 285
Map error differences, managing, 333
Map projections, 160–162
Margfit analysis technique, 356, 357*f*, 359
Marin County, California
    maximum likelihood algorithm classification, 297, 298*f*
    supervised image classification, 288*f*
    supervised vs. unsupervised classification, 306, 307*f*
    unsupervised classification, 281*f*
Masking (postclassification) change detection, 329, 330*t*
Maximum likelihood algorithm, 296, 296*f*, 297, 298*f*
Mercator projection, 161
Military applications, 4, 7
Minimum distance algorithm, 296, 296*f*, 297
Minimum mapping units (MMUs)
    and classification schemes, 180, 182*f*
    and image segmentation, 279
    and manual interpretation, 271

      and spatial resolution requirements, 71
      and thematic map accuracy assessment, 350
Minimum-maximum stretch, 121–122
Mining, 5
Misregistration, 332
Mississippi River Delta, 232f
Mixed caches, 377
MMUs. See Minimum mapping units
ModelBuilder, 246, 248
Moderate- and low-spatial-resolution imagery, 94–98, 95f, 96t, 97t
Moderate Resolution Imaging Spectroradiometer (MODIS)
      land surface reflection data, 262
      moderate- and low-spatial-resolution imagery, 98
      multitemporal image analysis, 240
      spectral resolution, 95f
      thermal imaging, 99
MODTRAN (MODerate resolution atmospheric TRANsmission), 156
Mosaic datasets, 129–132, 361–384
      about, 129
      adding rasters to, 368–369
      advanced tools, 370–372
      attribute table, 363–364
      block adjustment, 371
      catalog properties, 368
      client-defined processing, 380
      cloud considerations for image services, 381–382
      color correction, 371
      components, 129–130
      creating, 368
      download properties, 368
      dynamic mosaicking, 130, 366
      elevation data, 373–374
      footprints, 362–363, 363f
      from full-motion video, 374
      image mosaics and, 128, 128f
      image properties, 367
      imagery from aircraft and unmanned aerial systems, 373
      integrating satellite data with different levels of processing, 373
      managing image collections with, 361–374
      mosaic methods, 130
      mosaic operators for, 130, 131f
      on-the-fly processing with function chain, 370
      open standards, 381
      optimizing services for performance, 378–379
      overviews, 364, 365, 365f
      processing templates, 370
      properties, 366–368, 366f
      publishing as image services, 374–382
      pyramids, 364–365, 365f
      raster functions, 131, 131f, 369
      scaling to massive collections, 372–373
      scientific multidimensional raster management, 374
      seamline generation, 372
      structure, 362–368
      use cases, 372–374
Mosaicking, 175
Mount St. Helens, 241, 241f
MSS (Landsat Multispectral Scanner), 245
Multispectral data, 19f
Multispectral imagery, 52, 52f, 53, 53f
Multispectral Imagery Service, 92
Multispectral satellite systems, 35

Multispectral sensors, 51
Multitemporal image analysis, 240, 240*f*, 241, 241*f*, 318–328, 320*f*–326*f*
    continuous change analysis, 327–328
    image differencing, 322–323, 322*f*–325*f*, 325
    image-to-map comparisons, 328–330, 330*t*
    managing nonchange differences, 330–333
    manual interpretation of change from imagery, 319–322, 320*f*, 321*f*
    map-to-map comparisons, 317, 318*t*
    unsupervised/supervised classification, 325–327, 326*f*, 327t
MXD files, 270, 271*f*
Myrtle Beach, South Carolina, 316*f*

Nadir, 57–59, 60*f*
NAIP. See National Agriculture Imagery Program
National Aeronautics and Space Administration (NASA)
    DAACs, 80
    EO-1 program, 100
    first Landsat satellite, 66
    free access to imaging from, 64
    GEDI mission, 39
    Global Land Surveys datasets, 97
    Landsat Global Land Survey, 98
    public domain earth science data, 75
    SRTM, 46, 102, 224, 373–374
    thermal imagery, 99
National Agriculture Imagery Program (NAIP)
    and Esri World Imagery, 71
    HVH-resolution imagery, 85, 92*f*
    imagery web services, 134
    multitemporal image analysis, 240
    and NDVI, 248
    positional accuracy assessment of imagery, 341*f*
    pricing of imagery from, 64–65
    public domain imagery from, 75
    Sonoma Vegetation Mapping project, 9, 70
    spatial resolution, 56, 56*f*, 57*f*
National Archives, 94
National Center for Atmospheric Research (NCAR), 261
National Elevation Dataset (NED), 221, 226, 254
National Geospatial Agency of the Department of Defense, 65
National Geospatial-Intelligence Agency (NGA), 102
National Hydrography Dataset (NHD), 259
National Imagery and Mapping Agency, 102
National Land Cover Database, 186, 245, 300
National Map, 227
National Map Accuracy Standards (NMAS), 341
National Oceanic and Atmospheric Administration (NOAA)
    Climate Data Online, 80
    Coastal Change Analysis Program (See C-CAP)
    Digital Coast, 80
    free access to imaging from, 64
    Landsat program, 66–67
    public domain data, 75
National Park Service, 187
National Standard for Spatial Data Accuracy (NSSDA), 342, 344, 346–347
National Vegetation Classification (NVC), 186
    basis for new classification schemes, 187
    basis for Sonoma County vegetation map classification key, 189, 191–192
    vegetation hierarchy, 188*f*
Native display scale (NDS), 109
Natural disaster assessment, 6
Natural Resources Conservation Service (NRCS), 262–263

NCAR (National Center for Atmospheric Research), 261
NDS (native display scale), 109
NDVI. See Normalized Difference Vegetation Index
Nearest neighbor interpolation, 164, 164f, 165f
Near-infrared (NIR) light, 231, 243
    and change images, 320–321, 321f
    and NDVI, 247
NED (National Elevation Dataset), 221, 226, 254
Neighborhood, image filtering and, 124, 124f, 125, 125f
New Orleans, Louisiana, 232f
NGA (National Geospatial-Intelligence Agency), 102
NHD (National Hydrography Dataset), 259
NIR light. See Near-infrared (NIR) light
NIR/red ratio, 150
NMAS (National Map Accuracy Standards), 341
NOAA. See National Oceanic and Atmospheric Administration
NOAA Coastal Change Analysis Program. See C-CAP
NoData pixel value, 116, 117, 117f, 363
Nonimagery geospatial data, 389
Normalized burn ratio, 248
Normalized difference moisture index, 247
Normalized Difference Vegetation Index (NDVI), 246–248, 247f, 250, 251f
Normalized DSMs, 222. *See also* Digital height models (DHMs)
Normalized Vegetation Index/red ratio, 150
NovaSAR-S, 103
NRCS (Natural Resources Conservation Service), 262–263
NSSDA (National Standard for Spatial Data Accuracy), 342, 344, 346–347
NVC. See National Vegetation Classification

Object-based mapping, 275, 276f
Objects, supervised image classification and, 288
Off-nadir, 58, 59, 60f, 61f
OGC (Open Geospatial Consortium), 381
Oil and gas exploration, 6
OLI (Operational Land Manager), 99
Olympic Mountains, Washington, 331f
On-demand caching, 378
Open Geospatial Consortium (OGC), 381
Openings, 42
Operational Land Manager (OLI), 99
Oregon white oak (Quercus garryana), 302–303, 303t
Organizational characteristics of projects, 64–67
    access, 65–66
    pricing/licensing, 64–67
Orthorectification, 163, 172–175, 172f, 173f
Overall accuracy formula, 355
Overviews, mosaic dataset, 364, 365, 365f

Panchromatic and multispectral imagery
    from active sensors, 100–103
    passive (*See* Passive panchromatic and multispectral imagery)
Panchromatic sensors, 51
Parallax, 223
Parallelepiped algorithm, 296, 296f, 297
Passive panchromatic and multispectral imagery, 84–100
    growing supply of, 84f
    high- and very-high spatial resolution, 84–94
    hyperspectral imagery, 100
    moderate and low spatial resolution, 94–98
    satellites, comparison of, 86t–91t
    sources of, 85, 92–94
    thermal imagery, 98–99

Passive sensors
    active sensors vs., 33–35, 34f
    wavelengths sensed by, 35–36, 37f–38f
Passive systems, 62
Paz, 103
PCA (principal components analysis), 244, 245f
Pentagon (Washington, D.C.), 232, 233f
Percent clip stretch, 122
Percentile height metrics, 263
Per-pixel classification
    segmentation and object-oriented classification, 389
    in semiautomated image classification, 275–277
Pflugerville, Texas, 6
Photoelectric effect, 31
Photogrammetry, DEMs from, 222–224
Photo interpretation, 268
Pitch, 48, 49f
Pixels
    band ordering, 110, 112
    and digital array imaging, 33
    recommended sizes when creating overviews, 379f
    as sampling unit for reference data, 352–353
    in semiautomated image classification, 275, 276f
    storage formats, 110, 112
    supervised image classification, 295–297
    and supervised image classification, 287
    tiling, 110
Plainfield, Vermont, 81–83, 83f
Platforms, 7, 43–50
    agility of, 49–50
    altitude of, 47
    defined, 28
    piloted vs. unpiloted, 46
    power of, 50
    speed of, 47–48
    stability of, 48, 49, 49f
    types of, 44–45
Pleiades satellite imagery, 71
PNG format, 377
Point cloud, 13, 40, 100–101, 224, 263
Point density, 40–41
Points, 15, 16f
Polygons, 15, 16f
Positional map accuracy assessment, 340–347, 341f, 345f, 347t
    collecting reference data, 343–344, 345f
    computing descriptive statistics, 345–347, 347t
    initial considerations, 343
    with thematic map accuracy assessment, 352
Postclassification change detection (masking), 329, 330t. *See also* C-CAP (NOAA Coastal Change Analysis Program)
Precipitation, 261
Precision, accuracy vs., 338, 339f
Precision agriculture, 5
Pricing, 64–67, 388
Primary colors, 231
Prime Meridian, 161
Principal components analysis (PCA), 244, 245f
Processing templates, 370
Producer accuracy formula, 355–356
Projective transform, 170, 170f
Properties, mosaic dataset, 366–368, 366f
Push broom (along-track) scanners, 35, 38f

Pyramids. See Image pyramids
Python
    automating image-processing workflows with, 390
    and client-defined processing, 380
    and NDVI, 248
    Python Adapter function, 370
    and TCT, 246

Quercus agrifolia (coast live oaks), 235*f*
Quercus garryana (Oregon white oak), 302–303, 303*t*
Quercus lobata (valley oaks), 235*f*
Query filters, 130
QuickBird, 246

R (statistical package), 299, 301
Rabben, E. L., 242
Radar imagery, 38, 39*f*, 102–103, 103*f*
Radar images, 30
Radial basis function (RBF) kernel, 304
Radiance
    and atmospheric correction, 153
    reflectance vs., 148–150, 149*f*
Radiometric correction, 151–156
    atmospheric correction, 153–156, 154*f*, 155*f*
    sensor correction, 151–152, 152*f*
    sun angle and topographic correction, 152–153
Radiometric resolution, 54–55
Random error, 158
Random Forests (machine learning algorithm), 277, 301–304, 303*f*
Raster(s)
    adding to mosaic dataset, 368–369
    bands in, 18–20, 19*f*, 20*f*
    cell size of, 20f, 21
    scientific multidimensional raster management, 374
    types of, 17, 17*f*–18*f*
    vectors vs., 15, 16*f*
Raster algebra, 246, 248
Raster data (raster grids), 13, 14*f*, 15, 137–138, 138*f*
Raster functions, 131, 369
Raster products, 114
Raster pyramids, 365, 365*f*
Raster web services, 140, 140*f*
Ratio (derivative band), 243
Rational polynomial coefficients (RPCs), 113, 174–175
RBF kernel, 304
Rectangular arrays, 13
Redwood (*Sequoia sempervirens*), 228
Reference data
    ground truth vs., 349
    nonsampling factors, 350–352
    sampling factors, 352–354
    for thematic map accuracy assessment, 350–354
Reflectance, 29*f*
    and atmospheric correction, 153
    radiance vs., 148–150, 149*f*
Reflection, 147
Registration. See Image registration
Relative atmospheric correction, 154
Remote sensors, 4, 29–43, 30*f*
    active vs. passive, 33–35, 34*f*
    bodies of, 43
    correcting radiance to reflectance, 148–150, 149*f*

  defined, 28
  imaging surfaces of, 31–42, 31*f*, 32*f*
  lenses in, 42
  openings in, 42
Representational state transfer (REST) API, 380
Resampling, 163–167, 164*f*–167*f*
REST (representational state transfer) API, 380
Riparian Topography toolbox, 260
Roll, 48, 49*f*
Root mean square error (RMSE), 342–343, 345–346, 347*t*
RPCs (rational polynomial coefficients), 113, 174–175

Sampling units/schemes, 352–354
Satellite imagery. *See also* Passive panchromatic and multispectral imagery; *specific sources*, e.g.: Landsat imagery
  atmospheric correction, 153
    integrating for mosaic dataset, 373
Satellite Imaging Corporation, 79
Satellites, 7
SAVI (soil-adjusted vegetation index), 247
Scale, image, 107–109, 108*f*
Scaling, mosaic datasets, 372–373
Scan Line Corrector (SLC), 150–151, 151*f*
Scanners, 35, 159
Scientific multidimensional rasters, 374
Seamlines, 130, 372
Seasat radar mission, 102
Second-order polynomial transform, 171, 171*f*
Selection of imagery, 70–104
  accessibility, 75
  budgetary considerations, 76
  determination of smallest identifiable image, 71
  framework for, 70–76
  needs vs. availability, 70
  project area size/shape/accessibility, 73–74
  project time frame, 72, 72*f*
  sharing considerations, 75
  spatial/spectral accuracy requirements, 74–75
  types of features needed, 73
  for visualization vs. mapping, 71
Semiautomated image classification, 275–311, 276*f*
  algorithms for, 279, 280*f*
  combined approaches to, 306–309, 307*f*, 308*f*
  economies of scale and, 390
  history of, 276–277
  image segmentation, 278–279, 278*f*
  map validation and editing, 309–311, 310*f*
  supervised classification, 285–306
  unsupervised classification, 280, 281*f*, 282, 283*f*–284*f*, 284–285
senseFly eBee, 82
Sensor correction, 151–152, 152*f*
Sensors. *See* Remote sensors
Sentinel, 7, 63, 75, 240, 387
Sentinel-1B, 103
Sentinel-2, 95, 95*f*
Sequoia sempervirens (redwood), 228
Serving imagery. *See* Image services
Shadow, 234, 235, 235*f*
Shift and scale change, 169, 169*f*
Shuttle Imaging Radar (SIR), 102
Shuttle Radar Topography Mission (SRTM), 46, 102, 224, 373–374
Site-specific accuracy assessment, 348–349, 349*f*
6S (Second Simulation of the Satellite Signal in the Solar Spectrum), 156

SLC (Scan Line Corrector), 150–151, 151*f*
Slope and aspect, 225, 256, 257*f*
Small-scale imagery, 108
Smoothing (low-pass) filters, 126, 127*f*
Snow, 156–157
Soil-adjusted vegetation index (SAVI), 247
Soil data, 262–263
Solar insolation, 258
Solar insolation rasters, 226
Sonar images, 30
Sonoma County, California, Vegetation Mapping project
    bare-earth DEM, 255, 256*f*
    clip and email viewer, 383, 383*f*
    data for DEMs, 226
    fog in, 262
    HAR, 260, 261*f*
    hierarchical forest classification scheme, 185*f*
    imagery datasets used for, 70
    lidar-derived bare-earth hillshade, 228, 228*f*
    manual interpretation of vegetation types, 270, 271*f*
    map, 9, 11, 12*f*
    map classification key, 189, 191–218
    multispectral imagery, 19, 19*f*, 52*f*, 53*f*
    Random Forest importance matrix, 302–303, 303t
    stream centerlines, 259, 260*f*
    supervised image classification, 291*f*
    vineyard patterns, 234*f*
    wildfire history, 263, 264*f*
Sonoma County Agricultural Preservation and Open Space District, 9
Sources of imagery, 77–103
    active sensors, 100–103
    ArcGIS Online, 77
    commercial photogrammetric and remote sensing firms, 78
    commercial satellite companies, 79
    comparison of, 78t, 79t
    government agencies, 79–80
    passive panchromatic and multispectral imagery, 84–100
    unmanned aerial systems, 80–83, 83*f*
SPA (spectral pattern analysis), 249–251, 250*f*, 251*f*
Spain, radar platforms, 103
Spatial accuracy, importance of, 388
Spatial aliasing, 164
Spatial analysis, 382–383
Spatial Analyst
    flow accumulation rasters, 259
    machine learning algorithms, 299, 304
    raster calculation, 246
    Slope tool, 256
    solar insolation, 226, 258
Spatial autocorrelation, 344, 353–354
Spatial awareness, 8
Spatial information, mainstreaming of, 8
Spatial resolution, 21, 55–57, 56*f*–57*f*, 108–109
Spectral accuracy/calibration, importance of, 387–388
Spectral pattern analysis (SPA), 249–251, 250*f*, 251*f*
Spectral resolution, 51–54
Spinning filters, 36
SPOT (Satellite Pour l'Observation de la Terre), 59, 95–96, 224
Spurr, Stephen, 242, 274
SRTM. See Shuttle Radar Topography Mission
Standard deviation stretch, 122, 122*f*
Standard frame camera model (orthorectification), 173–174, 173*f*

Stereoscope, 274
Stereo viewing, 274
Story maps, 140, 140f
Stream centerlines, 259, 260f
Stretch. See Image stretch
Subtractive primary colors, 231
Sun angle, 152–153
Sun-synchronous platforms, 44, 62
Supervised image classification, 285–306, 307f
    machine learning techniques for, 298–306
    manually derived rulesets for, 292–293, 294f
    multitemporal, 327, 327t
    per-pixel classification, 276
    traditional techniques for, 295–298, 295f, 296f, 298f
    training sites, 285–292, 288f, 289f, 291f
Support vector machines (SVMs), 277, 304
Surfaces, imaging, 31–42
    active vs. passive energy sources for, 33–35, 34f
    and electromagnetic spectrum, 31–32, 31f, 32f
    film vs. digital array, 33
    wavelengths sensed by, 35–36, 37f–41f, 38–42
SVMs (support vector machines), 277, 304
Syncom, 44
Systematic error, 158

Tasseled-cap transformation (TCT), 245–246, 246f
Technology, for communicating results, 391
Teledyne Optech bathymetric lidar system, 13f
Teledyne Optech Titan system, 41–42
Temperature, vegetation mapping and, 261
Temporal resolution, 62
Texture (image element), 235–236, 236f–237f
Thematic maps. See also Digital elevation models (DEMs)
    accuracy assessment, 340, 348–359, 349f, 355f, 357f–358f, 359t
    defined, 24
    reference data collection, 350–354
    supervised vs. unsupervised classification, 306, 307f
    SVMs and, 304
    totally exhaustive classification schemes, 183
Thermal imagery, 98–99
Thermal Infrared Sensor (TIRS), 99
3D Elevation Program (3DEP), 226
3D transform, 172–175, 172f, 173f
3D viewing, 274
TIFF files, 113
Tile cache services
    about, 134, 375–376
    best practices: test before deploying, 378
    cache configuration, 376
    cache formats, 377
    imagery in visualization applications, 71
    image services, 375–378
    on-demand caching, 378
    tiling scheme, 376–377
TIRS (Thermal Infrared Sensor), 99
Tone, 231, 232f
Topographic correction, 152–153
Topographic displacement, 159, 160f
Topographic lidar, 40, 40f, 41
Training sites/samples, 285–292, 288f, 289f, 291f, 309–310
Transformations (derivative bands), 243–246
Transmittance, 147

Transverse Mercator coordinate system, 161, 162
Tree canopy, 257, 258*f*
Triangulation, 223
2D transforms, 168–171, 169*f*–171*f*

UAS. See Unmanned aerial systems
United Kingdom, radar platforms in, 103
United Nations, 186
US Department of Agriculture (USDA)
    Aerial Photography Field Office, 79–80, 93–94
    Farm Services Agency, 85
    National Agriculture Imagery Program (See National Agriculture Imagery Program)
    Natural Resources Conservation Service, 262–263
US Forest Service, 99, 315
US Geological Survey (USGS), 13
    Anderson classification scheme, 186
    data for DEMs, 227
    EO-1 program, 100
    EROS, 79, 85, 93–98
    Global Land Surveys datasets, 97
    Landsat Global Land Survey, 98
    Landsat imagery, 67
    Landsat Look Viewer, 141, 142*f*
    Landsat spectral data accuracy, 74–75
    Landstat imagery services, 66
    lidar specifications, 41
    NDEP, 342
    NHD flowlines, 259
    3DEP, 226
US National Satellite Land Remote Sensing Archive, 94
US National Vegetation Classification (USNVC). See National Vegetation Classification (NVC)
Universal transverse Mercator (UTM) coordinate system, 161–162
University of Maryland, 9n1
University of Vermont (UVM) UAS Team, 82
Unmanned aerial systems (UASs; drones), 7, 80–83
    adding imagery collections to mosaic dataset, 373
    HVH-resolution imagery, 85
    mapping woody debris in Great Brook, 81–83, 83*f*
    sensing platforms, 45, 46
    thermal imagery from, 99
Unsupervised image classification, 280, 281*f*, 282, 283*f*–284*f*, 284–285
    multitemporal, 325–327, 326*f*, 327t
    per-pixel classification, 276
Urban Observatory, 7
UTM (Universal transverse Mercator) coordinate system, 161–162

Validation of maps, 309–311, 310*f*
Valley oaks (*Quercus lobata*), 235*f*
Valtus, 92
Variation, in imagery and on ground, 25, 389
Vectors, rasters vs., 15, 16*f*
Vegetation indices, 246–248, 247*f*, 328
Vegetation maps, 11, 12*f*
    classification key, 189
    hydrology data and, 260
    precipitation and temperature, 261
Vernal pools, 255, 256*f*
Video, mosaic dataset creation from, 374
Viewing angle, 57–59, 60*f*–61*f*, 159
Viewsheds, 225
Volumetric analysis, 225–226

Washington, D.C., infrared imagery, 237*f*
Washington Monument (Washington, D.C.), 159, 160*f*
Water
    and BRDF, 150
    and supervised image classification, 287
Wavelet compression, 115–116
Weather satellites, 62
Web services, 132–141. *See also* Image services
    advantages of, 390–391
    in apps, 140–141, 142*f*
    in ArcGIS desktop, 135, 135*f*, 136, 137*f*
    in ArcGIS web maps, 139, 139*f*
    dynamic image services, 137–138, 138*f*
    Esri story maps, 140, 140*f*
    geoprocessing services, 134
    machine learning algorithms, 299
    for raster data, 133–134
WELD data, 98
Whisk broom scanners, 35
Wildfire history, 263, 264*f*
Windows, app development for, 141
World Elevation services, 227
World files, 113
WorldView satellite
    Esri World Imagery, 71
    HVH-resolution imagery, 84
    spatial resolution, 57
    viewing angle, 57, 58
World War I, 4, 337
World War II, 4, 337
Wurman, Richard Saul, 7

Yaw, 48, 49*f*

Zaatari refugee camp, 5